BC Science 6

Authors

Adrienne Mason
Professional Writer/Educator
Tofino, British Columbia

Karen Walsh
Peterson Road Elementary School
Langley, British Columbia

Karen Charleson
Professional Writer/Eductor
Hooksum Outdoor School
Hot Sprints Cove, British Columbia

Darcy Dobell
Professional Writer/Educator
Tofino, British Columbia

Tara Shepherd
Mount Crescent Elementary School
Maple Ridge, British Columbia

Eric Grace
Professional Writer/Educator
Victoria, British Columbia

Judy Gadicke
Adam Robertson Elementary School
Creston, British Columbia

Aboriginal Consultants

Dr. E. R. Atleo (Umeek of Ahousaht)
Co-founder, First Nations Studies Department
Malaspina University-College
Nanaimo, British Columbia

Dr. David Blades
Department of Curriculum Instruction
University of Victoria
Victoria, British Columbia

David Rattray
First Nation Education Centre
Fort St. John, British Columbia

Consultants

Sandy Wohl
Hugh Boyd Secondary School
Richmond, British Columbia

Dr. Douglas A. Roberts
University of Calgary
Calgary, Alberta

Herb Johnston
Department of Curriculum Studies
University of British Columbia
Vancouver, British Columbia

Assessment Consultant

Bruce McAskill
Hold Fast Consultants, Inc.
Victoria, British Columbia

Reading Consultant

Shirley Choo
Resource Teacher
Montroyal School
North Vancouver, British Columbia

Literacy Consultant

Gloria Gustafson
Coordinator, District Staff Development Team
Coquitlam School District
Coquitlam, British Columbia

Technology Consultant

Glen Holmes
Formerly with Claremont Secondary School
Victoria, British Columbia

Social Considerations Consultant

Nancy Josephson
School of Education
University of the Cariboo
Kamloops, British Columbia

Special Education/ At Risk Consultant

Elizabeth Sparling
Integration Support Teacher
Parkland Secondary School
Sidney, British Columbia

Differentiated Instruction Consultant

Jennifer Katz
Education Consultant
Richmond School District

Health and Safety Consultant

Ken P. Waldman, C.R.S.P.
Richmond, British Columbia

Advisory Panel

Elspeth Anderson
Crofton House School
Vancouver, British Columbia

Dr. Allan MacKinnon
Faculty of Education
Simon Fraser University
Burnaby, British Columbia

Darren Macmillan
Banting Middle School
Coquitlam, British Columbia

Dr. Jolie Mayer-Smith
Department of Curriculum Studies
University of British Columbia
Vancouver, British Columbia

Steve Williams
Shuswap Junior Secondary School
Salmon Arm, British Columbia

Josie U. Wilson
Ashcroft Secondary School
Ashcroft, British Columbia

Toronto Montréal Boston Burr Ridge, IL Dubuque, IA Madison, WI New York
San Francisco St. Louis Bangkok Bogotá Caracas Kuala Lumpur Lisbon London
Madrid Mexico City Milan New Delhi Santiago Seoul Singapore Sydney Taipei

COPIES OF THIS BOOK MAY BE OBTAINED BY CONTACTING:

McGraw-Hill Ryerson Ltd.

WEB SITE:
http://www.mcgrawhill.ca

E-MAIL:
orders@mcgrawhill.ca

TOLL-FREE FAX:
1-800-463-5885

TOLL-FREE CALL:
1-800-565-5758

OR BY MAILING YOUR ORDER TO:

McGraw-Hill Ryerson
Order Department
300 Water Street
Whitby, ON L1N 9B6

Please quote the ISBN and title when placing your order.

Student text ISBN:
0-07-095900-5

McGraw-Hill
Ryerson

McGraw-Hill Ryerson BC Science 6

ISBN 0-07-095900-5

www.mcgrawhill.ca

2 3 4 5 6 7 8 9 10 TCP 0 9 8 7 6

Printed and bound in Canada

Library and Archives Canada Cataloguing in Publication

BC science 6 / authors, Darcy Dobell, Eric Grace, Adrienne Mason; senior pedagogical consultant, Sandy Wohl; consultants, Karen Charleson, Douglas A. Roberts.

Includes index.

ISBN 0-07-095900-5

1. Science—Textbooks. I. Grace, Eric, 1948- II. Mason, Adrienne III. Title.
IV. Title: BC science six.

Q161.2.D62 2005 500 C2005-901161-0

ASSOCIATE PUBLISHER: Keith Richards
DEVELOPMENTAL EDITORS: Tricia Armstrong, Susan Girvan, Sara Goodchild, Lenore Latta
MANAGER, EDITORIAL SERVICES: Linda Allison
SUPERVISING EDITOR: Crystal Shortt
COPY EDITOR: Valerie Ahwee
PERMISSIONS EDITOR: Linda Tanaka
EDITORIAL ASSISTANT: Erin Hartley
EDITORIAL COORDINATOR: Valerie Janicki
MANAGER, PRODUCTION SERVICES: Yolanda Pigden
PRODUCTION COORDINATOR: Janie Deneau
SET-UP PHOTOGRAPHY: Erin Carr, Judy Gadicke
COVER DESIGN: ArtPlus Limited
ELECTRONIC PAGE MAKE-UP: ArtPlus Limited
COVER IMAGE: Courtesy of Ron Thiele

OUR COVER: The only place in the world where Kermode bears live is in the rainforests of coastal British Columbia, west of Terrace. Kermode bears are a unique type of black bear. It is also called the ghost, or spirit, bear. The Kermode is just one of the thousands of different types of animals that live in British Columbia. In Unit 1 of this textbook, you will find out more about the many species that live in this province. Why are there so many more species in British Columbia when compared to other provinces in Canada? Read Unit 1 to learn about the diversity of species and spaces in British Columbia.

Acknowledgements

The authors, editors, project manager, and publisher of *McGraw-Hill Ryerson BC Science 6* extend their heartfelt thanks to the students, elders, educators, consultants, and reviewers who devoted their time, energy, knowledge, and wisdom to the creation of this textbook. We thank the students of Judy Gadicke at Adam Robertson Elementary School in Creston for taking part in the student photos throughout the text. We extend respectful thanks to the elders who shared their knowledge and wisdom with us: Barbara Wilson of Haida Nation, Jean Issac of Tsay Keh Dene Nation, and Wilfred Jacobs of Ktunaxa Nation. We also thank Nancy Turner and Maxine Matlipi for their kindness in facilitating the interviews with Barbara Wilson and Jean Issac, respectively. Dr. Jane Watson, Taras Atleo, and Marc Fricker kindly shared their expertise with us for end-of-unit features. We thank them, and thank Dr. Richard Atleo for facilitating the interview with Taras Atleo.

Pedagogical Reviewers

Kevin Amboe
Elementary Science and Information Technology Helping Teacher
Curriculum and Instructional Services Centre
Surrey, British Columbia

Tony Biollo
Skaha Lake Middle School
Penticton, British Columbia

Catherine Dickie
Duncan Cran Elementary School
Fort St. John, British Columbia

Sharon English
Langley School District
Langley, British Columbia

Marc Garneau
Mathematics Helping Teacher
Curriculum and Instructional Services Centre
Surrey, British Columbia

Erica M. Hargreave
Science Educational Consultant
Ahimsa Nature Adventures
Richmond, British Columbia

Cameron Hill
Hartley Bay School
Hartley Bay, British Columbia

Dino Klarich
Lakeview Elementary School
Burnaby, British Columbia

Cathy Lambright
Thornhill Elementary School
Terrace, British Columbia

Arthur Loyie
Sunshine Hills Elementary School
Delta, British Columbia

Darren Macmillan
Banting Middle School
Coquitlam, British Columbia

Colin McTaggart
Chief Dan George Middle School
Abbotsford, British Columbia

M. J. Medenwalt
Mitchell Elementary School
Richmond, British Columbia

Georgina Oberle
Glendale Elementary School
Williams Lake, British Columbia

Chantel Parsons
Courtenay Middle School
Courtenay, British Columbia

John Price
Elsie Roy Elementary School
Vancouver, British Columbia

Kristin Visscher
Ellendale Elementary School
Surrey, British Columbia

Doreen Wilkins
Westwood Elementary
Prince George, British Columbia

Josie U. Wilson
Ashcroft Secondary School
Ashcroft, British Columbia

Sandra Wong
Kwayhquitlum Middle School
Port Coquitlam, British Columbia

Anita Woode
William Konkin Elementary School
Burns Lake, British Columbia

Anthony Yam
Lakeview Elementary School
Burnaby, British Columbia

Accuracy Reviewers

Dr. Wytze Brouwer
Department of Physics
University of Alberta
Edmonton, Alberta

Nancy J. Flood
Department of Biological Sciences
Thompson Rivers University
Kamloops, British Columbia

Steve Lang
Director, Canadian Space Resource Centre
Marc Garneau Collegiate Institute
Toronto, Ontario

Barry McGuire
Former Physics Teacher
Western Canada High School
Calgary, Alberta

Contents

Safety in Your Science Classroom viii
Science, Technology, and Society xii
Aboriginal Technology xviii

Unit 1 Diversity of Life 2

Chapter 1 A Closer Look at Living Things 4

Starting Point Activity 1-A: It's a Live One! 5

1.1 Living Things 6

At Home Activity 1-B: Diversity in British Columbia. 8
Conduct an Investigation 1-C: Get Growing 12

1.2 A Cell: The Basic Unit of Life 15

Find Out Activity 1-D: Kitchen Cells 18

1.3 Tools of Biology 23

Find Out Activity 1-E: Look at the Big Picture 24
Think & Link Investigation 1-F: Take a Closer Look. 25
Conduct an Investigation 1-G: Zoom In: Using a Microscope 27
Find Out Activity 1-H: Microscope Practice. 29

Chapter 1 At a Glance 31
Chapter 1 Review 32

Chapter 2 Living Things and Their Environment. 34

Starting Point Activity 2-A: Home, Sweet Home. 35

2.1 Climates and Biomes. 36

Think & Link Investigation 2-B: Climates in British Columbia 37
Find Out Activity 2-C: People in Biomes 41
Find Out Activity 2-D: Illustrate a Biome. 44

2.2 Adaptations 47

Find Out Activity 2-E: Picky Eaters 49
Find Out Activity 2-F: Animal Voices. 52
Problem-Solving Investigation 2-G: Camouflage Creature 55

Chapter 2 At a Glance 57
Chapter 2 Review 58

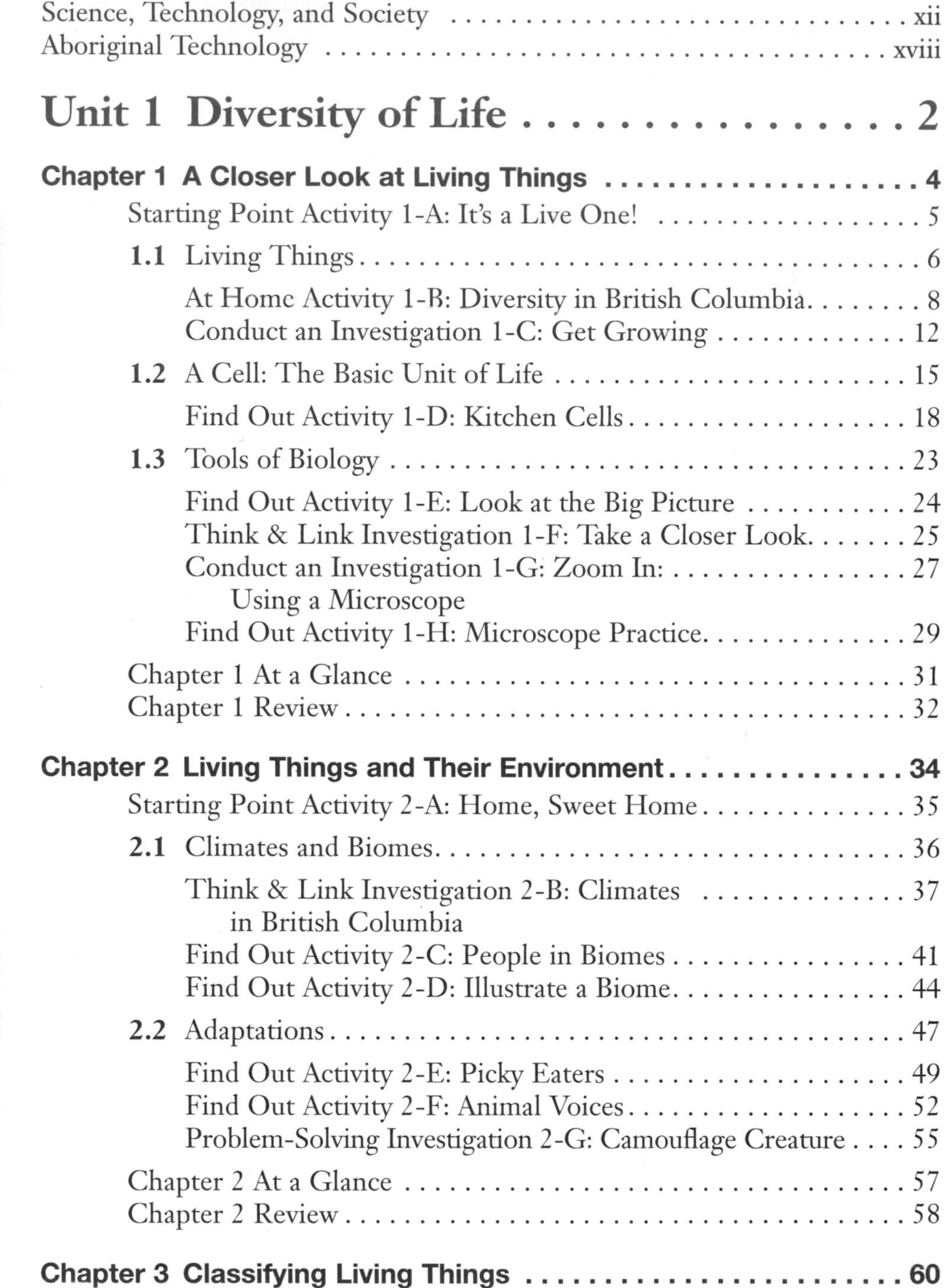

Chapter 3 Classifying Living Things 60

Starting Point Activity 3-A: Sort It Out!. 61

3.1 Grouping Living Things. 62

Find Out Activity 3-B: Identifying Canadian Cats 67
Find Out Activity 3-C: Create Your Own Classification Key 68

3.2 The Kingdoms of Life 70
Find Out Activity 3-D: Wanted: Bacteria That Make You Sick 72
Conduct an Investigation 3-E: Looking for Micro-organisms 74
Find Out Activity 3-F: The Function of Stems in Plants 76
Conduct an Investigation 3-G: Grow a Fungi Garden 78
Find Out Activity 3-H: Animal Collage 81
Find Out Activity 3-I: Who Am I? 83

Chapter 3 At a Glance 85
Chapter 3 Review 86

Unit 1 Ask an Elder: Barbara Wilson 88
Unit 1 Ask a Biologist: Jane Watson 90
Unit 1 Project 92

Unit 2 Electricity 94

Chapter 4 Electric Charges 96

Starting Point Activity 4-A: Balloon Buddies 97
4.1 Explaining Electricity 98
Find Out Activity 4-B: Make an Atomic Model 101
4.2 Static Electricity 104
Conduct an Investigation 4-C: Get Ready, ! Get Set, Charge 106
Find Out Activity 4-D: Make a Pepper Copier 110
At Home Activity 4-E: How Shocking! 113
4.3 Current Electricity 115
Conduct an Investigation 4-F: Lighten Up! 116
Find Out Investigation 4-G: Electric Lemon 120

Chapter 4 At a Glance 123
Chapter 4 Review 124

Chapter 5 Electricity at Work 126

Starting Point Activity 5-A: Light Your Way! 127
5.1 Series Circuits 128
Find Out Activity 5-B: Out of Control 130
Find Out Activity 5-C: It's the Only Way to Go 131
Conduct an Investigation 5-D: In Series 132
5.2 Parallel Circuits 136
At Home Activity 5-E: Current in Parallel and Series Circuits 137
Conduct an Investigation 5-F: On Parallel Tracks 138
Design Your Own Investigation 5-G: Build a Complex Electrical Circuit 140

5.3 Transforming Electrical Energy 143

Conduct an Investigation 5-H: What's the Big Attraction? 145

Chapter 5 At a Glance 147
Chapter 5 Review 148

Chapter 6 Power To You. 150

Starting Point Activity 6-A: Electric Lunch 151

6.1 Generating Electricity. 152

Find Out Activity 6-B: What's Your Source?. 153
Conduct an Investigation 6-C: Model Hydro-electric Power Station 155

6.2 Electricity Ahead. 160

Find Out Activity 6-D: Catch Some Rays 161
Problem-Solving Investigation 6-E: Production Teams 167

6.3 Bringing it Home 169

Find Out Activity 6-F: Blown Away 170
At Home Activity 6-G: Watts Up? 172
Problem-Solving Investigation 6-H: From Consuming to Conserving 173

Chapter 6 At a Glance 175
Chapter 6 Review 176

Unit 2 Ask an Elder: Jean Issac. 178
Unit 2 Ask an Electrician: Taras Atleo 180
Unit 2 Project. 182

Unit 3 Exploring Extreme Environments. . 184

Chapter 7 A Challenge to Explorers 186

Starting Point Activity 7-A: Explore Your Home Province. 187

7.1 What Are Extreme Environments? 188

Find Out Activity 7-B: Model an Extreme Environment . . . 189
At Home Activity 7-C: Eating Like an Astronaut 193
At Home Activity 7-D: Technology for Travel 195
Find Out Activity 7-E: Aboriginal Technology for Explorers 197

7.2 Observing the Sky. 199

Find Out Activity 7-F: A Clearer View 199
Conduct an Investigation 7-G: Gravity and Orbits. 202
Conduct an Investigation 7-H: Profile the Planets 204

7.3 Leaving Planet Earth 208

Find Out Activity 7-I: Racing Rockets 208
Conduct an Investigation 7-J: Working in a Spacesuit 212
Find Out Activity 7-K: Floating in Space 215

Think & Link Investigation 7-L: Space Report 218
Problem-Solving Investigation 7-M: A Visit to Venus...... 221

Chapter 7 At a Glance 223
Chapter 7 Review 224

Chapter 8 Technologies for Unknown Worlds 226

Starting Point Activity 8-A: Extremely Useful Technology...... 227

8.1 Rovers and Robots 228

Conduct an Investigation 8-B: Mission to an Unknown Planet 230
At Home Activity 8-C: Robots in Your Life 232
Conduct an Investigation 8-D: Communicating with a Robot on Mars 234
Find Out Activity 8-E: Robot in a Rabbit Hole 237

8.2 Satellites and Space Stations 240

Find Out Activity 8-F: Weather Forecasts from Space 243
Find Out Activity 8-G: Technology Timeline............ 246

8.3 Canadian Technologies 248

Problem-Solving Investigation 8-H: Build a Space Arm 250
Find Out Activity 8-I: Canadian Technologies 254
Find Out Activity 8-J: Extreme Game Board 255

8.4 Putting Technology to the Test...................... 257

Problem-Solving Investigation 8-K: Suited for Space...... 259
Think and Link Investigation 8-L: Using Technology Responsibly 261

Chapter 8 At a Glance 263
Chapter 8 Review 264

Unit 3 Ask an Elder: Wilfred Jacobs...................... 266
Unit 3 Ask a Robotics Instructor: Marc Fricker 268
Unit 3 Project.. 270

SkillPOWER 1 Using Graphic Organizers 272

SkillPOWER 2 Care and Use of the Microscope 276

SkillPOWER 3 The Metric System and Measurement 281

SkillPOWER 4 Organizing and Communicating Scientific Data ... 285

SkillPOWER 5 Designing and Conducting Experiments 291

SkillPOWER 6 Technological Problem Solving 295

Glossary... 298

Index 304

Credits... 307

Safety in Your Science Classroom

Become familiar with the following safety guidelines. It is up to you to follow them, and your teacher's instructions. Following these guidelines will make your activities and investigations in ***BC Science 6*** safe and enjoyable. Your teacher will inform you about any other special safety instructions that you need to follow in your school.

1. Working with your teacher . . .

- Listen carefully to any instructions your teacher gives you.
- Inform your teacher if you have any allergies, medical conditions, or other physical conditions that could affect your work in the science classroom. Tell your teacher if you wear contact lenses or a hearing aid.
- Obtain your teacher's approval before beginning any activity you have designed yourself.
- Know the location and proper use of the nearest fire extinguisher, fire exit, first-aid kit, and fire alarm.

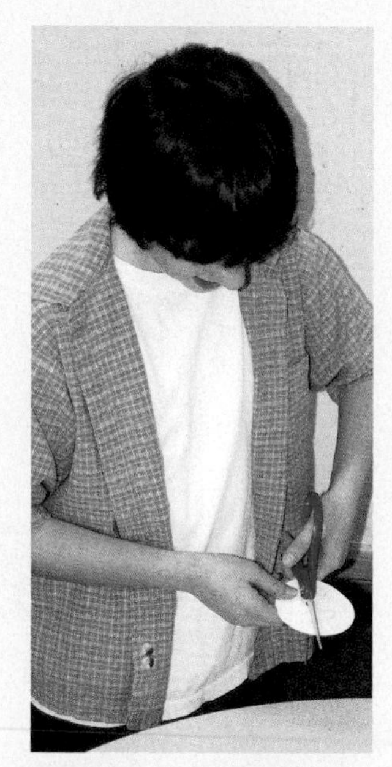

This symbol reminds you to take care when using sharp objects such as scissors.

2. Starting an activity or investigation . . .

- Before starting an activity or investigation, read all of it. If you do not understand how to do any step, ask your teacher for help.
- Be sure you have checked the safety symbols and have read and understood the safety precautions.
- Begin an activity or investigation only after your teacher tells you to begin.

3. Dressing for success in science . . .

- Tie back long hair. Do not wear scarves, ties, rings, or loose-fitting jewellery such as necklaces and bracelets.
- If you are directed to do so, wear protective clothing, such as a lab apron, gloves, or safety goggles.

4. Acting responsibly . . .

- Work carefully with a partner, and make sure that your work area is clear.
- Handle equipment and materials carefully.
- Make sure that stools and chairs are resting securely on the floor.
- If other students are doing something that you consider dangerous, report it to your teacher.

5. **Working in a science classroom . . .**
   - Make sure that you understand all the safety labels on school materials and materials you bring from home. Familiarize yourself with the WHMIS symbols, HHPS symbols, and the special safety symbols used in this book (see pages x–xi).
   - When carrying equipment for an activity or investigation, hold it carefully in front of your body. Carry only one object or container at a time.
   - Be aware of others during activities and investigations. Make room for students who are carrying equipment to their work stations.
   - Do not chew gum, eat, or drink in your science classroom.

6. **Working with sharp objects . . .**
   - Always cut away from yourself and others when using a knife or any other cutting tool.
   - Always keep the pointed end of scissors or any other sharp object facing away from yourself and others.
   - If you notice sharp or jagged edges on any equipment, take special care with it and report it to your teacher.
   - Ask your teacher to dispose of broken glass.

7. **Working with electrical equipment . . .**
   - Make sure that your hands are dry when touching electrical cords, plugs, or sockets.
   - Pull the plug, not the cord, when unplugging electrical equipment. (See the photographs on the right.)
   - Report damaged equipment or frayed cords to your teacher immediately.
   - Place electrical cords in places where people will not trip over them.

8. **Designing, constructing, and experimenting with structures and mechanisms . . .**
   - Use tools safely to cut, join, and shape objects.
   - Handle modelling clay, glues, and paints carefully. Wash your hands after using these materials.
   - Use special care when observing and working with objects in motion (for example, gears and pulleys, elevated objects, and objects that spin, swing, bounce, or vibrate). Keep your fingers safely away from moving objects.

This symbol reminds you to take care when using electrical equipment.

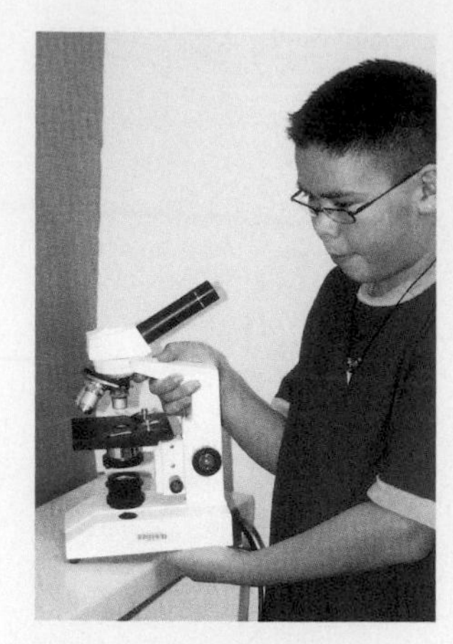

Carry equipment such as microscopes carefully at all times. SkillPower 2 describes how to carry and put away a microscope properly.

9. Working with living things . . .

On a field trip:

- Try to disturb the area as little as possible.
- If you move something, do it carefully and always replace it carefully.
- If you are asked to remove plant material, remove it gently and take as little as possible.

In the classroom:

- Treat living creatures with respect.
- Make sure that living creatures receive humane treatment while they are in your care.
- If possible, return living creatures to their natural environment when your observations are complete.

10. Cleaning up in the science classroom . . .

- Clean up any spills, according to your teacher's instructions.
- Clean equipment before you put it away.
- Dispose of materials as directed by your teacher. Never dispose of materials in a sink unless your teacher directs you to do this.
- Wash your hands thoroughly after doing an activity.

SAFETY SYMBOLS

The following safety symbols are used in the ***BC Science 6*** program to alert you to possible dangers. Be sure that you understand each symbol you see in an activity or investigation before you begin.

	Disposal Alert This symbol appears when care must be taken to dispose of materials properly.		**Electrical Safety** This symbol appears when care should be taken when using electrical equipment.
	Fire Safety This symbol appears when care should be taken around open flames.		**Clothing Protection Safety** A lab apron should be worn when this symbol appears.
	Thermal Safety This symbol appears as a reminder to use caution when handling hot objects.		**Skin Protection Safety** This symbol appears when chemicals might irritate the skin or when micro-organisms might transmit infection.
	Sharp Object Safety This symbol appears when a danger of cuts or punctures caused by the use of sharp objects exists.		**Eye Safety** This symbol appears when a danger to the eyes exists. Safety goggles should be worn when this symbol appears.

WHMIS (Workplace Hazardous Materials Information System)

Look carefully at the WHMIS (Workplace Hazardous Materials Information System) safety symbols that are shown below. The WHMIS symbols are used throughout Canada to identify the dangerous materials that are found in all workplaces, including schools. Make sure that you understand what these symbols mean. When you see these symbols on containers in your classroom, at home, or in a workplace, use safety precautions.

A Compressed Gas	**B** Flammable and Combustible Material
C Oxidizing Material	**E** Corrosive Material
D1 Poisonous and Infectious Material Causing Immediate and Serious Toxic Effects	**D2** Poisonous and Infectious Material Causing Other Toxic Effects
D3 Biohazardous Infectious Material	**F** Dangerously Reactive Material

HHPS (Hazardous Household Product Symbols)

You may have noticed the HHPS (Hazardous Household Product Symbols) on containers around your home. These symbols were developed to help people use household products safely.

▽ The triangle shape means that the *container* is dangerous.

⬡ The octagon (stop sign) shape means that the *contents* of a container are dangerous.

Symbol	Why is it dangerous?
Explosive	This *container* can explode if it is heated or punctured. **Examples:** spray paint, water repellant
Corrosive	This *product* will burn skin or eyes on contact. It will burn the throat and stomach if swallowed. **Examples:** toilet bowl cleaner, oven cleaner
Flammable	This *product*, or its fumes, will catch fire easily if it is near heat, flames, or sparks. **Examples:** glue, gasoline
Poison	Licking, eating, drinking, or sometimes smelling this *product* will cause illness or death. **Examples:** windshield washer fluid, furniture polish

Instant Practice

1. Find four of the ***BC Science 6*** safety symbols in activities or investigations in this textbook. Record the page number and title of the investigation or activity in which you found each symbol. What possible dangers are in the activity or investigation?
2. Ask your parent, guardian, or teacher to find an example of a container with a WHMIS symbol in the workplace.
   (a) What is inside the container?
   (b) Where is the container stored?
   (c) What dangers are associated with the contents of the container?
3. Ask your parent, guardian, or teacher to help you find an example of a container with an HHPS symbol at home or at school.
   (a) What is inside the container?
   (b) Where is the container stored?
   (c) What dangers are associated with the container or the contents of the container?

Science, Technology, and Society

Have you ever wondered how dolphins can spend so much time underwater without breathing? Have you ever thought about what causes a bolt of lightning? Have you ever gazed in awe at the night sky, and wondered what it would be like to visit a distant star or planet?

Dolphins need to breathe just like humans do, but dolphins can remain underwater for 15 minutes or more.

A bolt of lightning can explode a tree, shatter rock, and melt sand into glass.

Space is beautiful, but it is also cold and airless. Enormous distances separate planets, moons, and stars. Yet people have found ways to explore and find out about space.

What Is Science?

Nature inspires us to wonder. The questions people ask about nature are the starting point for science. **Science** is a way of thinking and asking questions about nature on Earth and far beyond Earth. Science is also a collection of knowledge and ideas that help people explain how nature works. Using science, we create explanations for the events we observe. And we explore new questions that arise from our explanations.

What Is Technology?

Scientists ask questions about nature to satisfy their curiosity. For example, a scientist might ask: "Is there a pattern to lightning strikes during a thunderstorm?" To investigate answers to this question, scientists might turn to technology for help. They might send a weather balloon (a type of technology) into the sky to monitor storm activity. Technology is not the same as science. How is it different?

Technology is the use of devices, methods, and scientific knowledge to solve practical problems. So technology can be an actual device, or it can be the way we use devices and knowledge. For example, a doghouse is a technology that solves the problem of providing shelter for an outdoor dog. The hammer you might use to build the doghouse is technology, too. So are materials you use, and how you decide to join them together.

Both of these doghouses solve the problem of keeping a dog comfortable and dry. How are the doghouses similar? How are they different? Practical problems usually have many possible technological solutions.

How Are Science and Technology Related?

Science and technology are different, but they are also related. Suppose you build a robotic fish or a group of robotic ants. Are you solving technological problems, or carrying out scientific investigations? The answer depends on the question you are asking and the reason you are asking it.

Technology Can Aid Scientific Inquiry

The photograph on the left shows RoboTuna. Researchers invented RoboTuna to help them answer questions about how fish swim. The tuna is one of the world's fastest swimmers. It can swim as fast as 80 km/h. Why can it move so quickly? The researchers created a technological device, RoboTuna, to help them find out. The scientific knowledge the researchers gain using RoboTuna also has a technological benefit. Researchers are using their new scientific knowledge of the tuna to help them build better submarines.

Is RoboTuna (a robotic tuna) an example of science in action or technology in action?

Science Can Aid Technological Problem Solving

These mini-robots work as a team like real ants. Are they science in action or technology in action?

This photograph shows real ants interacting with tiny robotic ants. Researchers invented them to solve a practical problem: How can robots work as a team to reach a goal? The researchers modelled the robots on real ants. Scientists have discovered that ants work together to do complex tasks without any one ant serving as "leader." The robots may benefit both the general public and scientists. The researchers hope to create colonies of robots to work in unsafe places such as disaster areas. The robots could also explore extreme environments such as the Moon and Mars.

The Processes of Science and Technology

The concept map on the left shows a process for answering scientific questions. The concept map on the right shows a process for solving technological problems. You will use these processes often in ***BC Science 6***, as well as in future studies. You can also find more information in SkillPower 5 (Designing and Conducting Experiments) and in SkillPower 6 (Technological Problem Solving).

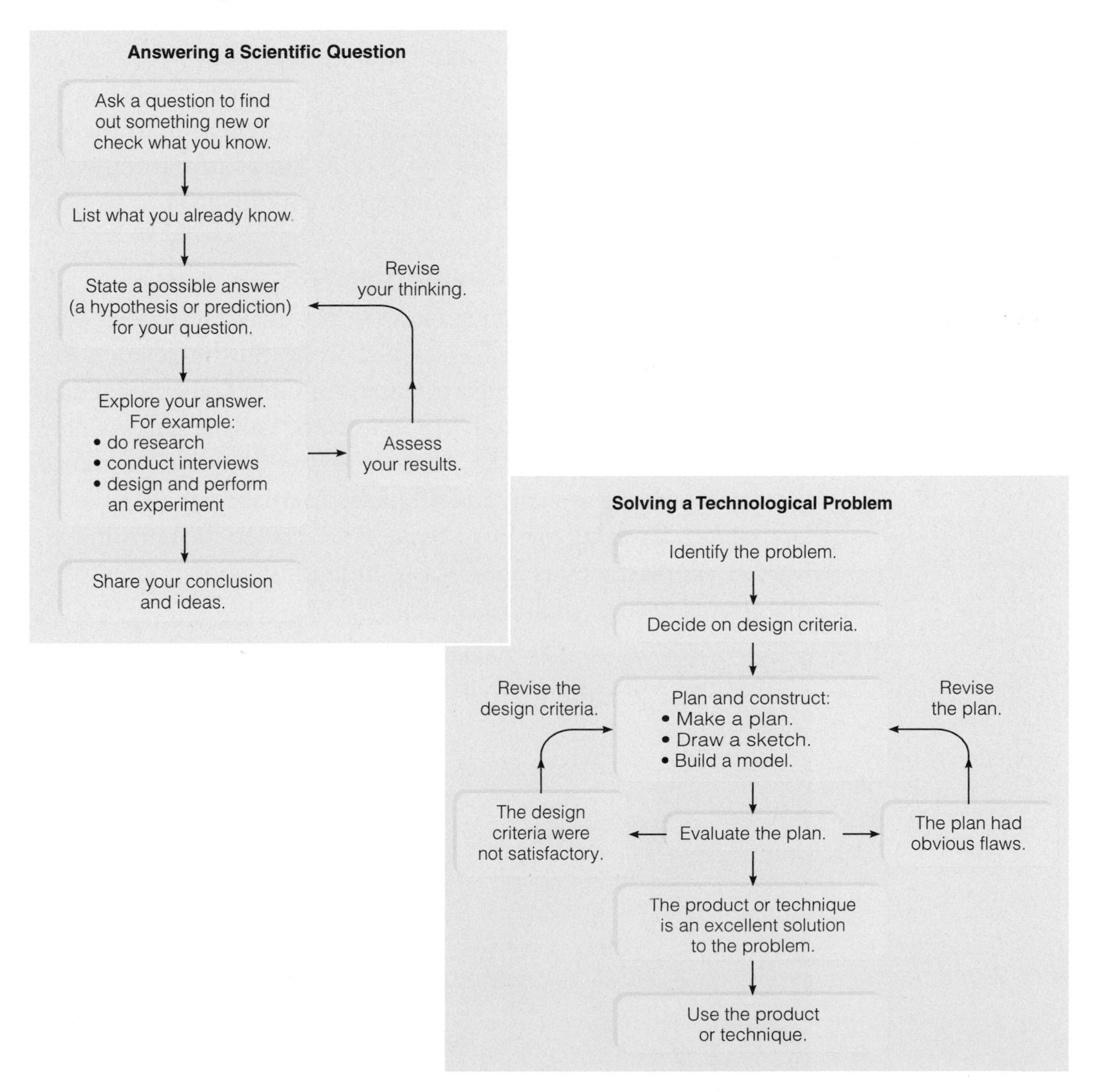

Science and Technology Affect Society and the Environment

All the people who live together in a certain place and at a certain time make up a **society**. The knowledge and devices that we create affect our society. Sometimes society is affected in a helpful way. For instance, plastic is a human invention. Scientists developed plastics using their understanding of substances. Plastics are cheap to produce, and they do not rust or fall apart easily. As a result, people can use plastics to make all kinds of useful products such as drink bottles, packaging, and building materials.

The invention of plastic has helped society, but it also has harmful effects. Too often, people throw away plastic in their garbage. Because plastic does not rust or fall apart naturally, tonnes and tonnes of plastic build up in landfill sites.

A Creative Solution: Paving the Way

A young student from Prince George, B.C. saw that there was a problem with plastic waiting to be solved. So she decided to solve it. After months of experimentation, Gina Gallant invented a new material to use for paving roads. She calls it PAR, which stands for PolyAggreRoad. PAR is a mixture that is 6 percent plastic, 6 percent asphalt, and 88 percent crushed rock (called aggregate). By using plastic to make PAR, society benefits in two ways. Less plastic ends up as waste in landfill, and an effective material is used to build roads.

Gina Gallant has earned attention from around the world for her PAR material. An inventor since grade 1, she created PAR when she was in grade 7. She continues to improve on her product.

Issues: Problems with Many Points of View

Some science and technology problems are easier to solve than others. For example, people may have many ideas about the best solution to problems such as pollution, how to help an endangered species, or how to provide enough electricity for everyone. These complex problems are called issues. The diagram below shows one example of the ways that science, technology, and society overlap in an issue.

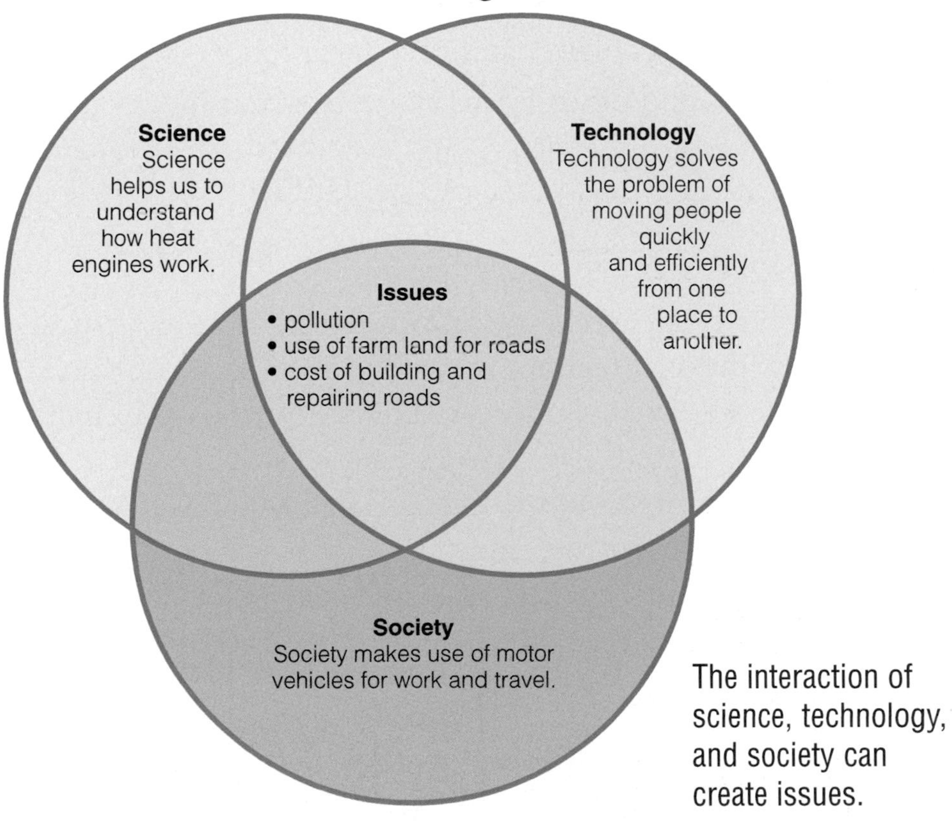

The interaction of science, technology, and society can create issues.

People often have many points of view about the best solution to an issue. There are usually both risks and benefits for any possible solution. As a result, issues are often very challenging to analyze. When issues involve science, technology, and society, you can compare the risks and benefits better if you understand the science and the technology. In ***BC Science 6*** you will have a chance to use your knowledge of science and technology to help you analyze issues.

Aboriginal Technologies

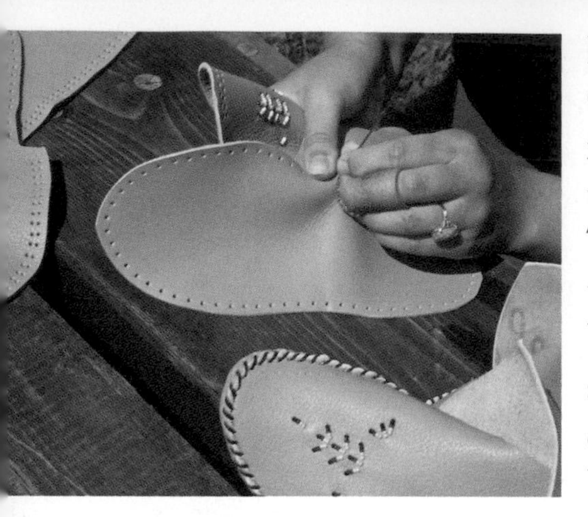

Tanned moose hide can be used to make long-lasting moccasins, clothes, snowshoes, and shelters.

Did you know that the Aboriginal Peoples of British Columbia developed and used many technologies in their daily lives? They made great cedar canoes for traveling over hundreds of kilometres of open ocean. They made tiny needles of bone and antler for sewing together and decorating moose-hide clothing. They peeled bark from living trees in ways that did not harm the trees, and prepared that bark to make products such as rope, clothing, and fine baskets. Some of the techologies created by Aboriginal Peoples include:

- tools for gathering, cooking, and preparing foods
- designs for boats and other forms of transportation
- techniques for creating works of art
- methods of harvesting foods and medicines
- methods of preparing materials for clothing and shelter

For example, the cedar canoe that you see below was made using a number of traditional technologies. The makers chose and felled a suitable log from the forest. They split the log in half using specially made wedges (A). Next, they shaped the outside of the log (B) and hollowed out the inside. Then the makers placed water and hot rocks inside the canoe. The canoe's sides could thus be steamed and stretched apart to make the boat wider (C). Finally, the finished boat was launched in a traditional ceremony (D).

This Nuu-chah-nulth canoe was made near Tofino on the west coast of Vancouver Island by Joe Martin of Tla-o-qui-aht and others. The Martins continue to use many of the same methods that their ancestors developed over many generations.

Aboriginal Technologies Today

Many Aboriginal technologies are still used today. In the northern interior of British Columbia, for example, large game animals such as moose and caribou are available only in certain seasons. Peoples such as the Sekani, Dunne-za, and Dakelh developed methods of drying meat so that large amounts could be preserved and eaten year round. The meat product was very nutritious and light enough to be carried on long trips. It remains an important food today.

This hunter is hanging moose meat to dry.

Other Aboriginal technologies are today being rediscovered. At Comox Bay on the east coast of Vancouver Island parts of a K'omoks fishing system are being mapped. Some of the stakes that supported the fences for this fishing system are over 1200 years old. Others are only 75 years old. They cover at least four square kilometres of beach.

The K'omoks People developed this complicated maze of fish fences to catch the types and sizes of fish they needed over many years. Fish that were too small, or the wrong species, could be easily released. Using this type of fishing technology, the K'omoks made sure that too many fish were never taken. This practice maintained a good relationship between fish and humans.

These two men are mapping some of the thousands of wooden stakes that were a part of a huge K'omoks fishing system.

Aboriginal Peoples of British Columbia

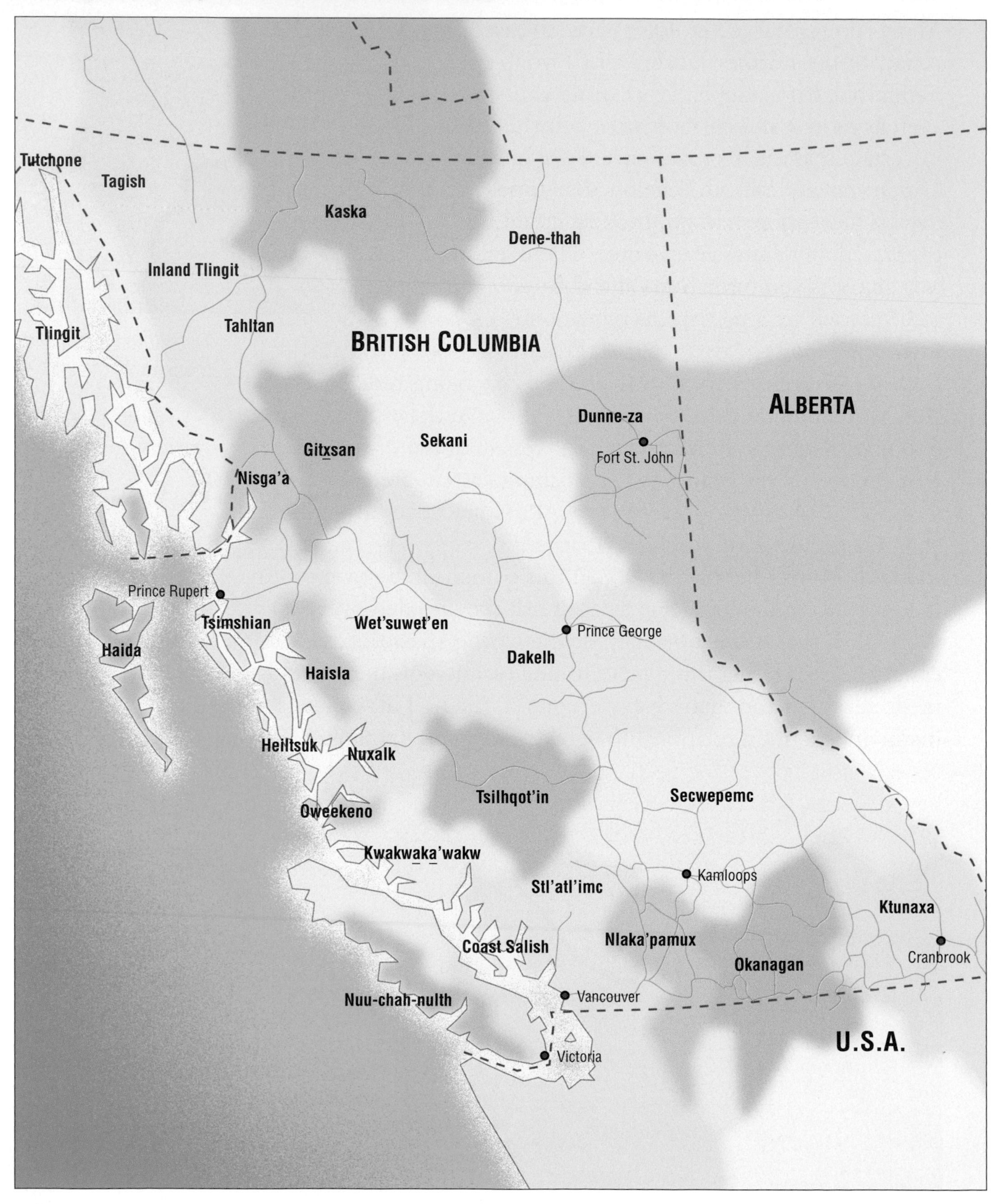

Different Technologies for Different Environments

Look at the map on the facing page. From hot, dry semi-desert areas, to wet, cool rainforest areas, British Columbia has many different environments. Over thousands of years, Aboriginal Peoples have discovered ways of using the natural materials of these environments. They have developed new technologies and improved old technologies to suit the different environments where they live. Aboriginal Peoples of interior British Columbia, where birch trees grow, for example, developed technologies using birch wood and bark. The baskets shown on the right are examples of useful products that interior Peoples developed from birch bark. Aboriginal Peoples of coastal British Columbia, where cedar trees grow, used cedar wood and bark. The bentwood box shown below is one example of a practical product that coastal Peoples made from cedar.

Dr. Mary Thomas of the Secwepemc People made these birch-bark baskets from the outer bark of paper birch, western red cedar root, and saskatoonberry wood.

Exchanging Technologies

Aboriginal Peoples also built a network of trade routes. They used these routes to trade goods and raw materials throughout the province and beyond. The Haida, for example, traded bentwood boxes. They traded for food products that did not exist on Haida Gwaii (the Queen Charlotte Islands), such as eulachon grease. (Eulachons are small, oily fish.) People also traded methods and ideas for technologies. For example, the Nuxalk People of the central coast of British Columbia learned how to make birch bark baskets and canoes from the Tsilqot'in and Dakelh Peoples of the central interior of British Columbia.

The Haida People of Haida Gwaii have long been known for their woodworking technologies. They traded bentwood boxes with the Tsimshian People of the mainland coast for eulachon grease and other animal products.

Spirituality and Aboriginal Technology

The Aboriginal Peoples of British Columbia used knowledge, experience, experimentation, and communication to develop their technologies. They also used spiritual methods. The Nuu-chah-nulth, for example, used a method of gaining knowledge, vision, and strength known as oosumich. Oosumich involved long periods of mental and physical preparation, and bathing in cold ocean, lake, or river waters.

Similarly, the Tahltan people of north-western British Columbia gained knowledge through sweat lodge ceremonies before going on hunts. Anyone who claimed to have gained knowledge through oosumich or a sweat lodge ceremony had to demonstrate this knowledge before it could be accepted.

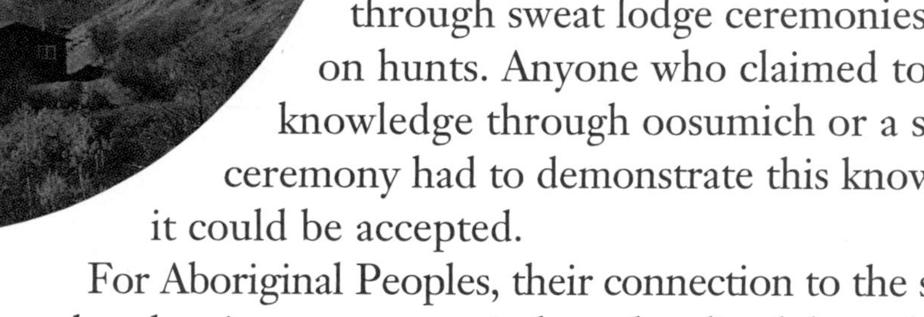

The Tahltan community of Telegraph Creek is shown here.

For Aboriginal Peoples, their connection to the spiritual world played an important part in how they lived, how they completed certain tasks, and how they solved problems using technology.

Snowshoes are an Aboriginal technology that many people use today.

Aboriginal Technologies in Everyday Life

Every time you go fishing on the ocean or the river; every time you wear a parka or leather mittens; every time you prepare certain foods or take certain medicines; every time you ride on a sled or in a canoe, you are using Aboriginal technologies.

There are other Aboriginal technologies that people are only beginning to understand today. Some examples are the K'omoks system of fish fences, or the Nuu-chah-nulth and Tahltan methods of connecting to the spiritual world. In the future, some of these methods may also become a part of the technologies once again used in peoples' everyday lives.

EXPLORING Further

Inspiration for New Technologies

How do people develop new technologies? A story from the Tahltan People tells of a hunter roasting a duck over a fire and noticing the shape of its breastbone. Planning to make a new pair of snowshoes, the hunter realized that the shape of the breastbone would work. It would shed the snow easily and not get stuck as it moved uphill. As a result of this hunter's observation, a new design of snowshoe came about.

How would you solve the problems of getting food, making a warm shelter, and finding a way to travel using only the natural materials near your home? Write down your ideas. Explain why you think each idea would work.

With the permission of your teacher or guardian, why not try out some of your ideas? You might find that you have thought of some of the same things that Aboriginal Peoples of your area tried many years ago.

The photographs below show some traditional Aboriginal technologies.

Mabel Joe, of the Nlaka'pamux People near Merritt, British Columbia, is shown here with a berry-drying mat she made from common reed grass.

Elder Sam Mitchell, of the Stl'atl'imc People near Lillooet, British Columbia, uses a root-digging stick he made from a mule deer antler. He is digging for mountain potato at Pavilion Mountain. The mountain potato was a staple food for many Interior Salish and Tsilhqot'in peoples of south central British Columbia.

Florence Davidson, a Haida elder, makes a hat out of spruce roots.

UNIT 1

Diversity of Life

Imagine you are an eagle soaring over British Columbia. You start your journey in the Rocky Mountains, above this meadow full of flowers. As you fly westward, you pass mountains and valleys, rivers and lakes, forests and grasslands. Finally you reach the coast of the Pacific Ocean and see waves sweeping the shore. Your flight has shown you that British Columbia has more different natural regions than most provinces in Canada. Each region has its own geography, climate, plants, and animals. This unit will introduce you to the variety of living things that make their homes in these different regions.

If you were a scientist, how would you learn more about the flowers growing in this meadow? One way would be to use a microscope to look at parts of the flowers and their seeds. This unit will show you how to use a microscope.

These flowers grow in a place that is covered with snow for many months of the year. When the snow melts, the flowers bloom quickly. Why are these plants able to live high in the mountains? Would these plants be able to grow in the dry grasslands in the province's interior? You will find out how living things are able to live in a particular environment. You will also learn how scientists classify all living things to help us understand the natural world.

Unit Contents

Chapter 1
A Closer Look at Living Things 4

Chapter 2
Living Things and Their Environment 34

Chapter 3
Classifying Living Things 60

Getting Ready...

- How does a tree living in northern British Columbia survive in a place that is cold and dry for many months of the year?
- A rose, a mushroom, and a wolf have some similar characteristics. Can you list what they are? Can you list the characteristics that make them different?
- What tools could you use to learn more about living things in the world around you?

CHAPTER 1

A Closer Look at

Getting Ready...

- How would you decide if something is alive?
- An ant and an oak tree are both alive. How are they alike?
- How would a magnifying glass help you learn about living things?

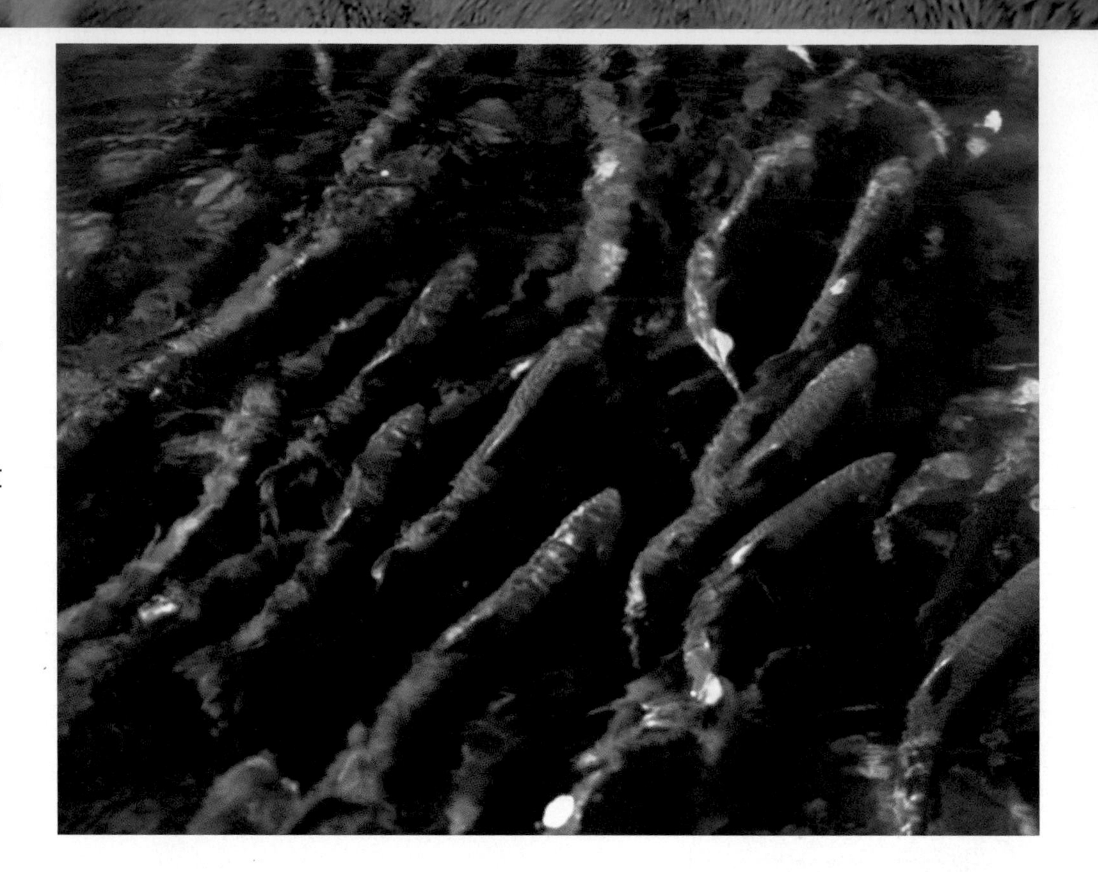

Each year, mature salmon leave the ocean and enter fresh-water streams. The salmon swim up stream, and many even travel far into the interior of British Columbia. At the end of their journey, the salmon spawn, and then they die. Animals, from eagles to bears to insects, eat the dying and dead salmon. Their bodies also supply food for even more living things in the streams and nearby forests.

A salmon is a living thing. It eats, reproduces, and moves. A stream moves, too. Is a stream a living thing? The characteristics that mean something is "alive" are what you are about to explore.

The salmon and all living things are made of cells, the basic (first) units of life. Most cells are so small that you cannot see them easily. What can you find out about cells? Are there differences between plant cells and animal cells? What tools could you use to find out more about cells? In this chapter you will take a closer look at cells, using tools of science.

Living Things

What You Will Learn

In this chapter, you will learn

- the main activities of living things
- about cells, the basic units of life
- how to use a microscope

Why It Is Important

- When science helps you understand the activities and needs of animals and plants, you are better able to see how all parts of nature—including ourselves—depend on one another.
- Microscopes are useful tools of science—they introduce you to a world of things you cannot see with just your eyes. You will use microscopes more as you advance in school, so it is important to learn how to use them correctly.

Skills You Will Use

In this chapter, you will

- design an experiment with a control and variables to see how changes in the environment affect the growth of a seed
- develop a model of a cell
- use a hand lens and a microscope to observe details you cannot see with just your eyes

A glacier moves slowly. Is it a living thing?

Starting Point **ACTIVITY 1-A**

It's a Live One!

How would you describe the differences between a living thing and a non-living thing?

What to Do

1. With a partner, list the things shown above under two headings: Living Things and Non-living Things.
2. Discuss how the things in the pictures are similar and how they are different.
3. Choose one thing that you have listed as living and one that you have listed as non-living. Prepare a Venn diagram that shows the similarities and differences between the two things.

What Did You Find Out?

1. Write a list of the details that tell you that living things are alive. Share your list with your classmates.
2. A glacier can move, and it can also grow larger (if temperatures are cool enough for snow to pile up). Is a glacier a living thing? Explain your answer using the details you listed in question 1.

Skill POWER

For tips on making a Venn diagram, turn to SkillPower 1.

Section 1.1 Living Things

Key Terms
cell
organism
photosynthesis
reproduction

Defining Living Things

As you will read in this section, in biology the differences between living and non-living things are clearly defined. However, the scientific method is only one way of defining living and non-living things.

The Aboriginal peoples of British Columbia define animals and plants as living things, and they also define natural features such as mountains, rocks, and streams as living things. For Aboriginal peoples, all parts of the natural environment are interconnected and related to one another. The Salish people of southwestern British Columbia, for example, know certain rock formations and mountains as their relatives.

Figure 1.1 Coyote Rocks are living parts of the Secwepemc world.

Many Aboriginal stories tell of people or animals changing form. Secwepemc stories from southcentral British Columbia, for example, tell of Coyote introducing salmon to the Secwepemc people and creating important fishing sites. They also tell of Coyote being foolish and not completing his work. They tell of the "Old One" transforming Coyote into the Coyote Rocks that can be seen today throughout Secwepemc territories. (See Figure 1.1) Both rocks and animals like Coyote are living, related parts of the Secwepemc world.

Would you define the rock crystals in Figure 1.2 as alive? In this section you will read about the ways that living things are defined in science.

Figure 1.2 This rock crystal candy "grows" from a sugary liquid as it cools, but it is not alive. What *is* the difference between living things and non-living things?

Living Things Are Made of Cells

Cells are the basic units of life. We call them the "basic units" because all living things (**organisms**) have at least one cell. Some organisms, like the bacteria in Figure 1.3, are made up of only one cell. This single cell can do everything that is needed to keep the organism alive. Other living things—like you—are made up of more than one cell and have many different kinds of cells. You will learn more about cells in Section 1.2.

Figure 1.3 This cluster of bacteria cells shows a group of organisms that have only one cell.

Living Things Exchange Gases

Living things use gases, including oxygen gas and carbon dioxide gas. When the whales in Figure 1.4 surface to breathe, their bodies get rid of carbon dioxide gas as waste and take in oxygen gas. Taking in oxygen gas and releasing carbon dioxide gas is known as respiration. All animals do this. During the day, plants use carbon dioxide gas to make food. They release oxygen gas as waste. This food-making process is called **photosynthesis** [foh-toh-SIN-thuh-sis]. Plants use oxygen gas, too. At night, when photosynthesis does not occur, plants respire. They take in oxygen gas and release carbon dioxide gas. Respiration and photosynthesis are both ways of exchanging gases.

Make a drawing that shows how oxygen gas and carbon dioxide gas each move in and out of plants and animals.

Figure 1.4 When a whale surfaces, it exhales carbon dioxide gas and inhales oxygen gas.

At Home **ACTIVITY 1-B**

Diversity in British Columbia

An incredible variety of organisms live in British Columbia. Over 35 000 kinds of insects, 454 types of birds, and 143 kinds of mammals call this province home. Let's take a close look at one animal found in the province.

What to Do

Choose one animal that lives in British Columbia. Conduct research to find out what the animal eats, how it catches or gathers food, where it lives, what enemies it has, and any other information that describes the way it lives.

What Did You Find Out?

Create a song, poem, play, poster, or computer presentation on the animal you have chosen. **Report** your findings to your class.

INTERNET CONNECT

www.mcgrawhill.ca/links/BCscience6

To learn more about the variety of life in British Columbia, go to the web site above. Click on **Web Links** to find out where to go next. You can use this web site to help you with Activity 1-B.

Living Things Use Energy

All living things need gases, water, and food to help make the energy they use to live. Animals eat food, while plants make food. Food provides living things with nutrients such as fats, proteins, and carbohydrates. Living things need nutrients and energy for growth, repair, and other activities.

Animals eat a variety of foods and gather it in many different ways. For example, an earthworm (which lives underground) finds its food in the soil, and a barnacle (which lives underwater), like the one in Figure 1.5, strains food out of the water. Some animals hunt and eat other animals. Some animals, like elk and bighorn sheep, eat only plants.

During photosynthesis, plants use water, carbon dioxide gas, and the energy of the Sun to make food. The food (sugar) is stored in the plant. The plants use some oxygen gas to break down the sugar to provide them with energy.

Figure 1.5 Barnacles use feathery, leg-like structures to filter food out of the ocean.

It is not always that easy to decide how some living things obtain the food they use to make energy. For example, many fungi, such as the mushroom shown in Figure 1.6, use decaying plants as a source of food energy.

Figure 1.6 Many kinds of mushrooms grow in British Columbia forests. They absorb nutrients from rotting plants in the soil.

Living Things Grow

All living things grow. Growth can mean an increase in size, just as you grow larger as you get older. Or growth can mean that living things create new cells to replace cells that are dead or damaged. Every day, skin cells come off your body as new skin cells grow. When you have a cut, your body heals itself and may create a scar made of new cells. The sea star in Figure 1.7 has lost an arm, but this injury did not kill it. Instead, new cells were formed at the end of the arm, and the sea star is growing a new one.

Figure 1.7 Living things, such as this sea star, can grow new cells to repair themselves.

How are plants and animals similar? How are they different?

Some animals can drop a body part when they are in danger, usually when a predator comes just a little too close. In British Columbia, young western skinks will drop their bright blue tails to confuse predators. Over time, a skink can grow a new tail.

Did You Know?

Most household dust is made up of dead human skin cells. You and everyone around you continually shed cells from the thin outer layer of your skin. Your entire outer layer of skin is completely replaced by new cells approximately every 28 days.

Living Things Reproduce

All living things die. Therefore all living things must reproduce so that their species will survive. **Reproduction** is how living things produce offspring.

Living things reproduce in many different ways. For example, one way in which the sea anemone in Figure 1.8 reproduces is by budding tiny anemones off its body.

Some animals, like birds and frogs, lay eggs, such as those shown in Figure 1.9. Others, like bears, horses, and humans, give birth to live young.

Figure 1.8 This sea anemone reproduces tiny copies of itself by budding. Can you see them?

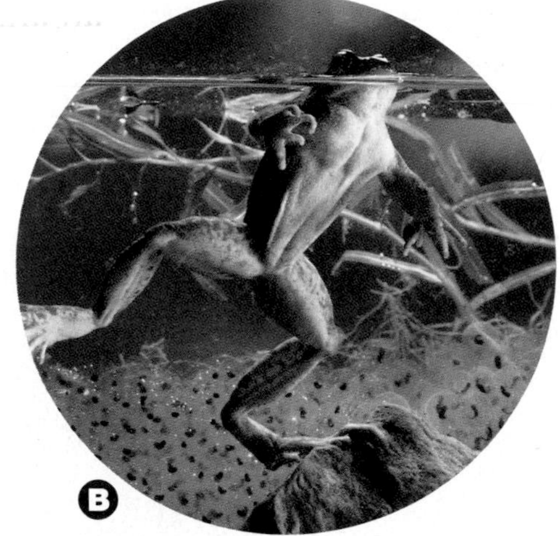

Figure 1.9 Birds (A) and some animals such as frogs (B) lay eggs when they reproduce.

Plants reproduce in different ways. They can produce spores or seeds, such as those shown in Figure 1.10.

What is the definition of the word "reproduction"? Use a dictionary to look up the prefix *re-* and the word "production."

Figure 1.10 Ferns reproduce with spores (A). Other plants, such as the tomato, make seeds (B).

Living Things Respond to Stimuli

Stimuli are situations or actions that cause a reaction. When you are cold, you shiver. The temperature makes you react. When a plant senses the Sun, it turns toward it. These are both ways in which living things respond (react) to stimuli.

Living things can react to stimuli such as touch, light, temperature, or sound. Sometimes an animal's brain will tell it to react. (If you burn your hand, your brain will tell you to pull your hand away from the heat.) Many reactions happen without a brain's instruction. A reaction can happen quickly or slowly. It can be easy to see—a flower like the Morning Glory in Figure 1.11 closes at night—or hard to see—an animal's heart beats faster when it is frightened.

Movement is not always a response to stimuli. A leaf moving in the wind is not reacting—the wind is moving it. Here, something outside the organism is causing the movement. However, a leaf on a plant that is turning toward the Sun is moving because the plant is reacting. Here, the organism is causing the movement.

See how seeds respond to changes in heat, water, and temperature in Investigation 1-C.

Stimuli is the plural of the word "stimulus." Use your dictionary to find other words related to stimuli.

Pause & Reflect

In question 2 of the Starting Point Activity, you were asked if a glacier is alive. (A glacier can move and grow larger. Is it a living thing?) Answer this question again in your science notebook.

Figure 1.11 Flowers like the Morning Glory open in daylight and close up tight when it gets dark.

CONDUCT AN

INVESTIGATION 1-C

SKILLCHECK

- Observing
- Controlling Variables
- Interpreting Data
- Communicating

Get Growing

In this investigation, you will change the temperature or the amounts of water or light that are available to plant seeds. How do you think changing these stimuli will affect how the seeds grow?

Questions

How does changing the temperature affect the growth of seeds? How does changing the amounts of water or light affect the growth of seeds?

Safety Precautions

Apparatus

nail
4 waterproof dishes or plates
waterproof felt marker
measuring cup
cardboard box
bucket

Materials

4 polystyrene cups or other small containers
masking tape
12 bean seeds
potting soil
water
ice

Procedure

Group Work

1. Use the nail to poke three or four small holes in the bottom of each cup or container.
2. Use the masking tape and felt marker to label your cups: Control 1, Control 2, Test 1, Test 2.
3. Fill each cup with potting soil and plant three seeds in each cup.

4. Place your cups labelled Control 1 and Control 2 near a sunny window in your classroom. Give each cup 60 mL of water every second day. These cups will get a day's worth of light and be at room temperature. The results in the test cups will be compared to the results in these control cups.
5. With your group, choose whether you will study the effect of light, water, or temperature on seed growth.
   - **If your group is studying the effect of light on growth:** Expose the cups labelled Test 1 and Test 2 to 1 h of light per day. For the rest of the day cover them with a cardboard box or put them in a dark place at the same temperature as the control cups. Give each cup 60 mL of water every second day.

- **If your group is studying the effect of temperature on growth:** Put the cups labelled Test 1 and Test 2 in a bucket of ice for 2 h every day. These cups should be exposed to the same amount of light as the control cups. Give each cup 60 mL of water every second day.
- **If your group is studying the effect of water on growth:** Decide whether you will give your cups labelled Test 1 and Test 2 more or less than 60 mL of water. Give each cup the amount of water you have chosen every second day. Keep these cups in the same location as the control cups.

6. Create a data chart in your notebook like the one shown below. Give your chart a title.

Date	Control 1	Control 2	Test 1	Test 2

7. When your group's seeds have sprouted, **observe** and **record** in the data table what the plants look like as they grow. Include sketches.

8. **Observe** and compare the results from your group with the results from the other groups.

Pause & Reflect

Think of what you know about the plants and climate of British Columbia. Do all plants need the same amounts of water or light? Do all plants grow in the same temperature? Explain your answer.

READING Check

The temperature and the amounts of water and light were variables in this investigation. What is a variable?

Analyze

1. Write a question for your experiment. For example, if you studied how the amount of water affected growth, a possible question might be "Can a bean seed grow with only 10 mL of water a week?" Write a question for the variables that you did not test as well. (For example, if you tested light, write questions for water and temperature.)
2. Which of the four groups (control, light, water, temperature) had the best growth? Why do you think this happened?
3. What factors other than water, light, and temperature might have affected the growth of the seeds?

Conclude and Apply

4. What is the purpose of having the control cups in your experiment?
5. In this experiment each group used four cups and 12 seeds. Why was it important to plant more than one seed?
6. Could the temperature group be placed in a freezer? Why or why not?
7. Based on your observations, answer the questions you wrote for question 1.

Extend Your Skills

8. Repeat the experiment using seeds from a different kind of plant and compare your results.

Skill POWER

To learn more about the use of *controls* and *variables* in experiments, turn to SkillPower 5.

Section 1.1 Summary

In this section, you learned to define living things. Living things are also called "organisms."

- There are different ways of defining living things. For example, Aboriginal peoples consider all parts of the natural environment to be interconnected and living.
- All living things are made of cells, exchange gases, use energy, grow, reproduce, and respond to stimuli.
- A cell is the basic unit of life.

Key Terms

cell
organism
photosynthesis
reproduction

Check Your Understanding

1. What are the differences between living things and non-living things?
2. **(a)** Name the two gases that living things exchange.
   (b) What is the difference between how animals exchange these gases and how plants exchange these gases?
3. **(a)** Explain the process called photosynthesis.
   (b) Name the three things that are necessary for photosynthesis to occur in plants.
4. **(a)** Describe three different ways that plants respond to stimuli.
   (b) Describe three different ways that animals respond to stimuli.
5. **Apply** Suppose you want to observe the effect of a change in stimuli. You set up an investigation with identical plants as shown in the chart below. Describe the changes you think will happen to each plant and explain why.

	Plant 1	Plant 2	Plant 3	Plant 4
sunlight	yes	no	yes	no
water	no	no	yes	yes

6. **Thinking Critically** Explain how a human being depends on energy from the Sun, air, water, and soil.

Section 1.2 A Cell: The Basic Unit of Life

Figure 1.12 An adult giant Pacific octopus is made of trillions of cells.

Figure 1.13 Each of these growing octopi began life as a single cell. When they hatch, they will have hundreds of cells.

Key Terms

organelles
nucleus
cytoplasm
mitochondria
vacuole
endoplasmic reticulum (ER)
cell membrane
cell wall
chloroplast
chlorophyll
unicellular
micro-organism
multicellular

The cluster of eggs in Figure 1.13 was laid by a female giant Pacific octopus. She glues up to 300 egg clusters on the ceiling of her den, using a sticky solution that comes from her mouth. Each cluster of eggs will produce up to 200 baby octopi.

A hatchling octopus weighs just 0.028 g at birth and is about the size of a grain of rice. An adult octopus like the one in Figure 1.12 can weigh up to 40 kg and stretch over 6 m in length. This huge animal is made of trillions of cells. What exactly is a cell? How do cells make life possible? These are some of the questions you will answer in this section.

DidYouKnow?

There are about 10 000 000 000 000 cells in your body. (That's ten trillion!)

What Is a Cell?

A cell is the basic unit of life because an individual cell carries out all of the life activities that you learned about in Section 1.1. For example, a cell grows and reproduces. It also uses energy, exchanges gases, and responds to stimuli. A cell has many different parts, called organelles, to do all of these jobs. These parts and the jobs they do are described on the next page.

Why is a cell called the basic unit of life?

INTERNET CONNECT

www.mcgrawhill.ca/links/BCscience6

Robert Hooke (1635–1703) was one of the first people to identify a cell. He saw a honeycomb-like pattern, which he called *cellulae*, meaning "little boxes," in a thin piece of cork. To see what Hooke saw, go to the web site above. Click on **Web Links** to find out where to go next.

Parts of Plant and Animal Cells

Organelles are the parts (structures) inside a cell that do specific jobs. In order to get all the jobs done, a cell must have different kinds of organelles. Most organelles are found in both plant cells (Figure 1.14) and animal cells (Figure 1.16). For example, most cells have a **nucleus**, the organelle that controls all of the activities in the cell. Other organelles are found only in a plant cell or only in an animal cell.

The onion cells shown magnified 100 times in Figure 1.15 are an example of plant cells. The human skin cells shown magnified 100 times in Figure 1.17 are an example of animal cells.

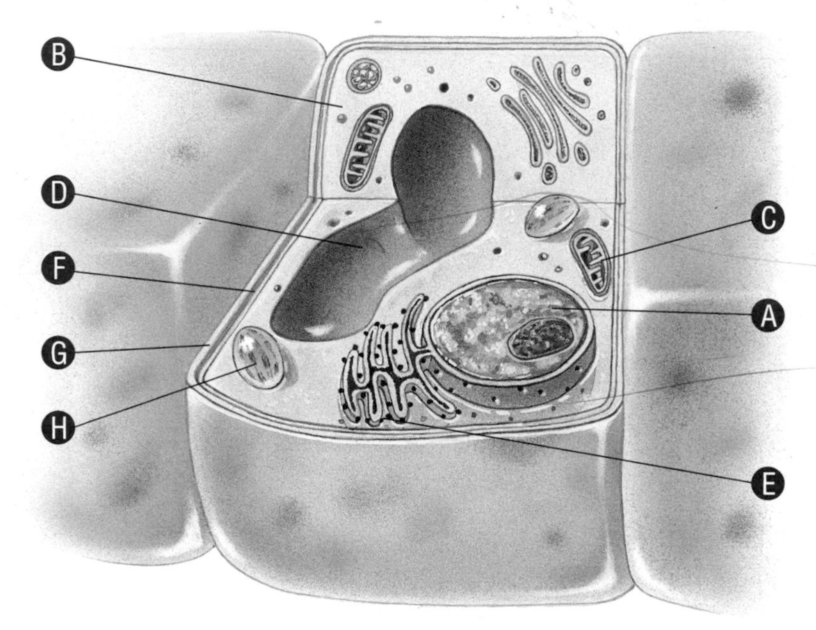

Figure 1.14 Parts of a typical plant cell

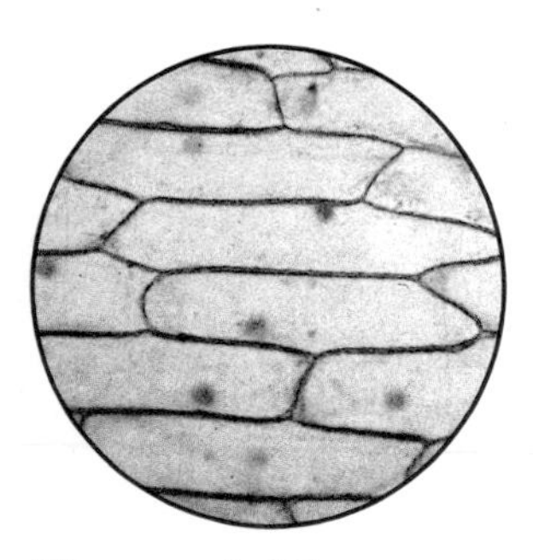

Figure 1.15 Plant cells (onion) enlarged 100×

Parts of a Cell

A. A **nucleus** controls the cell's activities. The nucleus is like the brain or control centre of a cell.

B. **Cytoplasm** holds the organelles in place. It is a clear, jelly-like material that surrounds the nucleus.

C. **Mitochondria** (singular: mitochondrion) are where the processes that produce energy for the cell occur. Cells have hundreds of mitochondria. These are the powerhouses of a cell.

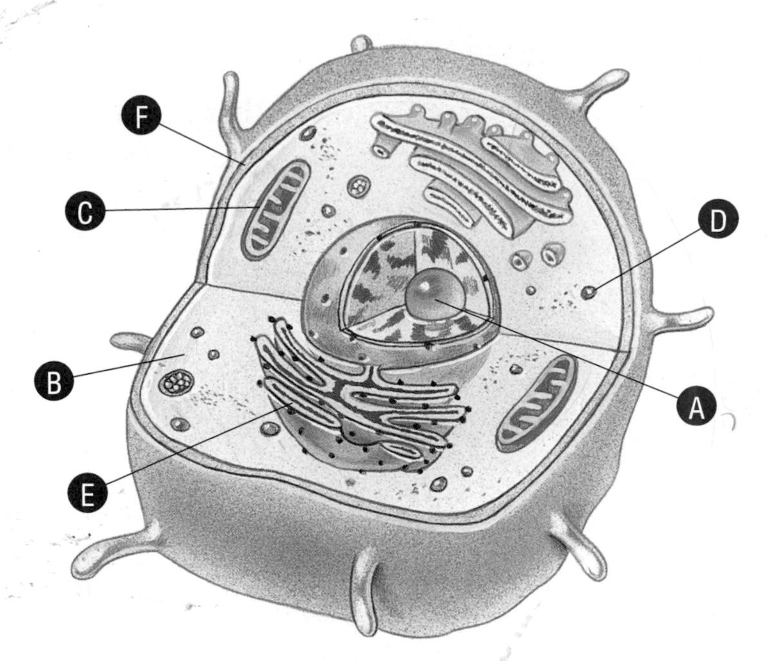

Figure 1.16 Parts of a typical animal cell

How can you tell a plant cell from an animal cell?

DidYouKnow?

When you eat celery, it makes a "crunch" sound. That is the sound of the cell walls breaking.

D. A **vacuole** stores water and food that the cell cannot use right away. It also stores wastes until the cell is ready to get rid of them. Plant cells usually have one large vacuole, while animal cells can have several vacuoles.

E. The **endoplasmic reticulum (ER)** transports food, water, and waste around and out of the cell.

F. The **cell membrane** controls what enters and leaves the cell. The membrane allows some things, such as food and water, through to the inside of the cell. It keeps out things that may be harmful to the cell. It also gives the cell shape and holds the cytoplasm.

G. The **cell wall** is a rigid structure that provides protection and support for a cell. It is found only in plant cells.

H. **Chloroplasts** contain a green colouring called **chlorophyll**. Chlorophyll traps energy from the Sun, which the plant uses to make food. It is found only in plant cells and gives plants their green colour.

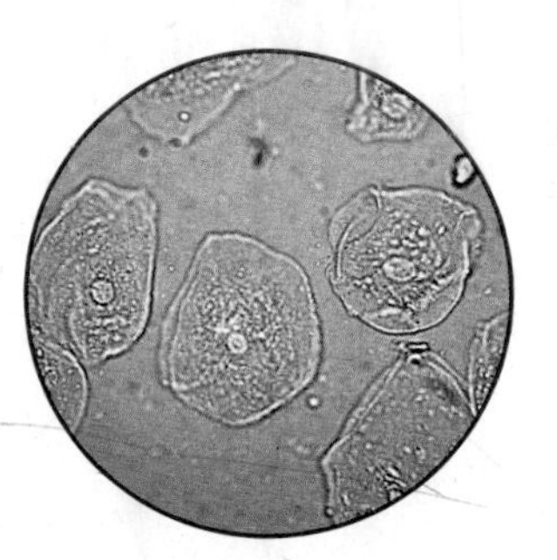

Figure 1.17 Animal cells (human skin cells) enlarged 100×

Try the activity on the next page to help you remember the parts of the cell and their functions.

Find Out **ACTIVITY 1-D**

Kitchen Cells

Scientists often use models to help them understand what they are studying. To help you understand the parts of a cell and the jobs (functions) they do, work in a group to make a model of a plant cell or an animal cell.

Safety Precautions

- Wash hands and work area thoroughly with soap and warm water before and after handling food.
- Never eat or drink anything in the science lab.

What You Need

Animal cell and plant cell:
piece of waxed paper (approx. 50 cm long)
small, self-sealing plastic bag
250 mL prepared gelatin
1 slice of banana
2 cherries
5 red grapes
1-cm wide strip of fruit leather

Plant cell only:
8 green grapes
500 mL plastic container

What to Do (Group Work)

1. Choose whether your group will make a plant cell or an animal cell.

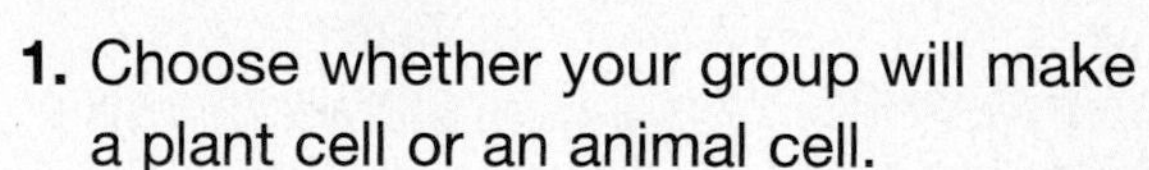

2. Cover your work area with the waxed paper and collect the materials needed to create your cell. Fold the fruit leather strip back and forth lengthwise so it is pleated like an accordion.
3. Your group will **develop a model** of a cell. Decide which ingredient will represent the following parts of the cell: nucleus, endoplasmic reticulum, mitochondria, cell wall (plant cell), cytoplasm, cell membrane, vacuoles, and chloroplast (plant cell).
4. In your notebook, create a data table like the one below. Give your data table a name.

Ingredient	Part of the Cell	Reason for Your Choice
plastic bag		
gelatin		
banana slice		

5. Help your group assemble the cell and complete your chart.
6. Draw all of the parts of your group's model cell. Label each organelle with its name and function.

What Did You Find Out?

1. Compare a model of a plant cell with a model of an animal cell. How are they the same? How are they different?
2. Compare your group's model with another group's model of the same type of cell. How are they the same? How are they different?

Skill POWER

For tips on making a data table, turn to SkillPower 4.

One Cell or Many?

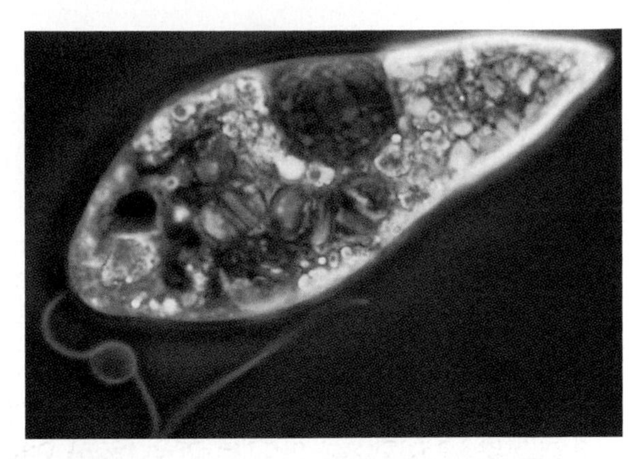

Figure 1.18 This colour-enhanced micrograph of *Euglena* shows a unicellular organism.

Some organisms are made up of only one cell. These are called single-celled, or **unicellular** (uni = one), organisms. Unicellular organisms are usually **micro-organisms**. These organisms are so small that you need to use a microscope to see them well. Bacteria (Figure 1.3, page 7) and the fresh-water *Euglena* (Figure 1.18) are examples of unicellular organisms.

Other organisms have many cells or are **multicellular** (multi = many). Although all the cells in a multicellular organism have a nucleus and the other organelles needed to function, they also have structures that help them do other jobs. A multicellular organism can do more complicated things than a unicellular organism can. This does not mean that multicellular organisms are always large. An ant and a giant red cedar tree are both multicellular organisms.

The Size of Cells

Cells grow to a size where they function best. When an organism grows larger, it usually does not grow larger cells. Instead, it adds more cells to its body.

Cells are measured in micrometres (μm). A micrometre is one-millionth of a metre. Most cells in plants and animals have a diameter between 10 and 50 μm. A cell of bacteria is much smaller. It is only 1–5 μm across. The sizes of typical plant, animal, and bacteria cells are compared in Figure 1.19.

READING Check

How do most multicellular organisms get bigger?

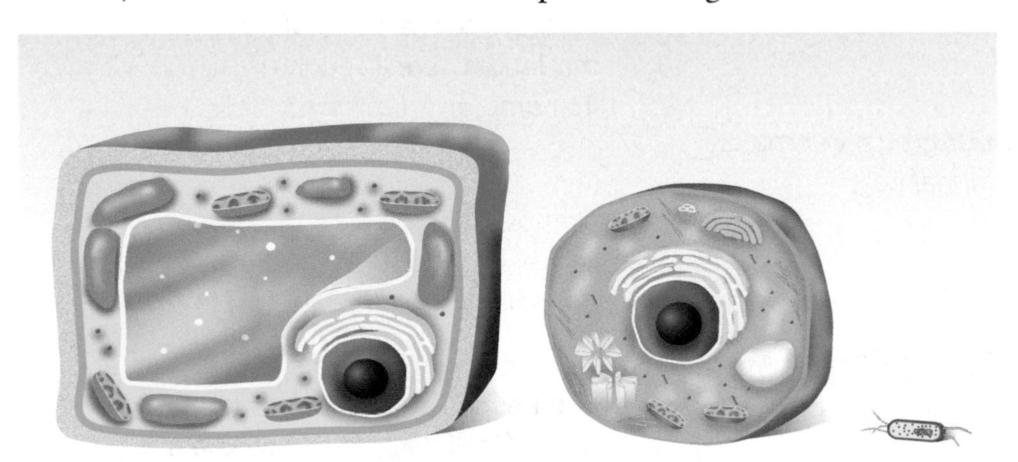

Figure 1.19 Comparison of the relative sizes of a plant cell, an animal cell, and a bacteria cell.

INTERNET CONNECT

www.mcgrawhill.ca/links/BCscience6

Take a virtual journey through a cell. Go to the above web site and click on **Web Links** to find out where to go next. You can zoom in, turn around, and check out different organelles inside a virtual cell.

Cell City

How is a cell like a city? Take a tour through this city in a cell to find out.

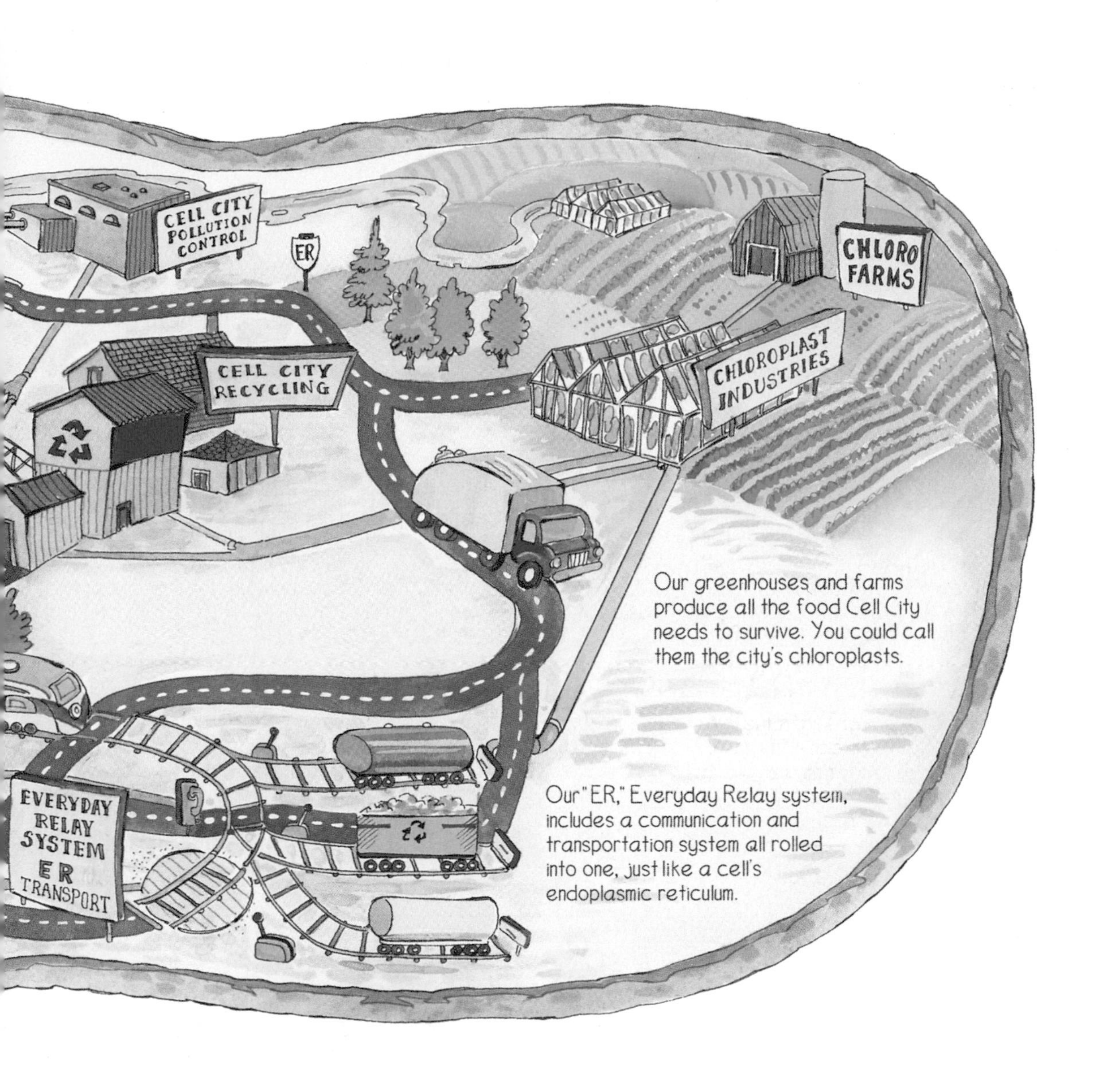

Cells are not always small. Some nerve cells can be as long as 60 cm. An ostrich egg, which can weigh almost 1 kg, is a single cell.

Does this cartoon model an animal cell or a plant cell? Explain your answer.

Section 1.2 Summary

All organisms—living things—are made of cells. Individual cells carry out all of the activities required for life, from consuming food to exchanging wastes and gases.

- Cells contain organelles—parts that do specific jobs for the cell.
- Cell walls and chloroplasts are found only in plant cells.
- Organisms can be unicellular or multicellular.
- The cells in large organisms are usually not much larger than the cells in small organisms. Instead, large organisms grow by making more cells.

Key Terms

organelles
nucleus
cytoplasm
mitochondria
vacuole
endoplasmic reticulum (ER)
cell membrane
cell wall
chloroplast
chlorophyll
unicellular
micro-organism
multicellular

Check Your Understanding

1. (a) Name one example of a unicellular organism and one example of a multicellular organism.
   (b) What are the differences between the cells of the two organisms?
2. Where is chlorophyll stored in plant cells?
3. Sketch the diagram below in your notebook and label the plant cell organelles: nucleus, chloroplast, vacuole. What are their functions?

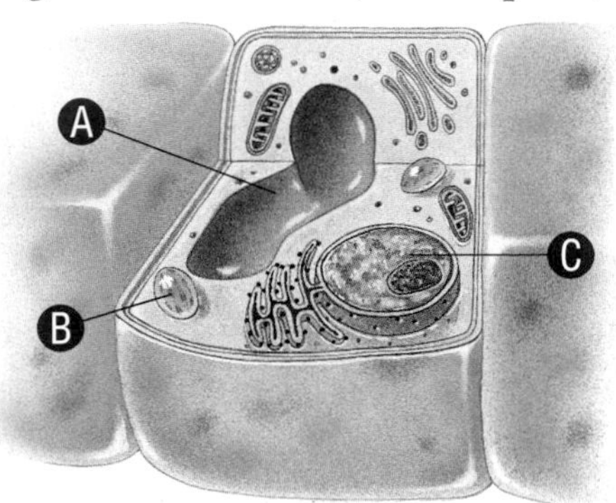

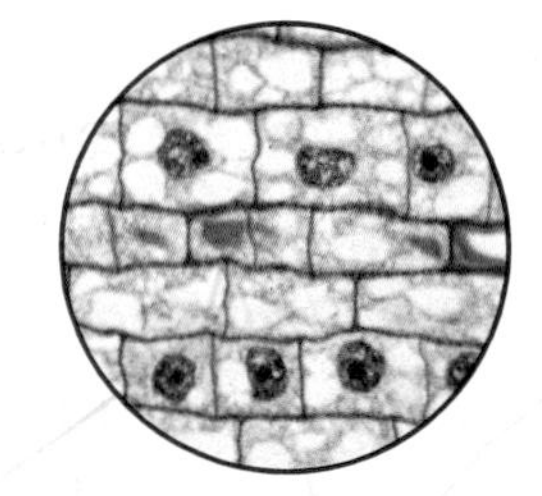

4. (a) What type of cells are shown in the micrograph opposite?
   (b) Explain your answer.
5. **Apply** Hamid wants to find out what happens to the cell walls in a plant if the plant does not get enough water. He knows that cell walls provide support for plant cells and give a plant its shape. What do you think will happen if he does not give the plant enough water? Start your sentence with "If."
6. **Thinking Critically** If a loaf of bread is left on the counter for a long time, mould will grow on its surface. What do you think will happen if another loaf of bread is stored in the same bread bag? Explain your answer by referring to the different ways that organisms reproduce.

Section 1.3 Tools of Biology

Figure 1.20 This pond is full of living organisms. Most of them are too small to see with just your eyes.

Key Terms
- magnify
- microscope
- slide
- cover slip

The water in the pond in Figure 1.20 is teeming with life that is too small to be seen with just your eyes. In fact, there are many living things around you that you cannot see at all. (Remember "micro-organisms"? Another way to describe them is to say that they're microscopic organisms.) Your eye can see only objects that are larger than 0.1 mm. Look at the circles shown in Figure 1.21. In some circles, you can see individual dots. Which circle looks like it has solid colour? Is the colour really "solid," or is it also made up of dots? For most of us to see separate dots, they must be more than 0.1 mm apart.

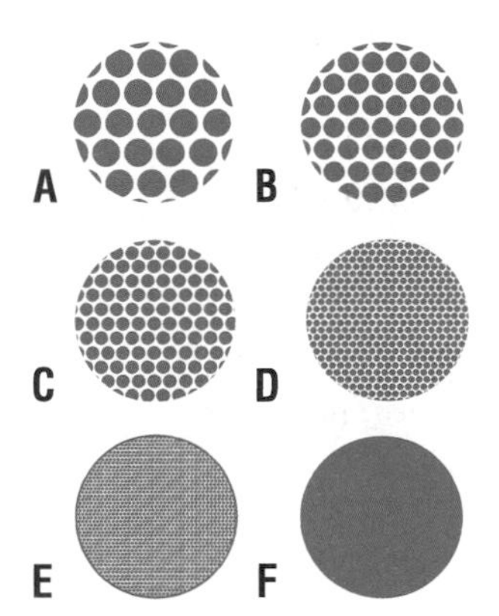

Figure 1.21 Are any of these circles solid, or are they all filled with dots?

Seeing a Microscopic World

In order to see microscopic organisms, you need to make them appear larger (**magnify**) them. A **microscope** is a tool that magnifies objects by bending light through a piece of curved glass (a lens). A lens can make an object appear many times larger, depending on how it is made. A hand lens (also called a magnifying glass) is a very simple type of microscope. Try out a hand lens in the next Find Out Activity.

What is a possible definition for "microscope"? Think of the words "micro" and "scope."

Find Out **ACTIVITY 1–E**

Look at the Big Picture

How can using a hand lens help you learn more about living things?

What You Need

hand lens
plant leaf
garden soil (approx. 25 mL)
layer of onion skin
pond water (approx. 50 mL)
petri dish
paper
pencil

What to Do

1. **Observe** the plant leaf with just your eyes (unaided).
2. Draw a circle in your notebook. Make a detailed drawing of what you see. Be sure your sketch shows size, colour, shape, and general appearance. For example, is it rough or smooth? Label this drawing "What I see with my unaided eye."
3. Use the hand lens to **observe** the plant leaf. Repeat step 2 but label your drawing "What I see using a hand lens."
4. Repeat steps 1–3 with the soil sample, the onion skin, and the pond water. Before viewing the pond water, pour a tiny amount in the petri dish. There should be a thin layer covering the bottom of the dish.

What Did You Find Out?

1. What did you see when you used the hand lens that you couldn't see with your unaided eye?
2. Describe how a hand lens can be a useful tool for learning about living things.
3. Were there things you couldn't do using the hand lens?
4. **(a)** Did you see any plants in your pond water sample? Explain how you know they are plants.
   (b) Did you see any animals in your pond water sample? Explain how you know they are animals.

THINK & LINK

INVESTIGATION 1-F

SKILLCHECK
- Observing
- Classifying
- Interpreting Observations
- Making Inferences

Take a Closer Look

In order to study organisms (or parts of organisms) that are too small to see with the unaided eye, scientists use microscopes. Use your knowledge about living things and non-living things and animal cells and plant cells to try to identify the objects in the photographs shown below, which were taken through a microscope.

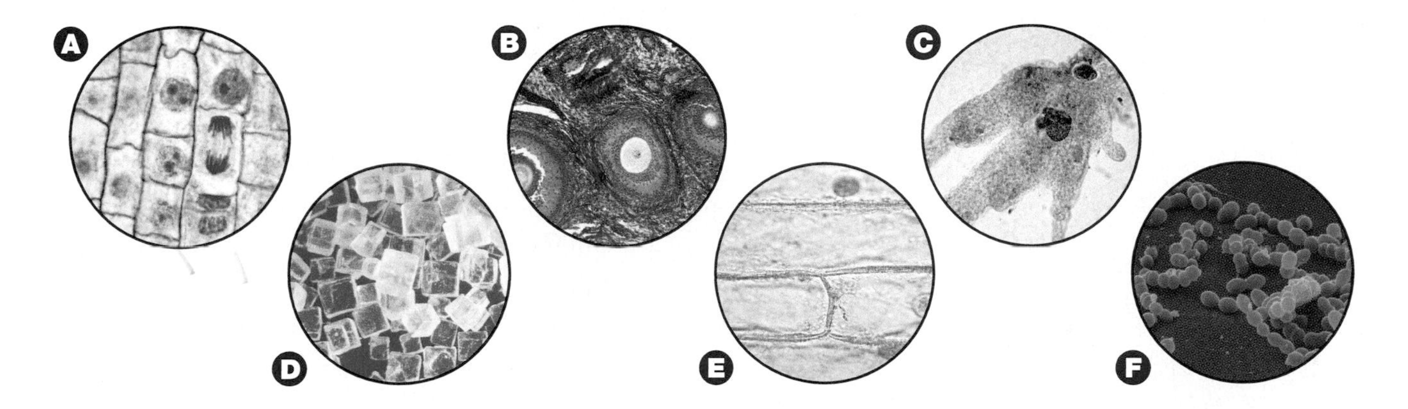

Think About It

What information can you use to determine whether you are looking at (a) a living thing or a non-living thing, and (b) a plant cell or an animal cell?

What to Do

1. Study the photographs. Try to determine:
   - **(a)** which photographs are of living things and which are of non-living things
   - **(b)** which photographs are of unicellular organisms and which are of multicellular organisms
   - **(c)** which photographs are of plants and which are of animals

Skill POWER

For tips on making Venn Diagrams, turn to SkillPower 1.

Analyze

1. Construct a Venn diagram to help you compare the photographs of living things and non-living things.
2. Construct a Venn diagram to help you compare the photograph of a plant cell with the photograph of an animal cell.

Conclude and Apply

3. Explain your answers to step 1 in "What to Do."
4. If the photographs had been black and white, would your answers have been the same? Explain.
5. Sketch the animal cell and label the organelles.
6. Sketch the plant cell and label the organelles.

Parts of a Microscope

The microscope shown is called a compound light microscope because it uses more than one lens, and it uses light to help you see the object. The different parts of the microscope are labelled in Figure 1.22.

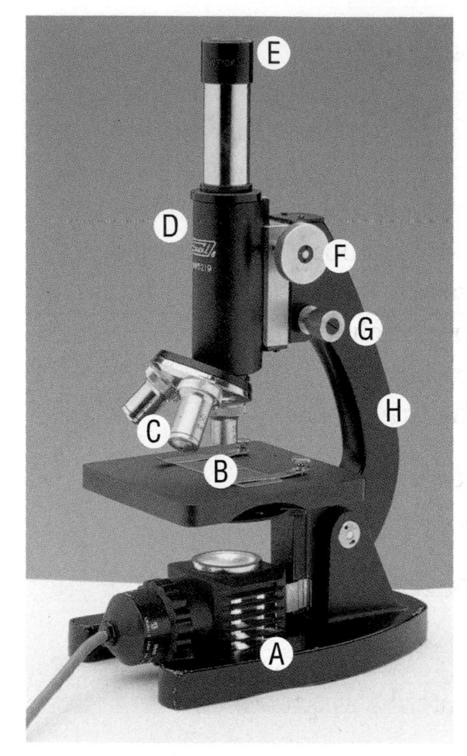

Figure 1.22 Microscope with an electric lamp

A. A **lamp** shines light through the object being viewed. Your microscope might have a mirror instead of a lamp.

B. The **stage** is a platform for the object you are viewing (often on a glass slide). An opening in the stage allows light to pass through the slide. Stage clips hold the slide in place.

C. **Objective lenses** produce a magnified image of the object. Some microscopes have low-, medium-, and high-power lenses.

D. A **tube** connects the objective lens to the eyepiece so you can see the image.

E. You look through the **eyepiece lens** to see the object. This lens magnifies the image produced by the objective lens.

F. The **coarse-adjustment knob** moves the tube (or the stage) up and down to bring the image roughly into focus. It is used with the low- and medium-power objective lenses.

G. The **fine-adjustment knob** brings the image into clearer focus. It is used with the medium- and high-power lenses.

H. The **arm** connects the stage and base to the tube. Use two hands to hold both the arm and the base of the microscope when carrying the instrument.

Pause& Reflect

Use the information here and in SkillPower 2 to make a list of the tips to protect a microscope from damage. Write the tips on an index card, and refer to them when you work with a microscope.

Objects viewed through a microscope are put on a glass microscope **slide**. If the object is wet, it may then be covered by a thin **cover slip** to hold the object in place and keep water off the lens, as shown in Figure 1.23. A prepared slide is a slide that has already been made for you.

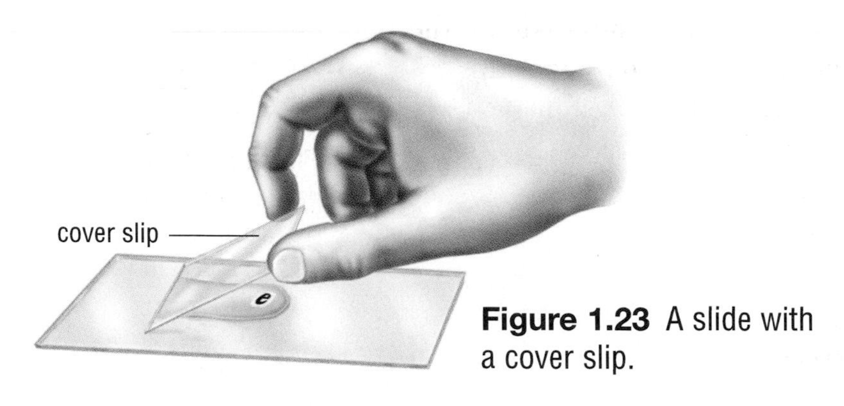

Figure 1.23 A slide with a cover slip.

CONDUCT AN

INVESTIGATION 1-G

SKILLCHECK
- Observing
- Communicating
- Interpreting Observations
- Problem Solving

Zoom in: Using a Microscope

Practise using a microscope with prepared slides.

Question

How does looking at an object through a microscope change what you see?

Safety Precautions

- If your microscope has an electrical cord, make sure your hands are dry when you plug in or disconnect the cord. Where possible, plug the cord in away from busy aisles so no one trips over it. Do NOT unplug using the cord—use the plug.
- Handle glass or plastic microscope slides and cover slips carefully so that they do not break and cause cuts or scratches.

Apparatus

microscope slides
microscope
prepared slide of onion skin
prepared slide of a cell from a human cheek

Materials

lens paper

Procedure

1. Turn to SkillPower 2 to review how to use a microscope.
2. Place the slide with the onion skin on the microscope stage. Use the stage clips to fasten down the slide.
3. Set your microscope to the low-power lens. Look through the eyepiece lens and move the slide until you can see the onion skin. Adjust the coarse-adjustment knob until the onion skin is in focus. Draw a circle in your notebook and draw what you see through the microscope. Label your drawing, including any organelles you can see. Give your drawing a title and note the magnification.
4. Change the lens to medium power and repeat step 3.
5. Observe the slide of the cheek cell on low power. Make a drawing and label any organelles you can see. Give your drawing a title and note the magnification.

Analyze

1. When scanning a microscope slide to find an object, would you use low or medium power? Why?
2. Could you see any organelles in the plant and animal cells? Why do you think you could not see all of the organelles in the cell?

READING Check

Photographs taken through a microscope are called "micrographs." Use a dictionary to define the word "graph."

Figure 1.24 Anton van Leeuwenhoek used a simple, homemade microscope to observe unicellular organisms and other objects.

Microscopes Then and Now

The first compound microscope was made in the 1590s by two lens makers in Holland, Hans and Zacharias Jansen. It magnified objects only nine times, but scientists and inventors in Europe immediately began work to improve the instrument.

In the 1670s, a Dutch amateur scientist named Anton van Leeuwenhoek (LAY-wen-hook) made a simple, hand-held microscope that had one lens (see Figure 1.24). He used this to magnify samples of blood and pond water.

Leeuwenhoek's single lens microscope was the best at that time, and it enabled him to see living cells of animals, including unicellular organisms, such as bacteria, that no one knew existed. Since that time, scientists have continued to improve the microscope and have used it to make discoveries that changed medicine, biology, chemistry, physics, and the development of new technology.

Today, most microscopes that use light rays can magnify objects 40 to 2000 times. There are also other, more powerful microscopes that are able to magnify objects over 2 000 000 times!

DidYouKnow?

Jewellers and watchmakers use a small microscope they hold at one eye to look at gems or the tiny parts of watches. This microscope is called a *loupe*.

All you need to create your own magnifying glass is a glass jar with curved sides and some water, just like this boy has here. This is the oldest type of magnifying glass. Two thousand years ago, early Romans used water-filled vessels to magnify small objects that they were carving.

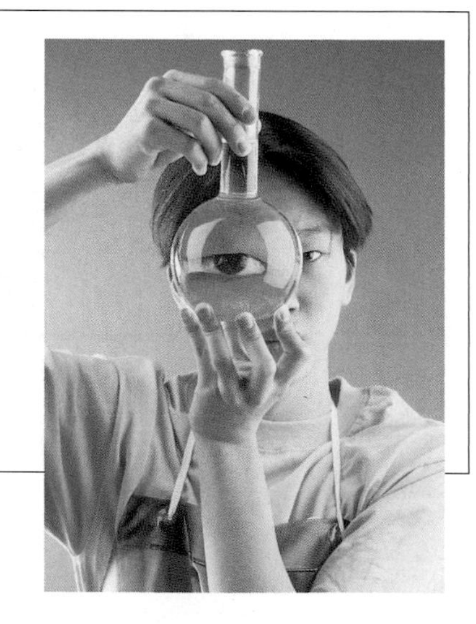

Find Out **ACTIVITY 1-H**

Microscope Practice

What You Need

small piece of newspaper, 6 cm × 6 cm
microscope

Safety Procedures 

- If your microscope has an electrical cord, make sure your hands are dry when you plug in or disconnect the cord of the microscope. Where possible, plug in away from busy aisles in the classroom so no one trips over the cord. Do NOT unplug using the cord; use the plug.

What to Do

1. Turn to SkillPower 2 to review how to use a microscope.
2. Find a three-letter word on the piece of newspaper containing the letters c, r, n, m, u, or a, with the letter e in the middle. **Observe** the letter e and draw what you see.
3. Place the paper on the microscope stage.

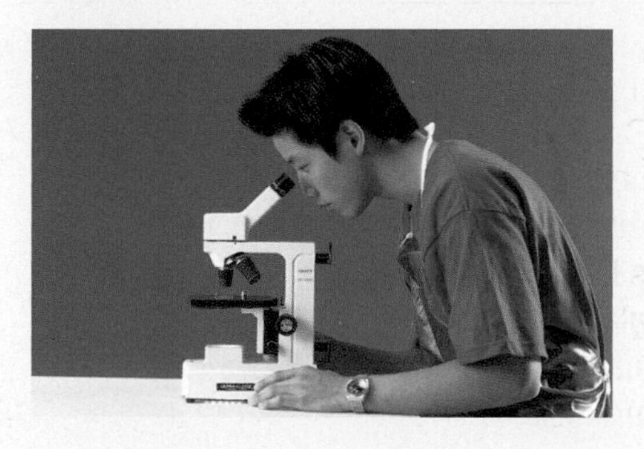

4. Set your microscope to the low-power lens.
   (a) Look through the eyepiece lens and move the paper until you can see the letter e. Adjust the coarse-adjustment knob until the letter is in focus. Draw a circle in your notebook and draw what you see through the microscope. Include the portions of any other letters you see. Note the length of the lines you see, the shape of the corners, and the distance between parts of the letters. Label your drawing with a title and the magnification.
   (b) Move the paper to the right. Record what happens to the letter.
   (c) Move the paper up and down. Record what happens to the letter.
5. Change the lens to medium power and repeat step 4 (a).

What Did You Find Out?

1. If you wanted to view the letter right way up, how would you position the paper on the microscope stage?
2. Compare the view of the letter when using the low-power lens with the view of the letter using the medium-power lens. What things are the same? What was different?

Section 1.3 Summary

Tools such as hand lenses and microscopes allow you to look at things that cannot be seen with the unaided eye. In this section, you learned about using tools to magnify images of objects.

- In order to see microscopic organisms or parts of organisms, such as cells, you need to produce an enlarged (magnified) image of the object you are looking at.
- You can use hand lenses and microscopes to magnify small objects.
- Most objects viewed through a microscope are put on a glass microscope slide and covered by a thin cover slip.

Key Terms

magnify
microscope
slide
cover slip

Check Your Understanding

1. What are the similarities and differences between a microscope and a hand lens?
2. When do you need to place a cover slip on a microscope slide before it is viewed? Why?
3. Make a sketch of the following photograph of a microscope in your notebook. Label the parts of the microscope (lamp, stage, objective lens, tube, eyepiece lens, coarse-adjustment knob, fine-adjustment knob, and arm). What does each of these parts do?

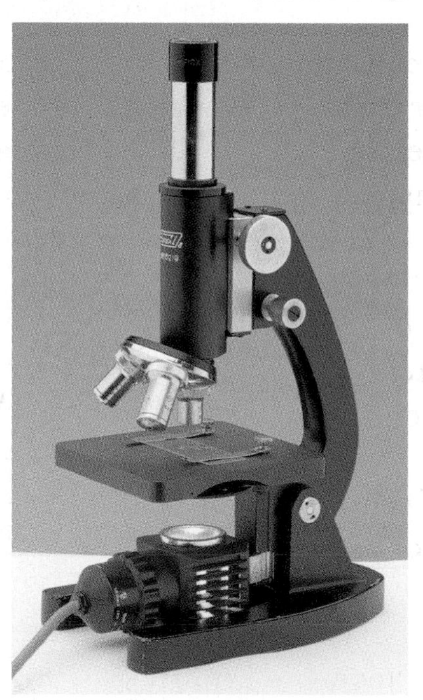

4. **Apply** You have been given a specimen of an unknown organism and need to determine what type of organism it is. What steps would you take with a microscope to find out whether the specimen was made up of plant cells or animal cells?

CHAPTER at a glance

Now that you have completed this chapter, try to do the following. If you cannot, go back to the sections indicated in brackets after each part.

(a) How would you tell if something is living or non-living? (1.1)

(b) What are the two gases that both plants and animals use or produce. (1.1)

(c) List five ways that organisms obtain food for energy. (1.1)

(d) List five ways that organisms reproduce. (1.1)

(e) Define the following: cell and photosynthesis. (1.1)

(f) What are some factors that affect the growth of seeds? (1.1)

(g) Name three differences between plant cells and animal cells. (1.2)

(h) Compare the size of plant and animal cells to the size of a bacterial cell. (1.2)

(i) Provide an example of a Venn diagram that could help you understand what you learned about cells. (Skill Power 1)

(j) Explain how and when a hand lens would be useful. (1.3)

(k) Describe how to store and move a microscope properly. (Skill Power 2)

Prepare Your Own Summary

Summarize Chapter 1 by doing one of the following. Use a graphic organizer (such as a concept map), produce a poster, or write a summary to include key chapter ideas. Here are a few ideas to use as a guide:

- List the characteristics of living things.
- Would you rather be unicellular or multicellular? Explain why. Express your reasons in a song, a poem, or a dialogue.
- What are the main structures of cells? Sketch a diagram of a plant cell or an animal cell and label the various organelles. Identify the main function(s) of each organelle and whether it is found in a plant cell, an animal cell, or both.
- Write a presentation to explain why you should have your own microscope. Include details about what you would use it for and how you would store it and take care of it.

CHAPTER 1 Review

Key Terms

cell
photosynthesis
organelles
cytoplasm
vacuole
cell membrane
chloroplast
unicellular
multicellular
microscope
cover slip
organism
reproduction
nucleus
mitochondria
endoplasmic reticulum (ER)
cell wall
chlorophyll
micro-organism
magnify
slide

Reviewing Key Terms

If you need to review, the section numbers show you where these terms were introduced.

1. Describe the difference between:
   (a) cell and organism (1.1)
   (b) cell membrane and cell wall (1.2)
   (c) nucleus and mitochondria (1.2)
   (d) unicellular and multicellular (1.2)
   (e) endoplasmic reticulum and vacuole (1.2)
   (f) chloroplast and chlorophyll (1.2)
   (g) slide and cover slip (1.3)

2. Copy the cell diagram shown below into your notebook and label the following animal cell organelles: endoplasmic reticulum, vacuole, mitochondria, nucleus, and cytoplasm. (1.2)

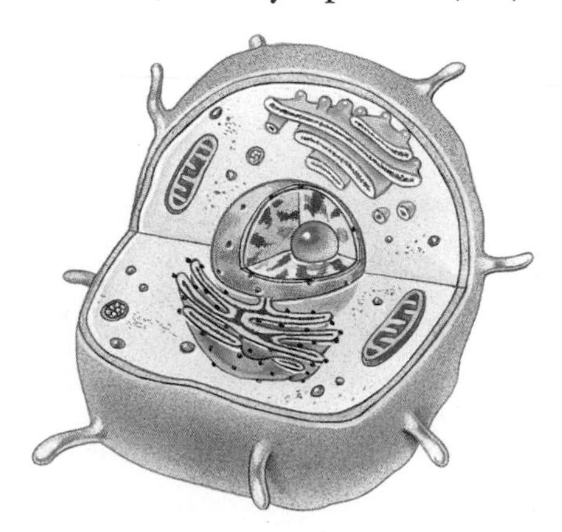

Understanding Key Ideas

Section numbers are provided if you need to review.

3. There is more than one way to divide living and non-living things. Describe the two ways that are included in this chapter.

4. How are the roles of the gases for plants similar to or different from the roles of same gases for animals? (1.1)

5. What is the role of the nucleus in the cell? (1.2)

6. Plant cells contain chloroplasts, which are organelles that contain chlorophyll. How do these organelles help plants? (1.2)

7. How does photosynthesis provide food for plants? (1.1)

8. Why is the cell called the basic unit of life? (1.1)

9. You are designing an investigation to find out what factors (light, water, temperature) affect a plant's rate of growth. How would you find out which factor has the strongest effect on the plant's growth? (1.1)

10. Unlike animal cells, plant cells have a rigid cell wall that lies outside the cell membrane. State the purpose of the cell wall and give two examples of what would happen to the plant if the cell wall did not exist. (1.2)

11. Which organelles do plants have that animals do not? What do these organelles do that helps plants? (1.2)

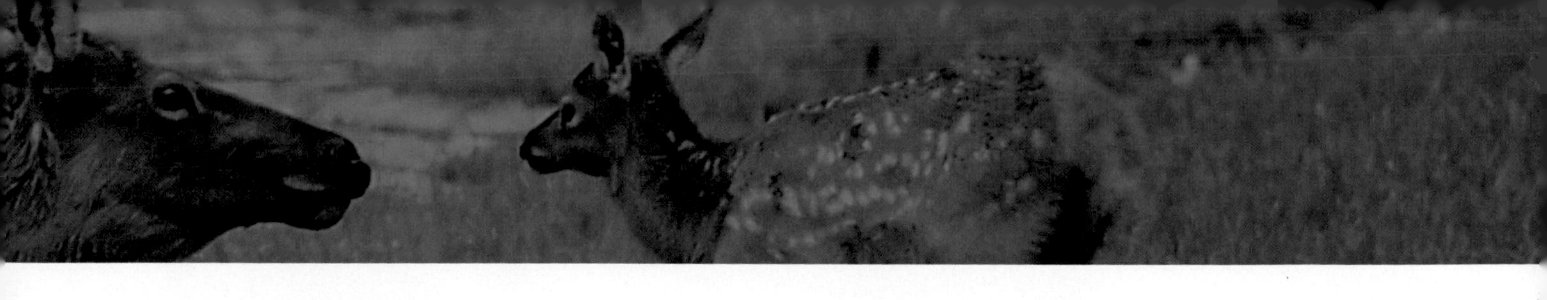

Developing Skills

12. You have been given a microscope slide with a sliver of a solid substance half the size of the end of your smallest finger. You need to look at it to tell if it is a sample of living or non-living material. Describe how you would set up the microscope and view the slide.

13. What would you use a hand lens for around your house?

14. You have made a model of a cell, and you have also seen an example of a city as a model for a cell.
(a) Choose another object or structure that could be a model for a cell and its organelles.
(b) Explain your model. Include a table showing which part of your model represents each organelle and why.

Problem Solving

15. Christina has been given slides of two cells by her teacher. She has been asked which one is a plant cell and which one is an animal cell. After she looks at the cells through a microscope, she sketches the cells. Based on her sketches, which cell is a plant cell and which is an animal cell? Explain what evidence you used to arrive at your answer.

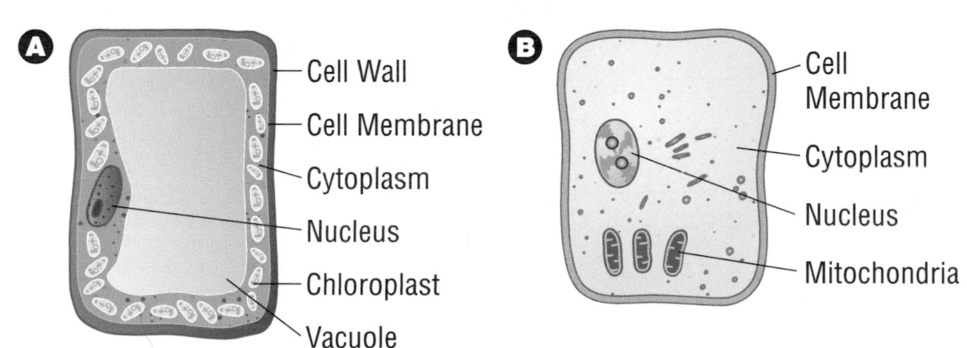

Critical Thinking

16. Cells are very small in size (10–50 μm).
(a) Why are cells so tiny?
(b) What would happen if a cell was as large as a basketball? Use what you know about the functions of cell organelles to support your answer.

Pause & Reflect

Go back to the beginning of this chapter on page 4 and check your original answers to the Getting Ready questions. How has your thinking changed? How would you answer those questions now that you have investigated the topics in this chapter?

CHAPTER

2 Living Things and

Getting Ready...

- Is a desert in Africa similar to a desert in Australia?
- Why do polar bears live in the Arctic but not in the desert?
- Why would an insect that does not sting look almost identical to an insect that *does* sting?

In winter, the feathers of ptarmigans are white. In summer, their feathers are brown, grey, and black. Why do you think their feathers change?

All living things have a place they call home. This ptarmigan [TAR-me-gen] lives high in the mountains of British Columbia. It is cold and windy there, and few plants and animals can survive. What makes ptarmigans well suited to life in this rugged place? A ptarmigan's winter feathers are white, but the colours are different in summer. How might the change in colour be a clue to how ptarmigans survive?

Each different natural area in the world has its own weather patterns, animals, and plants. For example, the animals and plants that live in a hot, dry desert are very different from those that live in a cool, wet rain forest. British Columbia has different types of natural areas, from tidal pools full of marine creatures to forests of pine and spruce to dry grasslands with sagebrush and rattlesnakes! In fact, British Columbia has the greatest variety of natural areas—and plants, animals, and other organisms—in all of Canada.

In this chapter, you will study how where a plant or animal lives affects how it looks and acts. You will also learn about the variety of natural areas in British Columbia and the plants and animals that live there.

Their Environment

What You Will Learn

In this chapter, you will learn

- how temperature and rainfall influence natural environments and the organisms that live there
- how Earth's natural regions are defined by their temperatures and amounts of rainfall as well as the plants and animals that live there
- how living things have certain features and behaviours that help them survive

Why It Is Important

- Do you live near an ocean, a grassland, or a forest? After completing this chapter, you will understand how the local climate affects the plants and animals that live around you.
- When you understand how weather patterns affect living things and where they live, you can better understand the problems caused by changes in temperature and rainfall.
- Understanding how populations of living things adapt to survive helps you understand how populations can change over time.

Skills You Will Use

In this chapter, you will

- compare the weather patterns of several places in British Columbia
- model ways that different kinds of birds all find enough to eat in the same small area
- develop a model of an organism that is able to hide in a particular environment

Could the animals and plants that live in northeastern British Columbia survive in the wet, cool forests on the coast?

Starting Point ACTIVITY 2-A

Home, Sweet Home

Have you ever wondered what it would be like to live in the Arctic, in a desert, or in the ocean? What challenges would you face?

What to Do Group Work

1. With a partner, write down the headings: Arctic, Desert, and Ocean. Using your own knowledge and library and/or Internet resources, list some of the plants or animals that live in each of these places.
2. Choose one organism from two different lists.
3. Draw a Venn diagram to compare and contrast what is similar and what is different about the two organisms you and your partner chose.
4. Repeat steps 2 and 3. Be sure to include an organism from the third list.

What Did You Find Out?

1. What characteristics might help each organism survive?
2. Could the organisms you chose survive in each others' homes? Why or why not?
3. Describe the environment (include information on the temperature and general amount of rain or snow) in the Arctic, the desert, and the ocean.
4. Write a paragraph, draw a picture, or act out a scene that describes or shows how you might survive in the Arctic, the desert, and the ocean.

Skill POWER

For tips on making a Venn diagram, turn to SkillPower 1.

Section 2.1 Climates and Biomes

Key Terms
- climate
- weather
- precipitation
- latitude
- altitude
- biome
- aquatic
- estuary

Why don't palm trees or cactus grow in northern British Columbia? Would you see caribou in Mexico? Plants, animals and other organisms live in climates where they can survive. The **climate** of an area is its average weather patterns according to records that have been kept for many years. Climate is not the same as weather. **Weather** means the local conditions that change from day to day, or even from hour to hour. Weather changes frequently; climate stays the same over a long time.

The climate of an area is defined by the range in temperatures and the amount of precipitation that falls each year. **Precipitation** is usually in the form of rain or snow.

How does climate affect organisms? Plants grow best under certain conditions. A plant that likes a dry, warm climate will not grow well where it is wet and cool. Plant-eating animals prefer certain food plants and live where those plants grow.

Look at the map and captions in Figure 2.1. Find two things that might determine the climate of the place where you live.

Is the climate the same all over British Columbia? Try Think & Link Investigation 2-B to find out.

Figure 2.1 This map shows the lines of latitude in British Columbia.

Latitude is the distance north or south of the equator. Latitude is measured in degrees north (N) or south (S). (For example, most of the British Columbia–United States border is at the latitude 49° N.) The amount of energy that a location receives from the Sun changes with latitude. The equator is the hottest area on our planet because it receives more direct sunshine than either the North or the South Pole. In British Columbia, the climate is colder in the north (which is at a higher latitude) than it is in the southern part of the province.

Altitude (elevation) is the height of a location above sea level. Sea level is the surface of the sea. For example, a place that is 10 m higher than the surface of the sea is said to be "10 m above sea level."

The higher the altitude of a place, the colder its climate. In locations at the same latitude, the climate on top of a mountain will be cooler than the climate at sea level.

INVESTIGATION 2-B

SKILLCHECK
- Communicating
- Making Inferences
- Interpreting Data
- Predicting

Climates in British Columbia

Think About It

British Columbia has a wide range of climates, from the wet, cool coastal rain forests to high, cold mountaintops and just about everything in between. In this investigation you will analyze the climates of five cities in British Columbia.

Table 1 Climate data for five British Columbia cities

	Dawson Creek	Cranbrook	Victoria	Kamloops	Prince Rupert
Latitude	55° 44' N	49° 30' N	48° 38' N	50° 42' N	54° 17' N
Altitude (m)	655	926	19	345	35
Average July temperature (°C)	15	18	16	21	13
Average January temperature (°C)	−15	−6	4	−4	1
Total annual rainfall (cm)	33	27	84	22	247
Total annual snowfall (cm)	17	13	4	8	13

What to Do

1. Plot a bar graph to show the annual rainfall and snowfall in each of the five cities. Make a red bar for rainfall and a blue bar for snowfall.
2. List the names of the cities from coolest to warmest in July.
3. List the names of the cities from coolest to warmest in January.

Analyze

1. Write one or more sentences describing the climate of each of the five cities. Include information on all of the categories listed in the first column of the table.
2. (a) How does latitude affect the climate of each of the five cities?
   (b) How does altitude affect the climate of each of the five cities?

Conclude and Apply

3. Where in British Columbia do you think the following animals would most likely live? Explain your answer.
   (a) A wolverine: Wolverines have thick, warm fur and large paws that can act like snowshoes. They have heavy claws, which are good for climbing trees and catching prey.
   (b) A rattlesnake: Rattlesnakes eat rodents, birds, and other warm-blooded animals. They like hot, dry environments.
   (c) A Canada goose: Canada geese prefer marshy areas where the water does not usually freeze. They eat grasses, berries, and seeds.

Skill POWER

For tips on making graphs, turn to SkillPower 4.

Land Biomes of the World

What does prairie in Saskatchewan have in common with grasslands in central Africa? They both have dry summers and large areas of grasses where animals such as antelope graze. Many widely separated land regions in the world have similar climates and similar (though not the same) types of plants and

Figure 2.2 Land biomes of the world

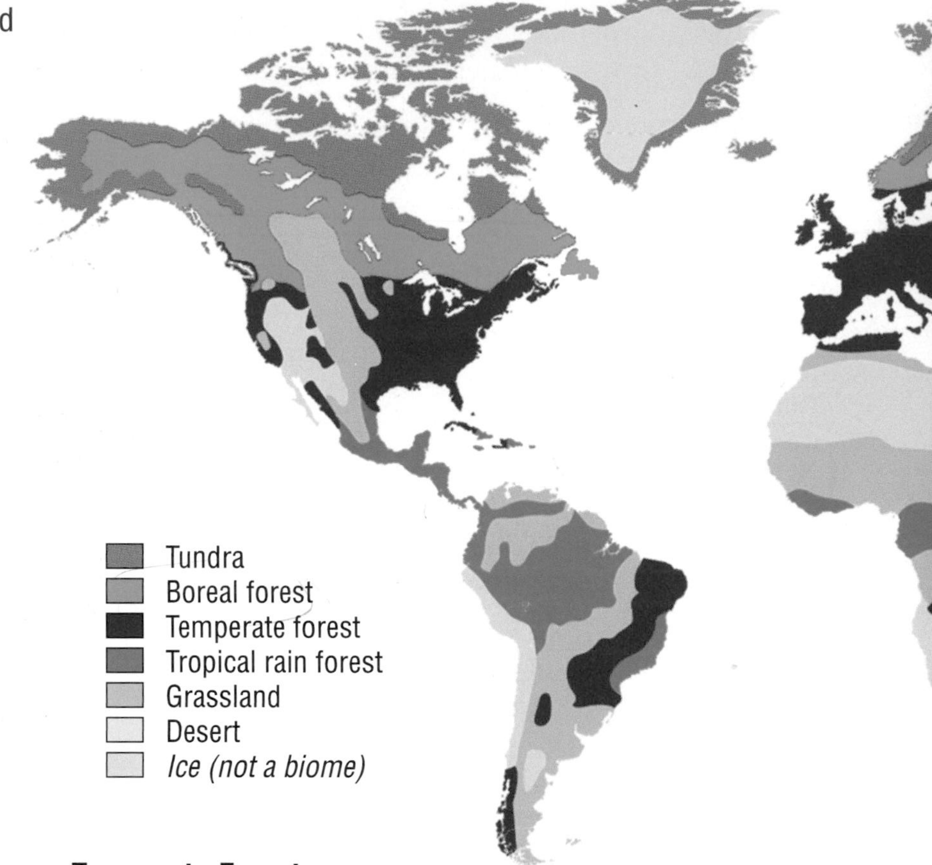

Tundra

Location: High northern latitudes such as the high Arctic region of Canada; also in high mountains at varying latitudes

Climate: Very cold, long winters; short and cool summers; average temperature about −12°C; 10–25 cm of precipitation a year

Plants: Grasses, wildflowers, mosses, small shrubs

Animals: Musk ox, caribou, polar bears, grizzly bears, Arctic foxes, weasels, snowshoe hares, owls, hawks, grouse, rodents, mosquitoes, black flies, and other insects

Boreal Forest (Taiga)

Location: Mid- to high latitudes, such as most of Canada, Russia, and Scandinavia (Sweden, Denmark, Norway)

Climate: Very cold winters, cool summers; about 50 cm of precipitation a year

Plants: Mostly spruce, fir, and other cone-bearing, evergreen trees

Animals: Rodents, snowshoe hares, moose, lynxs, least weasels, caribou, black bears, grizzly bears, wolves, many birds

Temperate Forest

Location: Mid-latitudes, such as most of the United States, United Kingdom, western Europe, Brazil, eastern China

Climate: Relatively mild summers and cold winters; usually have four distinct seasons; 75–150 cm of precipitation a year

Plants: Broad-leaved trees such as oak, beech, and maple

Animals: Wolves, deer, bears, and a variety of small mammals, birds, amphibians, reptiles, and insects

Tropical Rain Forest

Location: Near the equator, such as Central America, Indonesia, Costa Rica

Climate: Hot all year round; 200–600 cm of rain a year

Plants: Very wide diversity of vines, orchids, ferns, and trees

Animals: More species of insects, reptiles, and amphibians than in any other biome; monkeys, elephants, birds animals. These large land regions are called biomes. A **biome** has a distinct climate, soil, plants, and animals. Tundra, boreal forest (taiga), temperate forest, tropical rain forest, grassland, and desert are the six biomes most commonly referred to. These world biomes, their characteristics, and a few of the plants and animals that live in these biomes are shown in Figure 2.2.

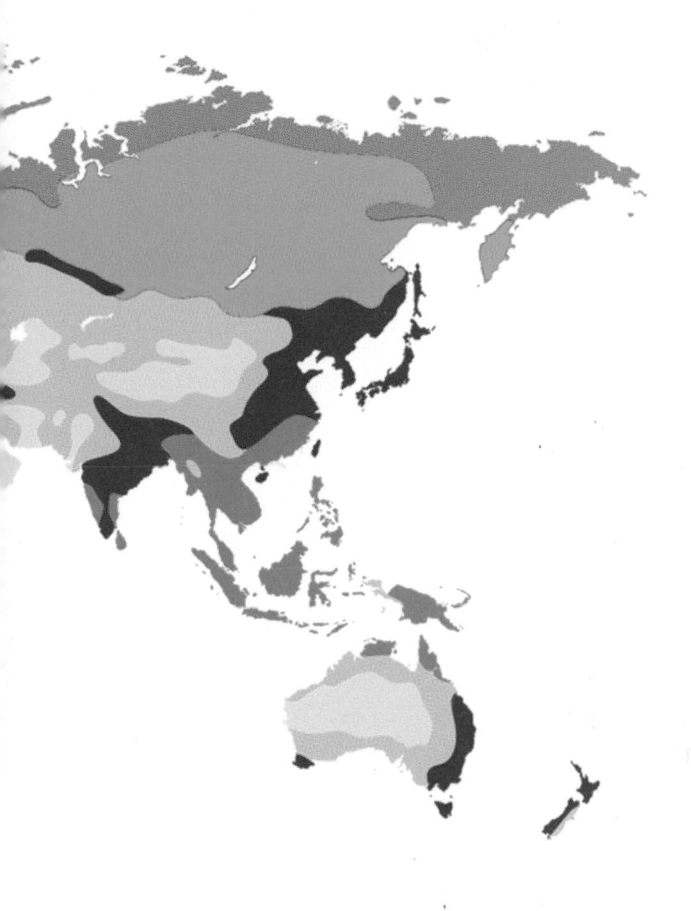

DidYouKnow?

When it is summer in the tundra biome only the top few centimetres of soil thaw out. Just below the surface is a layer of permafrost (permanently frozen soil). The permafrost prevents plants from growing deep root systems. This is why large plants do not grow in the tundra.

The tropical rain forest biome has the greatest diversity of plants and animals of all of the six world biomes. For example, one tree in a tropical rain forest might contain more types of ants than all the types of ants found in England, Ireland, Scotland, and Wales combined!

Grassland

Location: Mid-latitudes; interiors of continents such as central Canada and the United States, parts of China, and eastern Europe

Climate: Cool in winter, hot in summer; 25–75 cm of precipitation a year

Plants: Mostly grasses and small shrubs; some trees near water

Animals: Grazing animals that eat grasses; in North America: prairie dogs, foxes, snakes, insects, birds; in Africa: elephants, lions, zebras, and giraffes; in Australia: kangaroos

Desert

Location: Mid-latitudes, such as northern Africa, central Australia, and parts of China

Climate: Very hot days, cool nights; less than 25 cm of precipitation a year

Plants: Very few plants; cacti, grasses, shrubs, and some trees

Animals: Rodents, snakes, lizards, tortoises, insects, and some birds; in Africa, the desert is home to camels, gazelles, antelopes, snakes, lizards, and gerbils

Land Biomes in Canada

North America has all six land biomes, and four are found in Canada. As you can see in Figure 2.3, almost all of British Columbia is said to be boreal forest when you look at only the six biomes. But biomes are very broad categories. As you learned in Think & Link Investigation 2-B, there are many climates in British Columbia. Very different plants and animals live in these varied climates.

DidYouKnow?

Scientists have divided British Columbia into 14 *biogeoclimatic* zones. Each zone has a distinct combination of organisms **(bio)**, geography and landforms **(geo)**, and climate **(climatic)**.

What are the six major land biomes?

INTERNET CONNECT

www.mcgrawhill.ca/links/BCscience6

You can learn a lot about an area by finding out what kinds of trees grow there. Go to the web site above and click on **Web Links** to find out where to go next. Choose a type of tree and find out its ideal climate conditions.

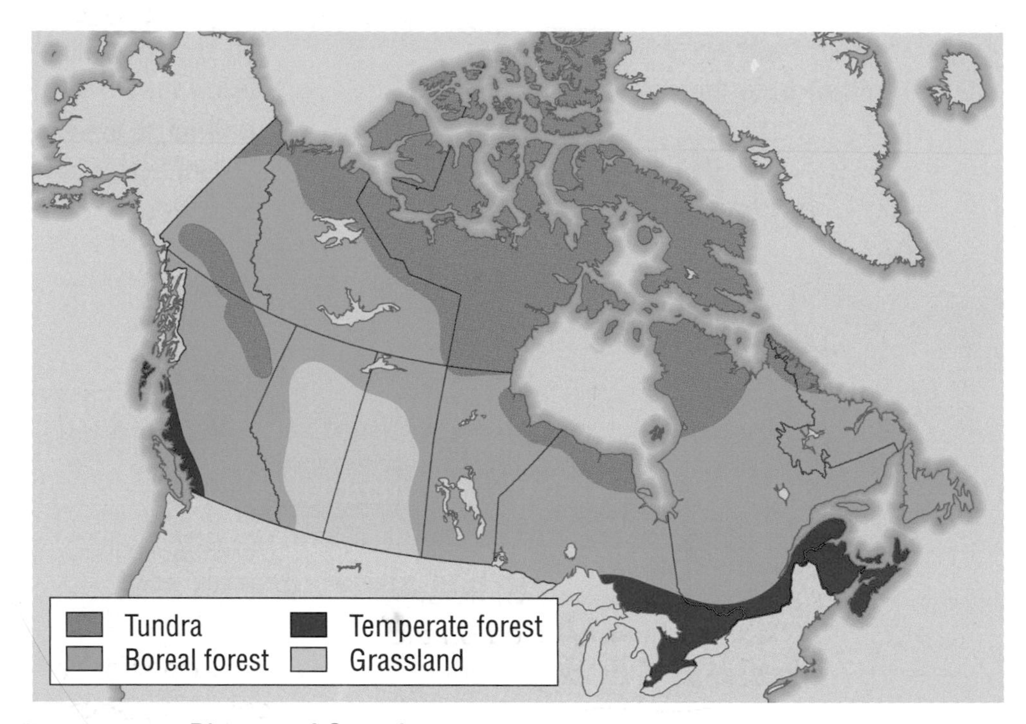

Figure 2.3 Biomes of Canada

Canada has a temperate forest on the West Coast of British Columbia. In some of the wettest parts of this forest it can rain as much as 5 m a year! Huge western red cedar, Douglas fir, and Sitka spruce trees grow in the mild, wet climate along the coastline.

A West Coast temperate forest is shown in Figure 2.4. This is where you can find Canada's largest trees.

Figure 2.4 A temperate forest grows on the West Coast of British Columbia.

British Columbia even has its own tiny piece of desert biome. The "pocket desert" near Osoyoos (see Figure 2.5) receives less than 4 cm of rain a year. Many parts of the Thompson River Valley and Okanagan Valley of British Columbia are semi-desert. These areas receive less than 25 cm of rain a year. You can find rattlesnakes, sage-brush, and other plants that live in dry conditions in these desert and semi-desert areas.

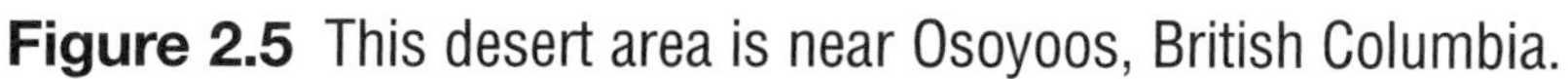

Figure 2.5 This desert area is near Osoyoos, British Columbia.

Find Out ACTIVITY 2–C

People in Biomes

For many thousands of years, humans have used the resources in their environment for food and clothing and also to build shelters. For example, the Interior Salish people built partially buried pit houses that kept them comfortable during cold, snowy winters. The Coast Salish people, who lived on the warmer, wetter coast, built houses out of planks from cedar trees. How did Aboriginal peoples use the plants and animals around them to help them live in particular environments?

What to Do (Group Work)

1. With a partner, choose a location in British Columbia. Research the climate and some of the plants and animals that live there.
2. Research how the Aboriginal people who lived in this climate used the local plants and animals to meet their needs. For example, how did they use plants and animals for food, clothing, shelter, and transportation?
3. Design and create a display that you and your partner can use to present your findings. Your display could be a written report, a poster, a role play, an oral or computer presentation, or a diorama.

What Did You Find Out?

1. How does an area's climate affect not only the plants and animals that live in a place but the people as well?

Extension

2. With another group, compare similar technologies from different parts of the province. For example, compare the methods of transportation that Aboriginal people developed in the far north with transportation methods developed in the southern part of the province.

READING Check

What are the main land biomes in Canada and in British Columbia?

Aquatic Biomes

There are also five **aquatic** (water) biomes. Aquatic biomes are defined by the amount of salt in the water and the water temperature. Fresh water contains little or no salt. There are two fresh-water biomes: rivers and streams, and lakes and ponds. The three salt-water biomes are **estuaries** (places where rivers flow into the sea), seashores, and coral reefs.

Rivers and Streams

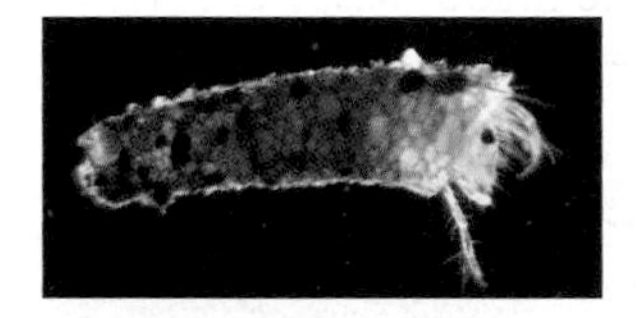

Figure 2.6 The larvae of the caddisfly cover their bodies in tubes made of sand or bits of plants.

Figure 2.7 Dippers feed on insects in fast-moving water.

Rivers and streams are fresh-water environments that flow. Rivers and streams differ in size and the speed at which the water flows. The section of the Columbia River of southeastern British Columbia shown here is slow and meandering. Many types of fish, including trout and salmon, live in fresh-water streams and rivers for at least part of their lives. Other animals living in such environments include insects such as caddisflies (Figure 2.6), river otters, and birds such as the dipper (Figure 2.7).

Canada's largest fish, the white sturgeon, lives in the big rivers of British Columbia. One caught in the Fraser River near New Westminster in 1897 weighed 630 kg and was about 6 m long! Overfishing, pollution, and dams built on rivers have made it hard for this huge fish to survive. Why do you think this is so?

What are the characteristics of at least two aquatic biomes?

Lakes and Ponds

Lakes or ponds form at a low place in the land that fills with rainwater or water from a stream. Water in lakes and ponds hardly moves at all. These aquatic biomes contain more plant growth than flowing-water environments do.

Ponds are usually small and shallow bodies of water. Lakes are larger and deeper than ponds. Lakes usually have more open water in the middle, and plants grow only on its edges. Animals that live in lakes and ponds include many types of insects, frogs, turtles, fish, and birds such as ducks. Atlin Lake, shown here, is the largest natural lake in British Columbia.

Estuaries

Estuaries are very important because the fresh water in rivers and streams brings nutrients into the ocean. Many ocean fish live in estuaries when they are young. Crabs, clams, oysters, and snails also live in estuaries. The Fraser River, British Columbia's largest river, creates a huge estuary where it flows into the ocean near Delta, British Columbia. This area provides valuable habitat for many bird species such as geese and shorebirds.

DidYouKnow?

The Fraser River is sometimes called the "mighty, muddy Fraser." Each year the river transports about 20 million t of sand, silt, and mud from the interior of British Columbia to the Strait of Georgia

Seashores

Seashores can be rocky, sandy, or muddy. Organisms that live on seashores need to tolerate changes in temperature, moisture, and salinity (saltiness), as well as the force of the waves. On sandy shores, such as Long Beach on Vancouver Island, the sand moves when waves wash over it. Organisms that live on sandy shores, such as clams, crabs, and worms, burrow into the sand. Organisms that can cling to rocks, such as sea stars, barnacles, and mussels, can withstand the crashing waves on rocky shores.

Coral Reefs

Coral reefs grow in shallow, warm waters. Coral reefs are made of living organisms—coral polyps—that live in colonies (groups). The coral polyps make a substance that hardens to make their skeletons. Thousands of these polyps join to form the reefs. Many types of corals, as well as hundreds of kinds of beautiful tropical fish, live in coral reefs such as the one shown above.

Find Out **ACTIVITY 2-D**

Illustrate a Biome

Aboriginal groups in British Columbia have ways of representing their history and environment with art and symbols. Carved totem poles and decorated curtains and dance shawls are some of the ways in which Aboriginal families portray their histories and the world around them. A "coat of arms," such as the one shown here for British Columbia, is another way to use pictures, symbols, and words to tell something about a place.

In this activity you will make your own representation of a biome.

What You Need

art supplies

What to Do

1. Select a biome from the ones described in this section. Do research to find out some of the plants and animals that live there. What is the climate like? You can look at books, use the Internet, watch a film, look at posters or paintings, or interview someone to help you find the information.
2. Choose a way to represent your biome visually. Use the art form of your choice, but include pictures that show parts of the biome.
3. Display your artwork and explain your creation to your classmates in a brief presentation.

Diversity of Life in British Columbia

The wide diversity (variety) of climates in British Columbia means that there are many different kinds of habitats for living things. In fact, British Columbia has more types of living things (species) than any other province in Canada. For example, of the 578 species of birds that live in Canada, 454 of them live in British Columbia.

For some species, British Columbia is home to most of the world's population. For example, almost all of the world's population of Cassin's auklet (a sea bird) nest on the British Columbia coast (Figure 2.8). Over half of the world's population of mountain goats (Figure 2.9) and about one quarter of the world's grizzly bears also live in British Columbia.

Pause & Reflect

National parks, provincial parks, and other areas such as British Columbia Ecological Reserves protect natural areas and the living things found there. Look on a map of British Columbia to locate a national or provincial park near you. In your science notebook, explain why such areas are important; include examples.

Figure 2.8 Cassin's auklets live most of their lives at sea but come ashore on the British Columbia coastline to nest and have their young.

Figure 2.9 British Columbia has a lot of the steep rock habitat that mountain goats prefer.

In British Columbia, there are also many species of organisms that are found *only* in this province. For example, the Newcombe's butterweed, shown in Figure 2.10A, grows only in Haida Gwaii (the Queen Charlotte Islands), and the Vancouver Island marmot, shown in Figure 2.10B, lives only on Vancouver Island.

The variety of habitats in British Columbia leads to a greater diversity of living things. But why? This is a question you will explore in the next section.

Figure 2.10 The Newcombe's butterweed (A) and the Vancouver Island marmot (B) are examples of organisms that you can't find anywhere else in the world!

Section 2.1 Summary

The climate of a region helps determine the types of plants and animals that live there. A biome is a large area with a distinct climate, soil, plants, and animals.

- Climate and weather differ. Weather changes frequently; climate stays the same over long periods of time.
- Tundra, boreal forest (taiga), temperate forest, tropical rain forest, grassland, and desert are the six land biomes most often described. The four biomes in Canada are tundra, boreal forest, temperate forest, and grasslands. Most of British Columbia is boreal forest.
- Land biomes are very broad categories. Scientists divide regions into more specific categories depending on the local climate and the living things found there.
- Aquatic biomes include ponds and lakes, rivers and streams, estuaries, seashores, and coral reefs.
- British Columbia has a great diversity of natural areas. Because of this, British Columbia has the largest variety of living things in Canada.

Key Terms

climate
weather
precipitation
latitude
altitude
biome
aquatic
estuaries

Check Your Understanding

1. **(a)** Is it easier to accurately predict weather or describe a climate? Explain your answer.
   (b) When would you use weather information? When would you use climate information? Explain your answer.
2. What are the six land biomes? Describe one climatic feature of each.
3. What are the five aquatic biomes? Describe one feature of each.
4. What characteristics are common for all aquatic biomes?
5. **Apply** A builder decides to put houses on a plot of land that includes a stream. The builder plans to change the path of the stream to make the land easier to build on. What impact, if any, will this have on the organisms that live on this land?
6. **Thinking Critically** The diagram provided shows elements of a pond biome. Is this diagram complete? Explain why or why not.

Section 2.2 Adaptations

Cactus in British Columbia? You bet! In very dry parts of the province, such as on the Gulf Islands near Victoria and in the dry interior of the Okanagan Valley, you can find the prickly pear cactus, which is shown in Figure 2.11. How do plants such as cactus survive in the hot, dry places they call home? The answer lies in the organism's adaptations. **Adaptations** are physical characteristics or behaviours that give an organism a better chance of surviving and reproducing in a particular environment. (A **behaviour** is the way an organism acts.)

Figure 2.11 The prickly pear cactus grows in a few very dry parts of British Columbia.

The thin, hard spines of the cactus are an adaptation. These spines are actually small, tough leaves. Because it has these spines, a cactus doesn't lose as much water as a plant with larger, softer leaves would. And the spines keep most animals from eating the cactus. How do spines help the cactus survive?

It is important to note that adaptations are *inherited* characteristics. An **inherited characteristic** is a characteristic that an organism is born with. For example, the ptarmigans shown at the beginning of this chapter have white feathers that blend in with the snow in winter, making it harder for predators to see them. In spring, they lose (moult) their winter feathers and grow darker feathers for summer. The colour of the ptarmigan's feathers is an inherited characteristic that makes them better able to survive.

Not all adaptations of behaviour are inherited. Looking both ways before you cross the street is a behaviour that helps you survive. But this behaviour is not inherited; it is learned. You were not born knowing to look before crossing a street.

Key Terms

adaptations
behaviour
inherited characteristic
physical adaptations
behavioural adaptations
hibernation
migration
colouration
camouflage
mimicry

INTERNET CONNECT

www.mcgrawhill.ca/links/BCscience6

To learn more about the adaptations of cactus and other plants that live in dry environments, go to the web site above and click on **Web links** to find out where to go next. Write and sketch your findings.

READING Check

Is wearing your seatbelt an adaptation that helps you survive? Explain your answer.

Physical Adaptations

Organisms have physical characteristics that make them suited to the biomes where they live. These are called **physical adaptations**. For example, the dense fur of wolves is a physical adaptation that helps them stay warm. The thick bark of a Douglas fir is also a physical adaptation. The tree's thick bark helps these trees survive forest fires. As shown in Figure 2.12, the outer bark may be burned, but the tree is not killed.

Figure 2.12 How does the thick bark of Douglas fir trees help them survive?

You have physical adaptations as well. For example, your skin has sweat glands that help you keep a constant body temperature no matter how hot or cold it is. When it is hot, sweat glands release water onto your skin. The water evaporates and, as a result, you become cooler. When it is cold, sweat glands do not release water, and you stay dry and warmer.

DidYouKnow?

The bark of Douglas fir trees in the interior of British Columbia is thicker than the bark of Douglas firs growing on the West Coast. This is because there are more forest fires in the interior than there are on the wet, cool coast, and thicker bark helps the trees survive a fire. In the interior, Douglas firs with thicker bark pass this characteristic to their offspring.

The plant moss campion (Figure 2.13) grows on high mountaintops where the climate is cold, dry, and windy. This plant has small leaves and grows very low to the ground. This size and shape traps heat, prevents wind damage, and reduces the amount of water that would be lost if the plant had larger leaves. Since they often grow in gravel or other soils with few nutrients, moss campion plants establish a good root system in the first few years of life. A long root, called a taproot, can reach 2 m into the ground. This helps the plant gather enough food and water from poor soils.

Animals that are active at night or live in very dark places often have physical adaptations that help them find food and escape predators in the dark. For example, owls, such as the great horned owl in Figure 2.14, have large eyes, a very flexible neck, and excellent hearing.

Learn more about physical adaptations in Find Out Activity 2-E.

Figure 2.13 How do the physical adaptations of the moss campion enable it to grow in the high mountains of British Columbia?

Figure 2.14 What physical adaptations of great horned owls make them well-adapted to hunting for food at night?

Find Out **ACTIVITY 2-E**

Picky Eaters

Many different types of birds can survive together in a small area because of their food choices. In this activity you will compare different tools and predict how the shape of a bird's beak influences what it eats.

A. bald eagle

B. pine grosbeak

C. northern flicker

D. robin

What You Need

scissors
1 jelly bean
4 sunflower seeds in the shell
1 gummy worm
1 chocolate chip
tongs
tweezers
clothespin

What to Do

1. Use the scissors to pick up each of the food items. Use the scissors to crack open a sunflower seed. **Record** your observations using sketches and notes.
2. Repeat step 1 using the tongs, tweezers, and clothespin.

What Did You Find Out?

1. Which tool best picked up the jellybean, the sunflower seed, the gummy worm, and the chocolate chip?
2. Which tool was the best at cracking a sunflower seed?
3. Do any of the tools resemble any of the beaks of the birds in the pictures?
4. What natural foods do you think each of the birds in the pictures usually eats?
5. Do you think all birds can use their beaks to eat any type of food? Explain your answer.
6. How do the differences in beak types help a variety of birds find enough food to survive in a geographical area that they share?

Behavioural Adaptations

Behavioural adaptations are habits and activities of organisms that are important for survival. For example, grizzly bears like the one in Figure 2.15 hibernate for part of the winter. **Hibernation** is a period of time when animals are much less active and use a lot less energy. During hibernation, the body temperature drops, and the heart rate and other body processes slow down. Hibernation is a behavioural adaptation that gives these animals a better chance of surviving winter when there isn't very much food available.

Figure 2.15 Grizzly bears that live in parts of British Columbia where it gets very cold hibernate for part of the winter. Hibernation is a behavioural adaptation.

Choose one plant and one animal. What is one adaptation that makes them well suited to the conditions in their environment?

Each spring, over 20 000 grey whales swim past British Columbia's West Coast as they travel from their winter feeding grounds in Mexico to their summer feeding grounds in Arctic waters. Other mammals, many birds, and insects such as monarch butterflies also take these long journeys (migrations) every year. **Migration** is the movement of animals from one region to another in response to a change in seasons. Animals often migrate when there are changes in temperature or the number of hours of daylight that mean less food will be available. Migration is another type of behavioural adaptation.

Behavioural adaptations are not always as dramatic as hibernation or migration. A snake moving under a rock to seek shade from the hot sun or a group of honeybees beating their wings to cool their hive are also behavioural adaptations.

Adaptations and Climate Change

The adaptations that organisms have made in order to live in their environments develop over many generations. Today, scientists are worried that our climate is changing the environment more quickly than plants and animals can adapt. For example, the warming of the climate in Canada's far north has resulted in thin sea ice. Polar bears need thick sea ice in order to go out to hunt for food. Scientists have reported that polar bears are thinner than they used to be, and that it is harder for their cubs to survive.

INTERNET CONNECT

www.mcgrawhill.ca/links/BCscience6

Will polar bears be able to adapt to the changing climate? Visit the web site above to see how warmer temperatures are affecting polar bears, birds, and other species. Click on **Web Links** to find out where to go next.

Adapting to Other Organisms: Symbiotic Relationships

Some organisms live closely with other organisms. Symbiosis is a relationship in which two species live closely together for a long time. Organisms adapt to these types of living relationships. For example, some organisms benefit from living with or on another organism. The tick in Figure 2.16 obtains its food from another animal such as a dog or a deer. In this relationship, the tick benefits, while the host animal is harmed (but usually not killed).

In symbiosis, there are also relationships where one organism benefits while the other is neither harmed nor benefits. The clownfish in Figure 2.17 lives amongst the tentacles of sea anemones. Since other fish avoid the stinging tentacles of anemones, the clownfish is protected.

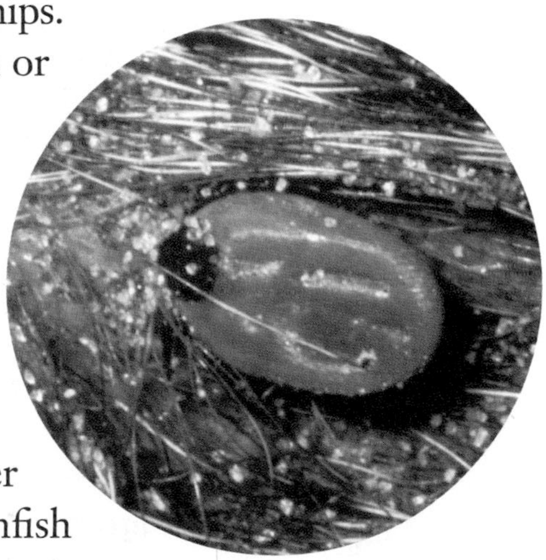

Figure 2.16 Ticks feed by sucking blood from host animals. This is a symbiotic relationship where one animal benefits while the other is harmed.

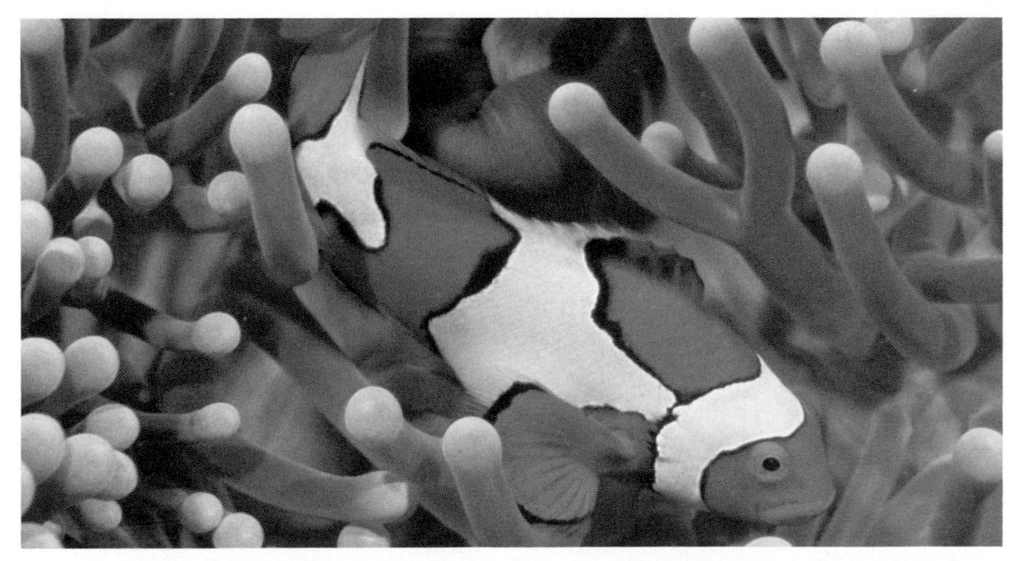

Figure 2.17 Clownfish are not harmed by the stinging tentacles of anemones. This means they can live amongst anemones, safe from predators.

Find Out **ACTIVITY 2–F**

Animal Voices

How can poetry help you learn about animal adaptations? To find out, read these poems from the book *Voices of the Wild: An Animal Sensagoria* by teacher and writer David Bouchard of British Columbia.

Poem 1

Though my ears are far from mammoth,
I can hear like others can't.
And you can show Them through your pictures,
You can paint me as I am,
Show my movements in the water
How I live and how I hear.

Tell Them of the sounds I make,
The far-off shrills, the whistling cries.
Tell them of the sounds I send
And how these sounds come back to me.

Tell Them how I hear an echo
Through the melon on my brow,
And with this locate my victim
And all life that swims with me.

Tell Them what I find I eat,
How I hunt and what I eat.
I feed on seals and birds and fish,
And walruses and massive whales.
I feed on anything I choose
And thus I'm known as "killer whale."

They should learn about my hearing
Show Them what you've come to know
How I keep in touch with others,
How I live and How I hear.

Poem 2

As I step out of my forest,
There are few who know I've left.
As I leave I move unnoticed
Like a breeze above the snow.

My tufted ears and wispy beard
Distinguish me from other cats.
Yet I am here because I know
You're one who seeks the truth.

They know me through my cousins
As sleek and swift and quick and strong
But have They ever noticed that my
Scent's as keen as any?

Behind my nose there lies a maze
Of bones like no one else here has.
I breathe in I draw the air
Across a bank of countless cells
And through this process I decide
To eat or leave disgusted.

So They think I'm finicky
About the food I choose to eat.
Please show Them that it is the smell
And not the taste that moves me.

As I leave I move unnoticed
Like a breeze above the snow.
And I am here because I know
You're one who seeks the truth.

continued

What to Do Group Work

1. **(a)** With a partner, read each of the poems. You could close your eyes as your partner reads the poem out loud a second time.
   (b) Try to picture in your mind the animal and where it is. Think about what the animal looks like and how it moves, smells, and hears, for example. Imagine you are the animal. What do you see, hear, smell, taste, and feel?
2. With your partner, discuss what animal you think is being described in each poem.
3. Draw a Venn diagram to compare and contrast the characteristics of each animal.
4. Draw a Venn diagram to compare and contrast the environment of each animal.
5. Draw a picture of each animal being described. Include the characteristics of the animal and its environment as described in the poem.

What Did You Find Out?

1. **(a)** What adaptations does each of the animals have that make it suited to live in its environment?
   (b) Would that animal survive without those characteristics? Explain your answer.
2. Use a chart with two columns to [illegible] your answers to the following ques[illegible]
   (a) Which of the adaptations are physical adaptations?
   (b) Which are behavioural adaptations?

Extension

3. Research one plant or animal that lives in British Columbia and find out its physical and/or behavioural adaptations. Describe the climate and biome in which your organism lives. Present your findings in a written report, a word-processed report, a poster, a diorama, a poem, or a song.

Skill POWER

For tips on drawing Venn diagrams, turn to SkillPower 1.

Pause& Reflect

Artists create dances, music, poetry, and other art works to share their knowledge and understanding of natural places. Through their art, these people share their love and appreciation for these places and the living things that call these places home. Look for examples of art that depict nature in books, in art galleries, and in peoples' homes!

e 2.18
lours
ttern of
hers and
haviour of
tern are
adaptations that
camouflage it.

Figure 2.19 This fish's colouration makes it hard to see on the sand.

Figure 2.20 The colouration of the non-stinging fly (B) mimics that of the stinging wasp (A). This is a type of camouflage called mimicry.

Figure 2.21 These thornbugs are hard to see because their shape mimics the thorns on the plant.

How is mimicry an adaptation that helps organisms survive?

Adaptations for Camouflage

The American bittern, shown in Figure 2.18, lives throughout British Columbia. This bird stands, still and silent by the edges of ponds, waiting for fish to swim by. The stripes on the underside of the bittern's neck and the colours of its feathers—its colouration—plus its ability to remain motionless make it less visible to its prey. **Colouration** is the colour and pattern of an animal's skin, fur, or feathers. Many organisms have adaptations that hide them. For example, the sand-like pattern and colour on the fish in Figure 2.19 make it blend in with the sand on the ocean floor where it lives. An organism's ability to blend into its surroundings is called **camouflage**. Camouflage hides animals while they wait for food to pass by or when predators are near.

Have you ever been stung by a bee or wasp? That experience would have taught you to avoid buzzing, black-and-yellow-striped insects such as the yellow-jacket wasp shown in Figure 2.20A. Animals that eat insects avoid wasps, too. They have been stung by insects with the bright colouration of the yellow-jacket wasps. There are insects that have the same colouration as yellow-jacket wasps, such as the fly in Figure 2.20B, but do not sting. However, *looking* like an animal that *does* sting has meant that insect-eating animals avoid eating the fly. Looking like something else—the surroundings or something unpleasant like an organism that stings or tastes bad—is called **mimicry**.

Other organisms camouflage themselves by mimicking the shape of something. For example, the thornbugs in Figure 2.21 stay close to the thorns on the stem of a plant. When thornbugs cluster, they look like thorns on a plant, and predators avoid them. Other insects mimic the shape of leaves. Some snakes mimic the colour and shape of vines.

PROBLEM-SOLVING

INVESTIGATION 2-G

SKILLCHECK
- Identify the Problem
- Decide on Design Criteria
- Plan and Construct
- Evaluate and Communicate

Camouflage Creature

Challenge

Create a 3-D model of a creature or object that will be camouflaged in your classroom.

Materials

containers of your choice such as milk jugs, food containers, and boxes
small items of your choice such as pins, buttons, string, pipe cleaners, and craft sticks
art supplies

Design Specifications

A. Your model must be at least 20 cm long.

B. Your model must be three-dimensional.

C. Your model must be camouflaged in its environment.

Plan and Construct

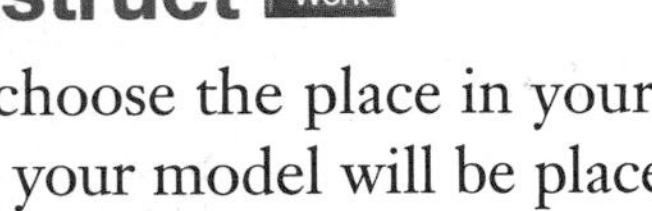

Group Work

1. With a partner, choose the place in your classroom where your model will be placed. For example, you might choose a bulletin board, a plant, or a piece of furniture.
2. Draw a labelled sketch of your model, indicating what materials you will use.
3. Obtain your teacher's approval and then construct your model.
4. Improve your model until you are satisfied with the way it looks and feel confident that it will be camouflaged in the location you selected.
5. Place your model in the location you chose.
6. Challenge your classmates to find your model.

Evaluate

1. How many students found your model?
2. Was your model well camouflaged in the location you chose? Why or why not?
3. What changes could you make to your model so that it is better camouflaged?
4. Where in your classroom would your model be the least camouflaged? Explain your answer.
5. Why is camouflage important for animals?

Section 2.2 Summary

Adaptations are physical characteristics or behaviours that increase an organism's chances of surviving in a particular environment.

- Adaptations are inherited characteristics.
- Physical adaptations are physical characteristics, such as the excellent vision of owls, which give organisms a better chance of surviving.
- Hibernation and migration are examples of behavioural adaptations.
- Camouflage refers to adaptations that help organisms hide from other organisms.
- Mimicry is a type of camouflage; mimicry allows organisms to gain protection by copying the colouration and/or shape of a harmful organism.

Key Terms

adaptations
behaviour
inherited characteristic
physical adaptations
behavioural adaptations
hibernation
migration
colouration
camouflage
mimicry

Check Your Understanding

1. How are leaves adapted to help plants survive?
2. What are the differences between physical and behavioural adaptations? Provide an example of each.
3. Why are some organisms, such as the cactus or the polar bear, able to survive in very harsh environments?
4. How are colouration and mimicry similar? How are they different?
5. Which of these is a physical adaptation: the wool on the sheep or wearing a wool sweater? Explain.
6. **Thinking Critically** Domestic dogs can be very small (like lap dogs) or very big (like Irish wolfhounds). Almost all domestic cats have a mass between 6 kg and 10 kg. Why do you think domestic cats are all about the same size while dogs can be big or small? Explain your answer.
7. **Thinking Critically** Like other birds that perch, robins have feet with three front toes, one long back toe, and a specialized tendon that automatically locks their back toes around a branch when they land. Describe two other adaptations that help birds survive.
8. **Apply** You are an ecologist studying wildlife populations. You notice lots of lynx tracks in a large section of forest isolated from human development. In a smaller section of the same type of forest close to human development, you see no signs of lynx.
   (a) What might be the reason for this?
   (b) How might you test your explanation?

CHAPTER at a glance

Now that you have completed this chapter, try to do the following. If you cannot, go back to the sections indicated in brackets.

(a) Explain how climate and weather are related. (2.1)

(b) If you know the climate of an area, you will be able to guess what kinds of animals you will find there. Is this true? Explain why. (2.1)

(c) Describe the characteristics of the six land biomes. (2.1)

(d) Describe the characteristics of the five aquatic biomes. (2.1)

(e) Explain why organisms on seashores need to be able to survive changes in temperature, moisture, and saltiness, as well as the force of wave action. (2.1)

(f) What are the differences between coral reefs and seashores? What are the similarities? (2.1)

(g) How do inherited characteristics enable organisms to survive? (2.2)

(h) Name two behaviours that help animals survive harsh winter conditions. Give an example of each and explain why it works best for that animal. (2.2)

(i) How does camouflage enable organisms to survive? (2.2)

Prepare Your Own Summary

Summarize Chapter 2 by doing one of the following. Use a graphic organizer (such as a concept map), produce a poster, or write a summary to include the key chapter ideas. Here are a few ideas to use as a guide:

- Describe how specific adaptations enable organisms to survive in their environments (for example, colouration, mimicry, camouflage, feeding, and other behaviours and physical characteristics).
- Describe how the land on Earth can be divided into six regions based on temperature, rainfall, plants, and animals.
- Choose an animal or plant that lives in your local environment. Describe how this organism's adaptations allow it to survive there.

CHAPTER 2 Review

Key Terms

climate (2.1)
weather (2.1)
precipitation (2.1)
latitude (2.1)
altitude (2.1)
biome (2.1)
aquatic (2.1)
estuary (2.1)
adaptations (2.2)
behaviour (2.2)
mimicry (2.2)
inherited characteristic (2.2)
physical adaptations (2.2)
behavioural adaptations (2.2)
hibernation (2.2)
migration (2.2)
colouration (2.2)
camouflage (2.2)

Reviewing Key Terms

If you need to review, the section numbers show you where these terms were introduced.

1. Describe the difference between:
   (a) climate and weather (2.1)
   (b) latitude and altitude (2.1)
   (c) camouflage and mimicry (2.2)
   (d) migration and hibernation (2.2)

2. In your notebook, match each description in column A with the correct term in column B. Use each description only once.

A	B
(a) biomes that differ by the amount of salt in the water and water temperature	• precipitation (2.1)
(b) the way an organism acts	• biome (2.1)
(c) the colour and pattern of an animal's skin, fur, or feathers	• aquatic (2.1)
(d) where a river flows into the ocean	• estuary (2.1)
(e) the amount of rain or snow that falls on a certain location	• behaviour (2.2)
(f) a trait that an organism is born with	• inherited characteristic (2.2)
	• camouflage (2.2)
	• colouration (2.2)

Understanding Key Ideas

3. Could a prairie dog from Saskatchewan survive in the grasslands in central Africa? Explain your answer.

4. Scientists have been able to teach certain animals to communicate using sign language. Would this skill be considered an adaptation? Explain your answer.

5. Humans use a variety of fabrics to make clothing. Explain why using fabrics is not a physical adaptation for humans.

6. A lynx is chasing a snowshoe hare through the snow. What adaptations enable each of these animals to survive winter?

7. What happens to organisms when a resource that they depend on becomes scarce? What might happen if the resource remains scarce?

8. Choose a creature that lives in the ocean. Identify two of its physical adaptations and two behavioural adaptations. Explain how each of these adaptations helps it survive in its environment.

Developing Skills

9. Choose a place in British Columbia. Imagine you are part of an Aboriginal community living there several hundred years ago. Using the materials available in that location (without modern technology), decide the following:
   - how you would make a shelter
   - how you would hunt, gather, and prepare food
   - how you would travel

   Make a diorama, a poster, or a computer presentation showing how you and your community would live in that environment.

Problem Solving

10. Every year Mr. Gardener tries to grow orchids in his garden in northern British Columbia. His orchids have never had any flowers. What could be preventing Mr. Gardener's orchids from flowering?

11. Aquaculture (fish farming) is a growing industry in Canada. In one region, an aquaculture company wants to raise exotic fish to be sold in pet stores. To save costs, the fish farmer wants to keep the fish in a net in a local lake instead of building a separate container. Some local sport fishers are worried that the exotic fish might escape and kill the native trout and bass populations or eat all their food and make it hard for them to survive. What characteristics should the exotic fish have that would make them unable to compete with the native fish populations? Give three examples.

Critical Thinking

12. Many people in Canada work very hard to preserve the tropical rain forest biome. Why would they work so hard to preserve a biome that does not even exist in Canada?

13. What adaptations would you look for in an edible plant to be grown on a spacecraft?

14. The colouration of female birds is not usually as bright as the colouration of male birds in the same species. Why do you think this is so? Give three examples of how this would help the birds survive.

Pause& Reflect

Go back to the beginning of this chapter on page 34 and check your original answers to the Getting Ready questions. How has your thinking changed? How would you answer these questions now that you have investigated the topics in this chapter?

CHAPTER

3 Classifying Living

Getting Ready...

- How is a daisy different from a dragonfly? How are they similar?
- How could you describe a turtle to another turtle expert so that she would know exactly what type of turtle you are talking about?
- Librarians sort books by topic. How do biologists sort living things?

This biologist is gathering information about the different types of organisms, such as insects and spiders, that live in the tops of the huge trees in our coastal rain forest. These forests, with large trees and thick growths of moss, are home to as many as 15 000 kinds of animals, mostly insects. At least 500 of these different kinds of organisms have never been identified by anyone before!

How do scientists organize and classify (group) all of the kinds of living things on Earth? How do they identify new types of organisms that they have not seen before? This is what you will be learning about in this chapter.

Have you ever seen an organism like the one shown on the opposite page? Do you think it is a plant or an animal? How would you decide if it was a plant or an animal? Plants and animals are two of the six major groups of types of organisms that biologists refer to when they classify living things. You will learn about these groups in this chapter.

Things

What You Will Learn

In this chapter, you will learn

- how scientists sort and classify living things
- the characteristics of six major classifications of living things
- similarities and differences between groups of animals, such as amphibians, reptiles, and fish

Why It Is Important

- All organisms are grouped because they share certain characteristics. In order to identify organisms, you should know what makes them similar to, and different from, other organisms.
- By learning how organisms are sorted and grouped, you are better able to find out more information about organisms.
- Classification is a way to organize and study the story of life on Earth.

Skills You Will Use

In this chapter, you will

- classify organisms into groups using their internal (inside) and external (outside) features
- use a classification key to identify and classify an organism
- use variables to test the best conditions for the growth of mould
- use a microscope to observe and classify organisms that live in pond water

Is this organism a plant or an animal? How do you know?

Starting Point ACTIVITY 3–A

Sort It Out!

Scientists have developed ways of grouping organisms to show how living things are related. What features could you use to classify a group of similar objects? Try this activity to find out.

What to Do

1. Describe each of the fruits shown above using only their *external* features.
2. Describe each of the fruits using their *internal* features.
3. Look at two of the fruits at a time. How are they similar? How are they different? Use a chart and/or a Venn diagram to **record** your observations.
4. Using your observations, sort the fruit into four groups of fruit that have similar features. Try to sort the fruit into three groups and then into two groups.
5. What features did you use to sort the groups of fruit? Explain your system to your classmates. Is there more than one way to sort the fruit?

Skill POWER

For tips on using Venn diagrams, turn to SkillPower 1.

Section 3.1 Grouping Living Things

Key Terms
- classification systems
- kingdom
- species

When you go into a grocery store, do you find lettuce, bread, and milk in the same section? Most grocery stores group similar types of foods together. You will probably find milk, yogurt, and cheese in the dairy department while lettuce, tomatoes, and apples are in the produce department. When you group similar items together, you are classifying them. To classify means to group objects, information, or even ideas based on the similarities they share. You use different **classification systems** (ways of grouping things) in your everyday life. For example, when you visit a library, a bookstore, or a department store, objects that are similar in some way or ways are grouped together. In another example, the students in your school are grouped according to their grade level.

There are several reasons for using classification systems. One reason is to put items in order so that you can find them. Librarians, for example, use a classification system to organize and label the thousands of books in a library. How hard would it be to find a book in a library if books were not organized in some way?

Scientists who study living things also use classification systems. Scientists show how types of organisms are similar by grouping them. There are over 1.5 million types of organisms in the world today. Grouping them is a way of organizing the information about all living things in one system. How would you classify the organisms in the tidal pool in Figure 3.1?

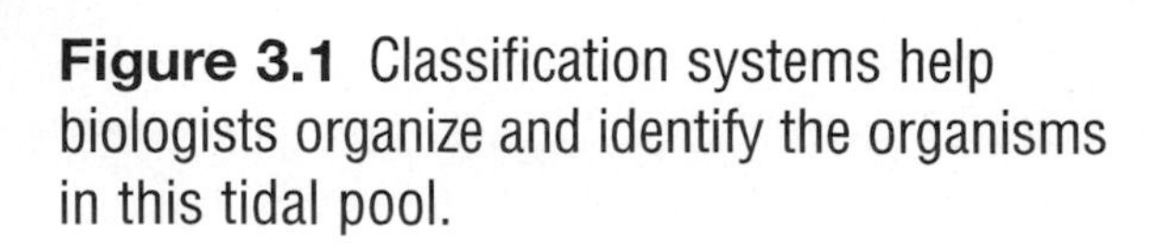

Figure 3.1 Classification systems help biologists organize and identify the organisms in this tidal pool.

How Are Organisms Classified?

More than 2000 years ago, Aristotle, a Greek philosopher, developed a system to classify living things. Aristotle sorted types of organisms into two categories: plants and animals. He called these two categories *kingdoms*. Today, the term **kingdom** is used for the largest groups of types of organisms. For example, all plants belong to Kingdom Plantae, and all animals belong to Kingdom Animalia.

In science, how is the term "kingdom" used?

Aristotle classified types of organisms by grouping them according to their outside appearance, such as whether an organism had fur or not. Then he used characteristics such as how it reproduced and where the organism lived to place it into a more specific group. Aristotle's system was one way to classify types of organisms, but it wasn't always useful. For example, how could you classify frogs using this system? Frogs spend part of their lives in the water and part on land. As scientists learned more about organisms, classifying them using Aristotle's system became more and more difficult. The invention of the microscope allowed scientists to see organisms that did not belong in either of Aristotle's two kingdoms.

New Classification and Naming Systems

In fact, scientists were developing new classification systems as their knowledge grew, but they were creating different systems. The variety of systems made it very hard for the scientists to share information. Sometimes the same type of organism was placed in different groups according to the different systems.

The naming of organisms was even more confusing. In many cases, the same type of organism had a different name in each country in which it lived. Some types of organisms might also be known by two or more names in the same country! For example, the seaweed in Figure 3.2 grows on the West Coast of North America. It is known by many names along the coast. Bladder wrack, popping weed, and rockweed are just a few of them.

Figure 3.2 Local communities along the west coast of North America, from Alaska to California, each have their own name for this seaweed.

DidYouKnow?

Names of plants often describe what a plant looks like. For example, the Ditidaht people on the west coast of Vancouver Island call the rockweed shown in Figure 3.2 on page 63 by a word that translates to "a lot of blown up things on the rocks."

INTERNET CONNECT

www.mcgrawhill.ca/links/BCscience6

Taxonomy is the study of scientific classification. Carolus Linnaeus is sometimes referred to as the "father of taxonomy." To find out more about Linnaeus and scientific classification, go to the web site above. Click on **Web Links** to find out where to go next.

In the 1730s, Carolus Linnaeus, a Swedish scientist, developed a way to classify organisms that scientists all over the world agreed to use. Linnaeus divided the large category of kingdom into smaller groups. First, each kingdom was divided into two or more phyla (singular, *phylum*). All members of a phylum share one or more important structures or characteristics. For example, the sunflower star and the salamander in Figure 3.3 are both in the same kingdom, Kingdom Animalia. However, salamanders have a backbone, but sea stars do not. All animals with backbones belong to the same phylum, *Chordata*. The sunflower star and other animals without backbones belong to other phyla.

Figure 3.3 The sunflower star and the western red-backed salamander are both in the same kingdom but are in different phyla.

Linnaeus then divided types of organisms in a phylum into a series of increasingly specific categories, or levels of organization, called *classes*, *orders*, *families*, *genera* (singular, *genus*), and *species* (singular, also *species*). As you go down the levels, the types of organisms in each level are more and more alike. At the bottom of the series, a genus is a group of organisms that share many similar characteristics. A **species** is the most specific level of classification for an organism. Organisms belonging to the same species can mate to produce healthy young. The example in Figure 3.4 shows the classification of dogs.

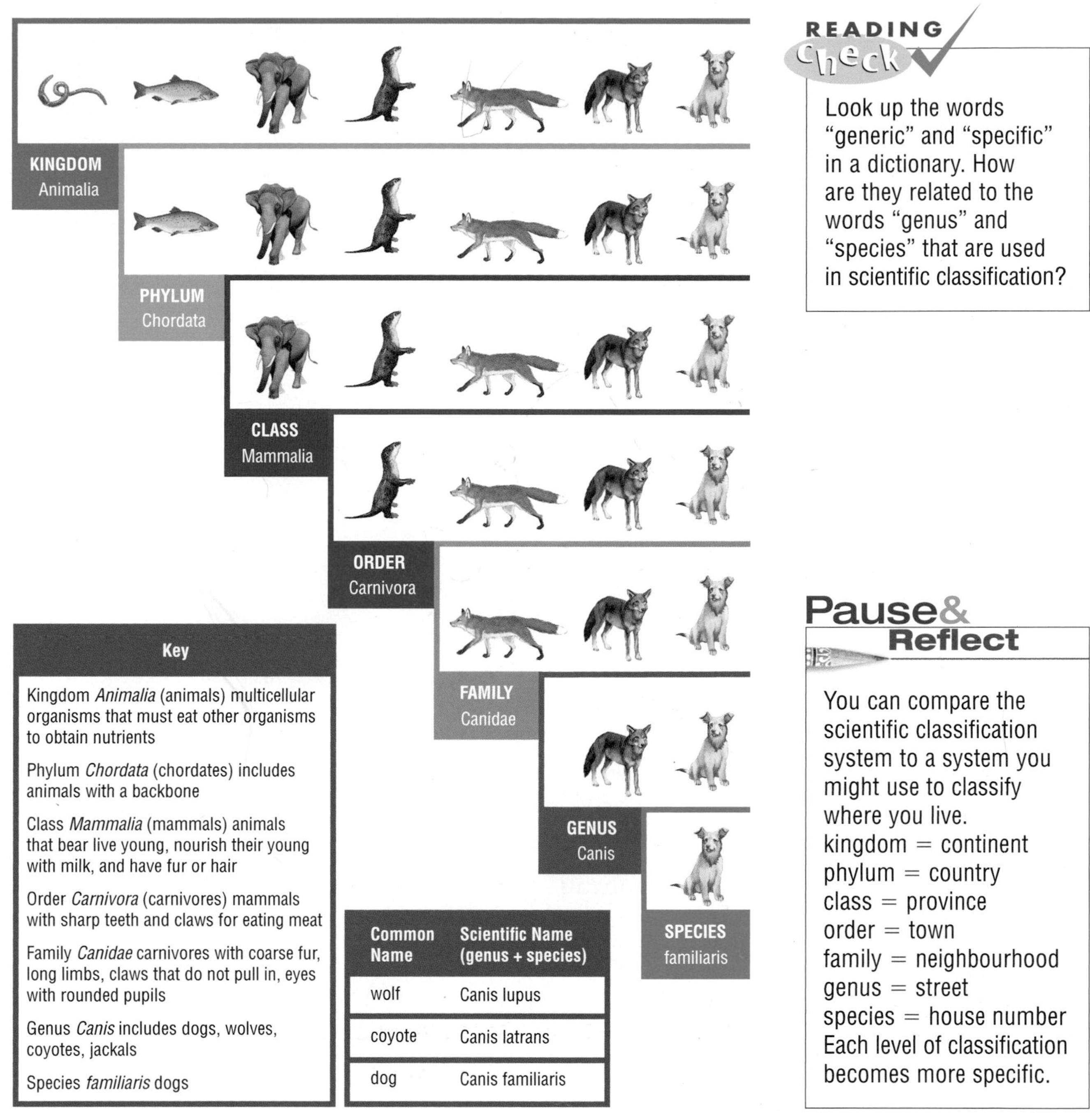

Common Name	Scientific Name (genus + species)
wolf	Canis lupus
coyote	Canis latrans
dog	Canis familiaris

Figure 3.4 Classification key for dogs

READING Check

Look up the words "generic" and "specific" in a dictionary. How are they related to the words "genus" and "species" that are used in scientific classification?

Pause& Reflect

You can compare the scientific classification system to a system you might use to classify where you live.
kingdom = continent
phylum = country
class = province
order = town
family = neighbourhood
genus = street
species = house number
Each level of classification becomes more specific.

Two-Part Scientific Names

To identify different organisms, Linnaeus gave each its own two-part scientific name. Scientists still use this system. The first part of an organism's scientific name is its genus, and the second is its species. If you look at the classification for dogs, you can see that the scientific name for dogs is *Canis familiaris*. The scientific name for humans is *Homo sapiens*. *Homo* is the genus for humans, and *sapiens* is the species.

Pause & Reflect

Think of all of the people you know with the same first name. For example, how many people named John do you know? How does using a system in which every person has at least two names help you identify one person from another?

No two types of organisms can have the same scientific name. For example, coyote in Figure 3.5(A) has the scientific name *Canis latrans*, and the wolf in Figure 3.5(B) has the scientific name *Canis lupus*. The coyote and the wolf (and dogs) are similar enough to be in the same genus, *Canis*, but they are placed in separate species because wolves don't tend to mate with coyotes.

People around the world can use the scientific name and know that they are all looking at or discussing the same species. This means that they can also share information about this species.

Figure 3.5 The coyote (A) and wolf (B) are both in the same kingdom, phylum, class, order, family, and genus. They are different species.

Other Classification Systems

While the classification system explained above is the system used by scientists, it is not the only way to classify living things. Different peoples around the world sort living things in their own ways and use different classification systems. Many of the Aboriginal peoples in Canada, for example, have classification systems that are more detailed than scientific classifications. For example, the Inuit of northern Canada do not have a word for mammals. Instead, they have a word for sea mammals and a word for land mammals. The coho salmon, shown in Figure 3.6, is considered a single species called *Oncorhynchus kisutch* by scientists. However, the Nuu-chah-nulth people of the west coast of Vancouver Island use one classification for when this fish lives in the ocean and one for when it lives in fresh water.

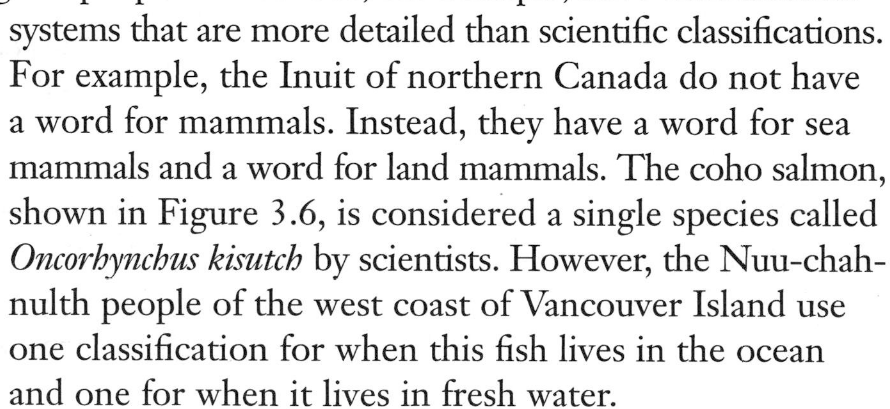

Figure 3.6 When the Nuu-chah-nulth people classify this coho salmon, its category depends on whether it lives in the sea or in fresh water.

Tools for Identifying Organisms

What would you do if you were asked to identify the plant shown in Figure 3.7? You could ask someone who knows a lot about plants. Or, you could use a plant field guide to help you find the plant's common names as well as its scientific name. Field guides have pictures and descriptions that enable you to identify living things. There are also field guides for non-living things such as rocks. Field guides often contain classification keys that give step-by-step instructions to help you to identify unknown organisms. Try to use a classification key, and then make your own, in the Find Out Activities 3-B and 3-C.

Figure 3.7 Tiger lily is a common name for this type of flower. Its scientific name is *Lilium columbianum*.

Identifying Canadian Cats

You can use classification keys to identify organisms. Try this activity to see how it's done.

Classification Key to Native Cats of North America
1. Tail length: **(a)** short, go to 2 **(b)** long, go to 3
2. Ear tufts: **(a)** long ear tufts tipped with black; lynx, *Lynx canadensis* **(b)** short ear tufts; bobcat, *Lynx rufus*
3. Coat: **(a)** plain coloured, go to 4 **(b)** patterned, go to 5
4. Coat colour: **(a)** yellowish to tan on back; white to beige on belly; cougar, *Puma concolor* **(b)** all brown or black; jaguarundi, *Herpailurus yaguarondi*

What to Do

1. Start at step 1. Choose either (a) or (b) to find either the scientific name of a species or directions to go to the next step.

What Did You Find Out?

1. What are the scientific names of the two cats?
2. Why is it important to begin with the first step instead of in the middle of a classification key?

Now that you've used a classification key, try to make your own in the next Find Out Activity.

Find Out ACTIVITY 3–C

Create Your Own Classification Key

Classification keys are also called dichotomous keys. *Dicho* means split in two. In every step, you have two choices.

What You Need

a collection of buttons, seeds, or pasta of different sizes and shapes
a large sheet of paper
ruler

What to Do

Group Work

1. Cover your work area with a large sheet of paper and copy the following table onto the sheet. Leave lots of room to write in each space in your chart.

A				B			
C		D		E		F	
G	H	I	J	K	L	M	N

2. With a partner, decide how to divide the collection of objects into two groups: A and B. All of the items must belong in either group. For example, a collection of buttons might be divided into groups of buttons with four holes and groups that have some number of holes other than four. If this is how you divided the items, your divisions would be:
   A = Buttons with four holes
   B = Buttons with less than four holes
3. Record your reasons for grouping the items on your chart and place the items where they belong on the chart.
4. After you have divided the collection into two groups, divide the first group (A) into two more groups. For example, if group A contains buttons with four holes, group C might contain four-holed buttons made of metal and group D might contain four-holed buttons made of a material other than metal.
5. Divide group B into two more groups.
6. Continue to divide the groups until each item has its own place on the chart. Make up a two-part name for each item based on the bottom two levels.
7. When you have completed your chart, put the items back into a pile and ask a classmate to use your chart to classify them.

What Did You Find Out?

1. Is it possible to create more than one dichotomous key for classifying and identifying the same group of objects? Explain your answer.
2. When two people use the same dichotomous key to identify the same object, is it possible for each of them to have different final answers?
3. Why are classification tools such as dichotomous keys useful?

Section 3.1 Summary

We use different classification systems to sort and organize many things, including books in a library or items in a grocery store. In science, organisms are sorted according to similar characteristics.

- Different types of organisms are sorted by kingdom, phylum, class, order, family, genus, and species. Kingdom is the broadest category, and species is the most specific.
- All organisms have a two-part scientific name, which includes both their genus and their species names.
- No two types of organisms share the same genus and species names. Each type of organism is placed in its own species. Several types of organisms can share the same genus name.
- You can identify organisms using classification keys.
- Aboriginal peoples in Canada have their own systems for classifying living things.

Check Your Understanding

Key Terms

classification systems
kingdom
species

1. What are the seven levels used in the scientific classification system?
2. What kinds of information do scientists use to classify organisms?
3. Why do scientific names have two parts?
4. (a) What is a dichotomous key?
   (b) Why is it useful? Give an example.
5. (a) What is a field guide?
   (b) How is it used?
6. **Apply** Create a classification system to do each of the following:
   (a) organize your desk
   (b) organize your drawers or closet
   (c) plan a meal
   (d) decide what clothes to take on a trip
7. **Thinking Critically** What are some examples of everyday words that name groups or classes of things? Think about subjects you study in school such as grammar, math, and social studies. What problems would there be if words such as "noun" and "fraction" did not exist?

Section 3.2 The Kingdoms of Life

Key Terms
- Monera
- Protista
- Fungi
- Plantae
- Animalia
- invertebrates
- vertebrates
- fish
- amphibians
- reptiles
- birds
- mammals

One classification system scientists commonly use today groups organisms into six kingdoms: animals, plants, fungi, protists, "true" bacteria, and ancient bacteria. Types of organisms are classified in a kingdom based on several characteristics, such as whether an organism is multicellular or unicellular or whether an organism can make its own food or not. The six kingdoms and some of their characteristics are shown in Figure 3.8.

Kingdoms of Bacteria (Monera)

Bacteria (singular, *bacterium*) are unicellular micro-organisms. A cell of bacteria does not have a nucleus. These organisms can be in one of three shapes: rods (shown in Figure 3.9), spirals, and spheres. Most bacteria do not make their own food. Instead, they break down (decompose) other living and once-living things.

Kingdom Monera once contained all bacteria. Scientists have now separated Monera into two kingdoms: ancient bacteria and "true" bacteria.

Ancient Bacteria	True Bacteria	Protista	Fungi	Plantae	Animalia
• Ancient bacteria are unicellular micro-organisms.	• True bacteria are unicellular micro-organisms.	• Some are unicellular; others are multicellular.	• Most are multicellular.	• All are multicellular.	• All are multicellular.
• Cells do not have a nucleus.	• Cells do not have a nucleus.	• Cells have a nucleus.	• Cells have a nucleus and a cell wall.	• Cells have a nucleus and a cell wall.	• Cells have a nucleus.
• Some ancient bacteria make their own food; others obtain food from the bodies of other organisms.	• Some true bacteria make their own food; most obtain food from the bodies of other organisms.	• Some use photosynthesis to make their own food; others obtain food from other organisms.	• All obtain food from other organisms.	• All use photosynthesis to make their own food.	• All must eat other organisms (such as plants or other animals) to obtain food.
• They can live in extreme environments, often without oxygen.					

Figure 3.8 The six kingdoms of life. The names come from the ancient language of Latin.

Ancient Bacteria

Ancient bacteria can survive in the kind of conditions that were found on Earth long ago. (These bacteria are placed in the Kingdom Archaebacteria. *Archae* is a Greek word that means ancient.) Ancient bacteria are often found in hot sulfur springs, muddy environments such as mudflats and swamps, and places deep in the ocean where lava and hot water seep through cracks in the ocean floor. These extreme environments have little or no oxygen.

Off the west coast of British Columbia there are several hot vents in the sea floor where the water is over 300°C, and archaebacteria thrive there. Some of the larger organisms that live there, such as giant crabs, feed on the ancient bacteria that form thick mats around the vents.

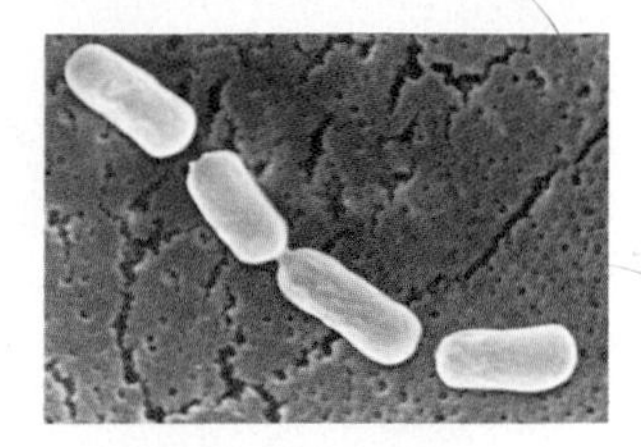

Figure 3.9 These rod-shaped bacteria (magnified 15 000) cause food poisoning.

One type of ancient bacteria lives in the digestive system of cattle. It enables the cattle to digest the tough parts of grasses, which they eat but can't digest without the bacteria.

True Bacteria

Bacteria that are referred to as "true" bacteria include bacteria that live all around us—in the soil, in the air, in the water, and in and on our bodies and other organisms. These bacteria are grouped in the Kingdom Eubacteria. Most of these bacteria do not live in the harsh environments that ancient bacteria live in, but they live everywhere else on Earth.

Most true bacteria do not make their own food. Instead, they absorb nutrients from other living or once-living organisms. One type of true bacteria, the blue-green bacteria, *does* make its own food, however. These bacteria use photosynthesis to make food, just as plants do. Blue-green bacteria often live together in long chains in ponds or lakes. They can cover the pond in a green, slimy growth as shown in Figure 3.10.

Learn more about true bacteria in Find Out Activity 3-D.

Figure 3.10 This pond is covered in a growth of blue-green bacteria.

DidYouKnow?

The six-kingdom system is just one way to group organisms. Before Linnaeus's time only two kingdoms (plants and animals) were used. Now, most biologists use five or six kingdoms. However, some biologists group life into three "domains" instead! Biology changes as new types of organisms are discovered. So the way we classify organisms can change, too.

Find Out ACTIVITY 3-D

Wanted: Bacteria That Make You Sick

The kingdom of true bacteria includes some bacteria that are very harmful to people or other living things. In this activity you will research a disease caused by bacteria and create a "Wanted" poster for that bacteria.

What You Need

poster-size paper
art supplies

What to Do

1. Select one of the following diseases caused by bacteria: meningitis, tetanus, diptheria, Rocky Mountain spotted fever, botulism, scarlet fever, or tuberculosis.
2. Use library or Internet resources to find as much of the following information as possible:
   - a diagram or picture of the bacterium as seen under a microscope
   - a written description of the size, shape, and colour of the bacterium
   - a description of how the bacterium causes the disease and how the disease is spread
   - physical effects of the disease
   - common victims of the disease
   - locations in the world where the disease is most likely to occur
   - the most effective way(s) to prevent this disease
   - any other information of your choice
3. Create a bacterium "wanted" poster that will tell your classmates everything they need to know in order to protect themselves from the disease it causes.

Good Bacteria Around You

Very few bacteria actually cause serious illnesses. In fact, most bacteria are helpful, not harmful. A shovelful of soil contains billions of bacteria! They decompose dead organisms and recycle nutrients needed for the survival of living things. Bacteria are in the foods you eat (such as cheese, yogurt, and even vinegar). Some of the bacteria that live in your body help you digest your food.

Kingdom Protista

Look at the organisms in Figure 3.11. Although these organisms look quite different from each other, they are all in the same kingdom. The Kingdom **Protista** has the biggest variety of members, called protists, of all of the kingdoms of life. Protists can be unicellular or multicellar organisms that usually live in moist or wet environments. Their cells have a nucleus. Some protists make their own food using photosynthesis, while others do not. Some, like the Euglena shown in Figure 1.14 in Chapter 1, can do both. Some protists are microscopic; others, such as giant seaweeds, can be very large.

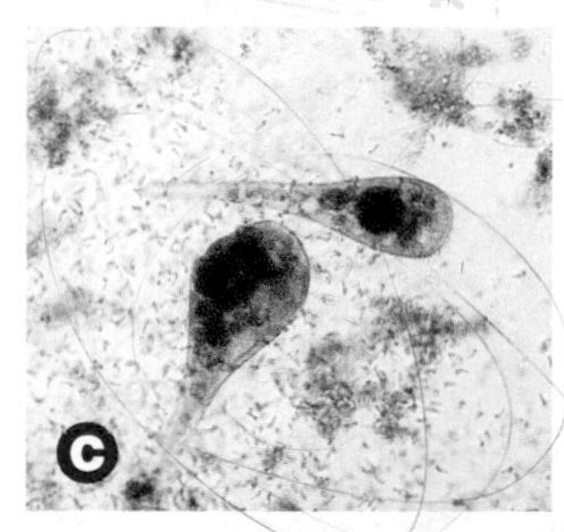

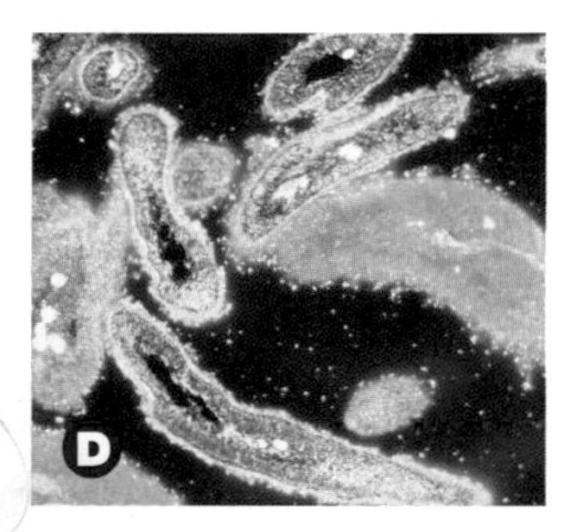

Figure 3.11 This slime mould (A), seaweed (B), amoeba (C), and paramecia (D) are all types of protists. Amoebas (C) and paramecia (D) move around to find food, such as bacteria.

The organisms in Figure 3.11 (C) and (D) are protists that, like animals, move and capture and eat food. These protists live in water, soil, or in both living and dead organisms and use different ways to move around as they pursue food. Some have many tiny hairs that they use to propel themselves. Others use one or two long "whips" to move. Others move by extending a flexible "foot."

To look at some protists, try Conduct an Investigation 3-E.

The brightly coloured slime mould in Figure 3.11(A) forms a delicate, web-like structure on the surface of its food. It obtains nutrients by decomposing its food. Slime moulds live on decaying logs or dead leaves in cool, moist, shady forests. Although slime moulds might look like fungi, they are different in many ways. For one thing, slime moulds can move. As they creep along, they feed on bacteria and decaying plants and animals.

CONDUCT AN

INVESTIGATION 3-E

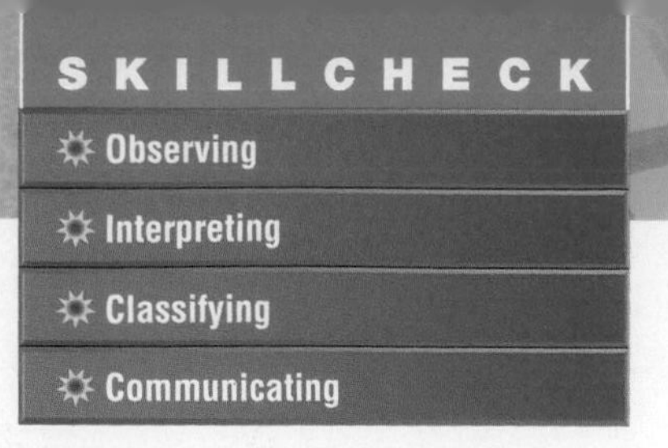

Looking for Micro-organisms

In this investigation, you will look for micro-organisms, such as protists, in a sample of pond water. To increase the number of micro-organisms in the pond water, you will make a hay infusion by adding hay or grass to the water to provide food for the organisms living there.

Question

What micro-organisms live in pond water?

Safety Precautions

- If your microscope has an electrical cord, make sure your hands are dry when you plug in or disconnect it.
- Always hold onto the *plug* to remove an electrical cord from the socket. *Never* pull on the cord.
- Handle microscope slides carefully so that they do not break or cause cuts or scratches.
- Do not put your hands on your face or near your mouth during this investigation.
- Wash your hands with soap and water after handling the pond water sample.

Apparatus

a large jar
spoon
medicine dropper
microscope
microscope slides
cover slips
pond life identification guide

Materials

pond water
grass or hay
a plastic drinking straw

Skill POWER

For tips on using a microscope and making a wet mount, go to SkillPower 2.

Procedure

1. To make the hay infusion, put the pond water in the large jar and add a small handful of grass or hay and a few grains of yeast. Cover the jar with a lid (not too tight) and keep it at room temperature for several days.
2. **Observe** how the water looks each day and **record** your observations. For example, note the colour of the water and whether you can see anything moving.
3. After one week, use the medicine dropper to remove a few drops of water from the surface. Prepare a wet mount of the sample.
4. View the slide under low power on the microscope. Use drawings and words to describe what you see.
5. Put your thumb over the end of the straw and put it into the jar. When the other end of the straw is at the bottom of the jar, release your thumb. Water will move up the straw. Put your thumb back on the straw and remove the water sample.
6. Put a drop of pond water on a clean slide by touching the end of the straw on the slide. Do not release your thumb; you will release far too much water. Prepare a slide from a sample from the bottom of the jar. Repeat step 4.
7. Repeat step 5, but remove a sample from the middle of the jar.
8. Make a slide from a sample from the middle of the jar. Repeat step 4.

Analyze

1. Describe the daily changes you observed in the jar of pond water.
2. **(a)** How many different micro-organisms did you observe?
   (b) Describe the size, colour, and shape of each organism. Also describe any movement you saw.
3. Why do you think it was important to sample different levels in the jar?
4. Use the pond life identification guide to identify some, or all, of the organisms you observed.

Conclude and Apply

5. **(a)** Which pond water sample (from the top, middle, or bottom of the jar) contained the most micro-organisms?
   (b) Why do you think the samples were different?
6. Predict what you might see in a sample of pond water one month after you started a hay infusion.

Extend Your Skills

7. Continue to feed the hay infusion by adding more pond water and hay or grass each week. **Record** your observations after taking new samples and viewing them under a microscope.

Kingdom Plantae

If you walked through the mountaintop meadow shown in the unit opener, you would be surrounded by wildflowers and other plants. Wildflowers, grasses, trees, and mosses are all members of the Kingdom **Plantae**. All plants are multicellular and use photosynthesis to obtain food. Plants have roots or root-like structures that help hold them in the ground and enable them to absorb water. Plants grow in almost all parts of the world.

Organisms in the Kingdom Plantae can be divided into two large groups, as shown in Figure 3.12 on the next page. The first group contains plants such as trees, grasses, ferns, and wildflowers. The steps these plants have structures that are like tubes. Water absorbed by the plant's roots moves up one set of tubes the leaves. At the same time, food (sugar) made in the leaves is moving through another set of tubes, which lead to all of the other parts of the plant. These tubes are called *vascular tissues*, and these plants are called vascular plants. You will see how a plant's vascular tissues work in Find Out Activity 3-F.

Pause & Reflect

How are vascular tissues like the veins and arteries in your body?

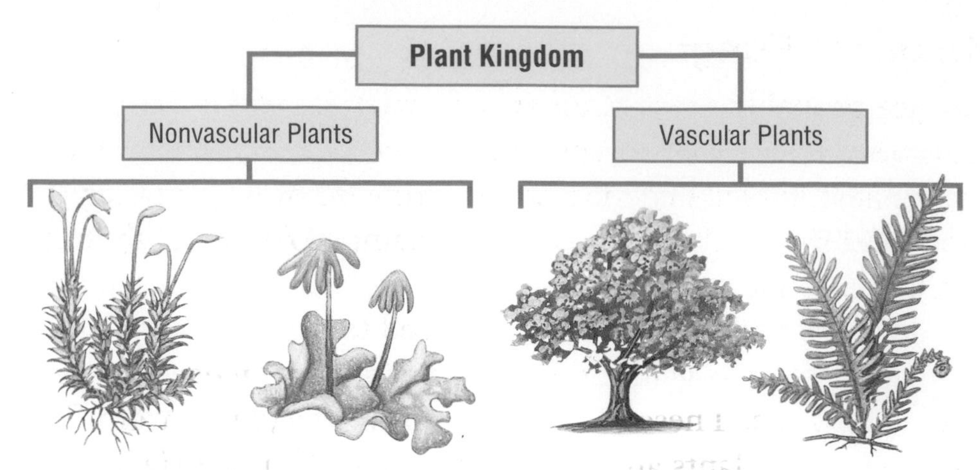

Figure 3.12 Plants can either be vascular or non-vascular. By examining other characteristics of plants, biologists have divided Kingdom Plantae into 11 different groups.

Mosses and other simple plants, such as the *liverwort* shown in Figure 3.12 do not have vascular tissue. They are called non-vascular plants. These types of plants are usually very small and absorb water directly through their cells walls.

Find Out ACTIVITY 3–F

The Function of Stems in Plants

In this activity you will observe the function of a stem and describe the role the stem plays in plant growth.

Safety Precautions

What You Need

1 fresh white carnation
scissors
500 mL container
water
food colouring

What to Do

1. Fill the container with 400 mL water. Add four or five drops of food colouring.
2. Use the scissors to trim the bottom of the carnation stem. Make a 5 cm-long cut up the length of the stem. Turn the stem a quarter turn and make another 5 cm-long cut up the length of the stem. The bottom 5 cm of the stem should now be split into four "strips."
3. Quickly place the cut carnation stem in the coloured water.
4. Record your observations after 5 min, after 30 min, and after 1 h.

What Did You Find Out?

1. How were your observations after 1 hr different from your observations after 5 min?
2. What are the functions of the stem?

Kingdom Fungi

You are probably more familiar with fungi (singular *fungus*) than you realize. The mushrooms in your soup or on your pizza are a type of fungi. The moulds that grow on bread or other foods and the mildew that grows on damp objects like shower curtains are also fungi.

Fungi obtain the nutrients they need to survive by absorbing them from other organisms. Often they do this by decomposing dead organisms. These fungi break down everything from food scraps to dead plants and animals and return the nutrients of the dead organisms to the soil. Other fungi are parasites. This means that they absorb nutrients from living things, usually in amounts small enough that the organisms providing the nutrients are not harmed.

Most fungi are multicellular, although there are some that are single-celled. All fungi cells have a nucleus and a cell wall. The mushrooms shown in Figure 3.13 are only one small part of a fungus. The rest of the organism is a mass of thread-like tubes that grow through the soil or tree trunk. These tubes are the part of the fungus that absorbs nutrients. The mushroom is the part of the fungus used for reproduction. It contains the *spores* of the fungus. (You saw spores on a fern in Chapter 1.) The mushroom releases tiny spores when it is time for the fungus to reproduce. Take a closer look at a fungi in Conduct an Investigation 3-G.

DidYouKnow?

Many fungi are very useful. For example, the mould *Penicillium notatum* produces a chemical that kills bacteria. It is used in a medicine called penicillin, which is used to fight bacteria that make people ill.

Figure 3.13 The visible part of this fungus called "chicken-of-the-woods" is just one part of the organism. Chicken-of-the-woods grows in British Columbia's forests.

CONDUCT AN

INVESTIGATION 3-G

SKILLCHECK
- Observing
- Communicating
- Measuring and Reporting
- Predicting

Grow a Fungi Garden

In this investigation you will grow mould in a variety of conditions.

Question

What conditions are necessary for the growth of a fungus we call bread mould?

Safety Precautions

- If you have any allergies to any type of mould, inform your teacher before the class begins this investigation.
- Wash your hands carefully after completing each part of this investigation.
- Once the plastic bags are sealed, do not open them again.
- After the investigation, all plastic bags must be put into the garbage. They should not be washed and reused.

Materials

permanent marker
1 slice of bread
plastic knife
4 new, resealable plastic bags
spray bottle filled with water
duct tape

Procedure

1. Use the marker to label each bag with your name. Label one of your four bags "light," one "dark," one "dark, moist," and the fourth, "light, moist."

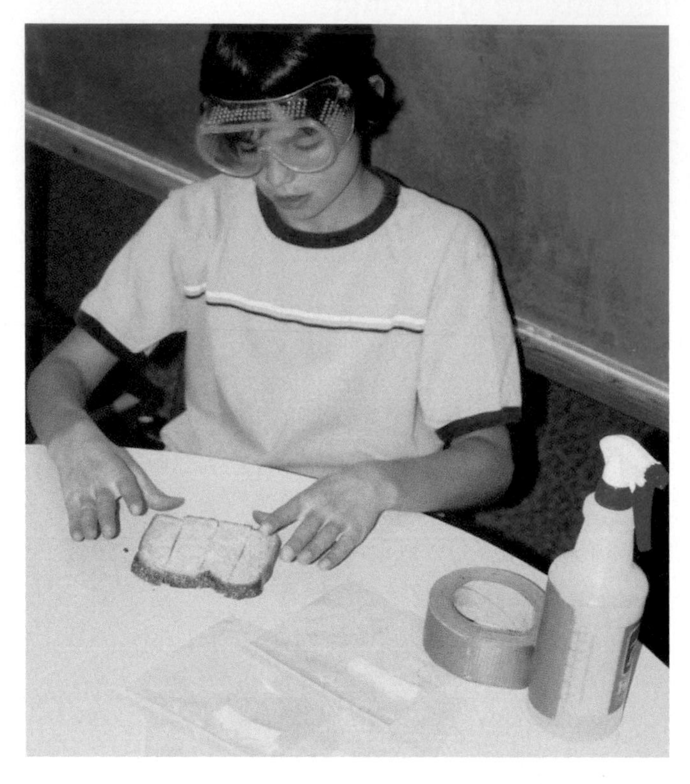

2. Take the slice of bread and wipe it on dusty surfaces around the classroom.
3. Cut the bread into four equal-sized pieces using the plastic knife.
4. Dampen two pieces of bread with the spray bottle. The bread should be moist, not soggy. Place one piece of damp bread in the bag marked "dark, moist." Seal the bag and cover the opening with duct tape. Place the bag in a drawer or cupboard. The location you place the bag in should remain closed except when you make your observations.
5. Place the second piece of dampened bread in the bag marked "light, moist" and seal it as you did in step 4. Place the bag in a location where it is exposed to indirect light.

6. Put the third piece of bread in the bag labelled "light" and seal it as you did in step 4. Place the bag in a location where it is exposed to indirect light.

7. Put the last piece of bread in the bag labelled "dark" and seal it as you did in step 4. Place the bag in a drawer or cupboard. The location you place the bag in should remain closed except when you make your observations.

8. Wash your hands thoroughly with soap and water.

9. Copy the table below into your notebook. **Observe** the bread slices daily and **record** your observations in the data table.

Day	Dry Bread Exposed to Light	Dry Bread in the Dark	Moist Bread Exposed to Light	Moist Bread in the Dark
1				
2				
3				
4				
5				

Analyze

1. Which bread sample had the most mould growth?
2. Which bread sample had the least mould growth?
3. Which bread sample was the first to show mould growth?
4. Compare your observations with the observations of other groups in your class. How are they similar? How are they different?

Conclude and Apply

5. Are you more likely to find mould and other fungi in areas that receive indirect light or in shady areas? Explain your answer.
6. If you were a mushroom grower, what conditions would you need to grow your mushrooms?
7. Why did you wipe the bread on surfaces in your classroom before beginning this experiment?

Extend Your Skills

8. Research some of the kinds of fungi that are found in British Columbia. Write a report describing species of three fungi. Include information on where they live.

Did You Know?

There are over 10 000 types of fungi in British Columbia. Some people harvest and eat mushrooms. However, since many mushrooms are poisonous, you should *never* eat mushrooms that you are not familiar with.

Kingdom Animalia

The animals shown in Figure 3.14 are all found in British Columbia. The body characteristics of all of these animals are quite different, yet they all belong to Kingdom **Animalia**. What makes an animal an animal? What are the characteristics that all animals share?

Animals are multicellular. Their cells all have a nucleus but do not have the protective walls found around the cells of plants and fungi. All animals need to eat plants or other animals to obtain food. Animals can move from place to place to find food, shelter, mates, and to escape from enemies.

As is true of the other kingdoms, Kingdom Animalia is further divided into phyla, class, order, family, genus, and species. Some of the phyla in Kingdom Animalia are shown in Figure 3.15.

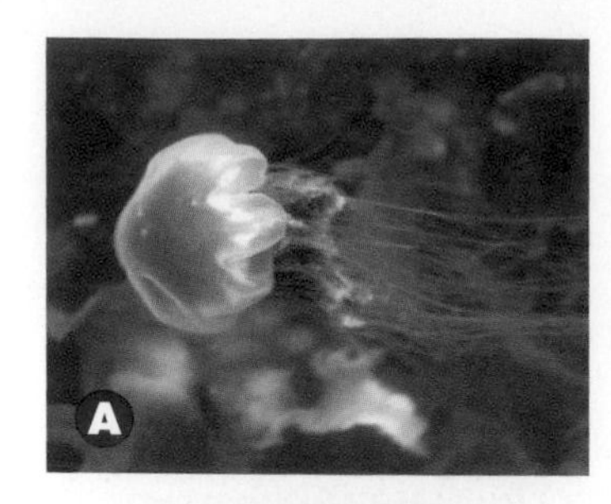

Figure 3.14 The lion's mane jellyfish (A), tufted puffin (B), and banana slug (C) are all animals that live in British Columbia. What characteristics do all animals share?

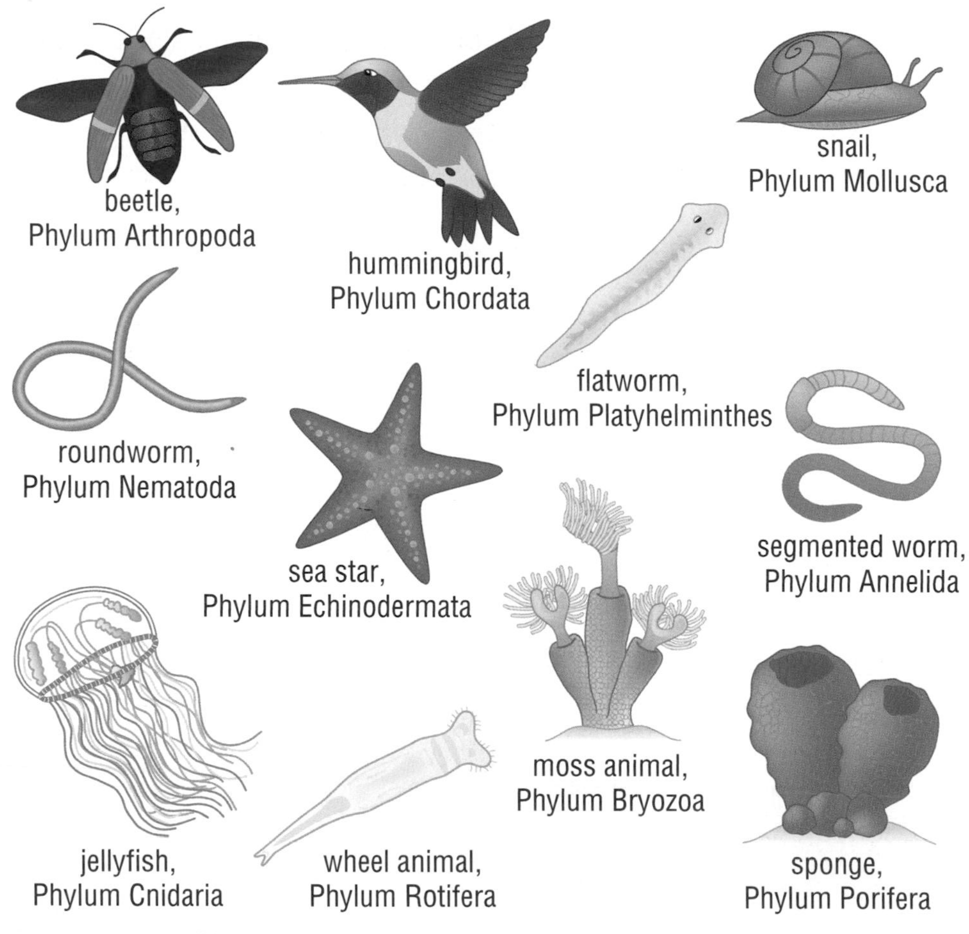

Figure 3.15 These are eleven of the approximately 35 phyla in Kingdom Animalia.

Invertebrates: Animals without Backbones

The earthworm and the snake in Figure 3.16 share the characteristics of all animals. As well, they both have long, thin bodies without any arms or legs. But there are also important differences—one is spineless! Animals with spines (backbones), such as the snake, are called **vertebrates**. Earthworms, which do not have backbones or any other bones, are called **invertebrates**.

Since most invertebrates are small, they do not need large structures to support their bodies. However, some invertebrates do have rigid body parts. For example, insects and spiders have a hard covering on the outside of their bodies. Snails have shells, and sponges have tiny glass or bone-like spines in their body. Invertebrates are much more common than vertebrates. In fact, most of the members of the animal kingdom are invertebrates.

Figure 3.16 An earthworm (A) is an invertebrate. The rubber boa snake (B) is a vertebrate.

Vertebrates: Animals with Backbones

The backbone of vertebrates is part of the internal skeleton that supports their bodies. Most vertebrates have two sets of paired limbs, such as fins, arms, or legs. Compare invertebrates and vertebrates in Find Out Activity 3-H.

Find Out **ACTIVITY 3–H**

Animal Collage

In this activity, you will classify pictures of animals as invertebrates or vertebrates.

What You Need

old nature and outdoor magazines
glue
poster paper
scissors

What to Do

1. Cut out pictures of animals.
2. Divide your poster paper in half and label one half "Vertebrates" and the other half "Invertebrates."
3. Sort your pictures into the two groups and glue them under the correct label.

What Did You Find Out?

1. How many different types of invertebrates and vertebrates did you find?
2. **(a)** What characteristics do invertebrates share?
   (b) What characteristics do vertebrates share?
3. Which side of your poster has more species of animals? Explain why this might be so.

Classes of Vertebrates

Vertebrates are divided into classes according to various features that are shared by some but not by others. The classes include three classes of **fish**, **amphibians**, **reptiles**, **birds**, and **mammals**. The characteristics of each class are shown below.

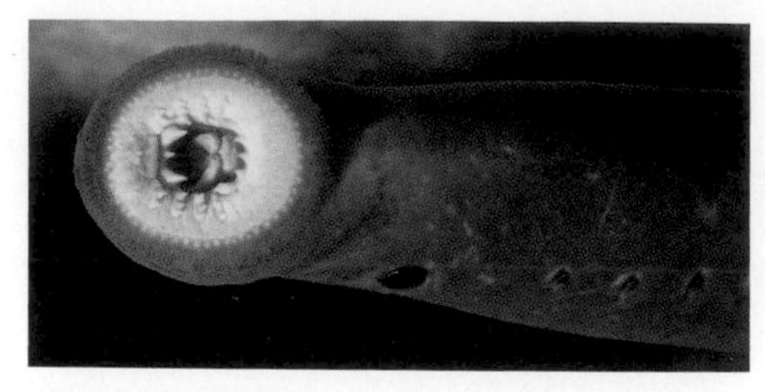

Jawless Fish
Example: lampreys
- live in water
- do not have scales on skin
- no jaw; small teeth in a round, sucking-type mouth
- soft skeletons made of cartilage
- lay eggs without shells
- gills gather oxygen from water

Cartilagenous Fish
Example: sharks
- live in water
- have scales on skin
- have jaws and soft skeleton made of cartilage
- most lay eggs without shells
- gills gather oxygen from water

Bony Fish
Example: trout
- live in water
- have scales on skin
- have jaws and skeletons made of bone
- lay eggs without shells
- gills gather oxygen from water

Amphibians
Example: frogs
- most live in water when young and on land as adults
- have smooth skin
- have jaws and skeletons made of bone
- lay eggs without shells, in water
- gills gather oxygen from water when young; as adults, small lungs and skin gather oxygen from air

Reptiles
Example: turtles
- live in water and/or on land
- have scales on dry skin
- have jaws and skeletons made of bone
- most lay eggs with soft shells, on land
- lungs gather oxygen from air

Birds
Examples: herons
- most live on land and most fly
- have feathers on skin
- have jaws and skeletons made of hollow bones
- lay eggs with hard shells, on land
- lungs gather oxygen from air

Mammals
Example: squirrels
- can live on land or in water
- have hair or fur on skin
- have jaws and skeleton made of bone
- most give birth to live young and feed them mother's milk
- lungs gather oxygen from air

Find Out **ACTIVITY 3-1**

Who Am I?

In this game you will review the kingdoms of life to try to classify different organisms.

What You Need

spare outdoor and nature magazines
scissors
glue
7 index cards

What to Do

1. Choose one organism from each of the six kingdoms, plus at least one organism from a class of vertebrate animal.
2. Research the organisms you chose using a library or the Internet.
3. On the front of an index card, either draw a picture of one of the organisms you chose or glue a picture from a magazine if one is available.
4. On the back of the same index card, neatly include the following information in this order:
   - describe the characteristics of the organism
   - kingdom
   - phylum (if known)
   - class (if known)
   - common name
5. Repeat steps 3 and 4 for the other organisms you chose.
6. Your teacher will collect the cards and redistribute them throughout the class.
7. Play the game as follows:
   - Find a partner.
   - Show your partner the picture of the organism.
   - Ask your partner: "In what kingdom do I belong?"

   - If your partner does not know the kingdom, give him or her clues (the characteristics) one at a time.
   - After your partner has determined the kingdom into which the organism is classified, see if he or she knows the phylum, the class, and the common name.
8. You could also play the game by reading off the clues without showing your partner the picture of the organism. See if your partner can guess the name of the organism just by hearing the clues.

What Did You Find Out?

1. Draw a classification table to help you review the characteristics of each kingdom. A sample table with some of the columns filled in is shown below.

Kingdom	Unicellular or multicellular?	Nucleus in Cell?	
Ancient Bacteria			
True Bacteria			
Protista			
Fungi			
Plantae			
Animalia			

Create your own table in your notebook. Give your table a title.

Section 3.2 Summary

Scientists group living things into one of six kingdoms: Ancient Bacteria, True Bacteria, Protista, Fungi, Plantae, and Animalia.

- All bacteria in the old Kingdom Monera have only one cell, which does not have a nucleus. Most bacteria do not make their own food; they decompose other living or once-living things.
- Ancient bacteria are found in extreme environments.
- Organisms in Kingdom Protista can be single-celled or multicellular, and their cells have a nucleus. Protists can make their own food by photosynthesis, or they can eat food, or they can absorb food into their bodies by decomposing other organisms.
- Most, but not all, fungi are multicellular. The cells of fungi have a nucleus and cell walls. Fungi obtain nutrients by absorbing them from other living things, often by decomposing these other organisms.
- Plants are multicellular organisms. Plant cells have a nucleus and cell walls. Plants use photosynthesis to make their own food.
- Animals are multicellular organisms. Animal cells contain a nucleus but do not have cell walls. Animals obtain nutrients by eating and digesting other organisms.
- Animals can be divided into two main groups: those with backbones (vertebrates) and those without backbones (invertebrates). There are many more invertebrates than vertebrates.
- Classes of vertebrate animals include jawless fish, cartilagenous fish, bony fish, amphibians, reptiles, birds, and mammals.

Key Terms

Monera
Protista
Fungi
Plantae
Animalia
invertebrates
vertebrates
fish
amphibians
reptiles
birds
mammals

Check Your Understanding

1. What are the six kingdoms that are commonly used to classify living things?
2. **(a)** How do fungi obtain their food energy?
   (b) How is this different from how plants obtain their energy?
3. What is the difference between a vertebrate and an invertebrate? Give an example of each.
4. **(a)** What are the classes of vertebrates?
   (b) What are the characteristics of each of these classes?
5. **Thinking Critically** Linnaeus's system of classification uses common physical characteristics. Create your own system of classification using some other idea—behaviour or habitat, for example. In your system, what organisms would be classed together that are not classed together in Linnaeus's system?

CHAPTER at a glance

Now that you have completed this chapter, try to do the following. If you cannot, go back to the sections indicated in brackets after each part.

(a) Identify the seven levels used to classify organisms. (3.1)

(b) Describe the main characteristics of the six kingdoms. (3.2)

(c) Describe the characteristics of the different classes of vertebrates. (3.2)

(d) Describe what a field guide is and how to use it. (3.1)

(e) Describe the characteristics of different protists that make them more like plants, more like animals, or more like fungi. (3.2)

(f) Explain why scientists do not consider plants to be fungi. (3.2)

(g) Why are bacteria and fungi called "nature's recyclers"? Give one example of each. (3.2)

(h) Explain why scientists might have decided to use the classification system that Linneaus developed. (3.1)

(i) Explain why Aboriginal peoples developed their own system of classifying organisms. (3.1)

Prepare Your Own Summary

Summarize Chapter 3 by doing one of the following. Use a graphic organizer (such as a concept map), create a poster, or write a summary to include the key chapter ideas. Here are a few ideas to use as a guide:

- Use Venn diagrams to compare the different kingdoms.
- Describe how you would classify a newly discovered organism.
- Explain why scientists have developed a classification system for organisms.
- What are the strengths and weaknesses of using two-part scientific names to identify organisms?

CHAPTER

3 Review

Key Terms

classification systems
kingdom
species
Monera
Protista
Fungi
Plantae
Animalia
invertebrates
vertebrates
fish
amphibians
reptiles
birds
mammals

Reviewing Key Terms

If you need to review, the section numbers show you where these terms were introduced.

1. Describe the difference between:
   (a) genus and species (3.1)
   (b) vertebrate and invertebrate (3.2)
   (c) Protista and Fungi (3.2)

2. In your notebook, match each description in column A with the correct kingdom in column B. Use each kingdom only once. (3.2)

A	B
(a) unicellular organisms with no nucleus	• Animalia
(b) multicellular organisms that use sunlight to make their own food	• Protista
(c) organisms that cannot make their own food	• Monera
(d) organisms that can be like plants, like animals, or like fungi	• Fungi
(e) multicellular organisms that eat and digest other organisms	• Plantae

Understanding Key Ideas

Section numbers are provided if you need to review.

3. What is the proper order of classification? (3.1)

4. Fill out the table below with at least 2 examples for each group. (3.2)

	Example 1	Example 2
Ancient bacteria		
True bacteria		
Protists		
Fungi		
Plantae		
Animalia		

5. **(a)** Why does each pair of animals on the chart have the same first name? (3.1)

Scientific Name	Common Name
Felis concolor	Mountain lion
Felis domesticus	House cat
Panthera leo	Lion
Panthera tigris	Tiger
Canis lupus	Wolf
Canis latrans	Coyote

   (b) Would you expect an animal with the name *Rania concolor* to look similar to a mountain lion? Why or why not? (3.1)

6. When an organism is named using the two-part naming system, what two classification levels are used? (3.1)

7. Which of the following would you expect to be most closely related? (3.1)
   (a) two species in the same genus
   (b) two species in different phyla
   (c) two species in the same class
   Explain your answer.

Developing Skills

8. A white-footed mouse has the scientific name *Preomyscus leucopus*. It is in the same class as lions and humans. It is a rodent that belongs to the family Cricetidae. Copy the table below for the classification of the white-footed mouse into your notebook. Use the information given here and in the chapter to fill it in.

Kingdom	
Phylum	
Class	
Order	
Family	
Genus	
Species	

9. Invent an imaginary organism. Write a description of it that will enable someone else to classify your organism into one of the six kingdoms described in this chapter.

Problem Solving

10. **(a)** Explain how scientists name a species. Give or use an example in your answer.
    (b) Why are species names important in scientific work?

Critical Thinking

11. We classify people in many ways—for example, by gender, race, religion, physical appearance, ethnic origin, profession, lifestyle, and so on. How can classifying human beings be helpful? How can it be harmful?

12. "Classification systems are tools, not rules." Explain what scientists mean by this statement when they talk about classifying newly discovered organisms.

Pause& Reflect

Check your original answers to the Getting Ready questions on page 60 at the beginning of this chapter. How has your thinking changed? How would you answer those questions now that you have investigated the topics in this chapter?

Ask an Elder

Barbara Wilson

Barbara Wilson is a Haida and an expert in the conservation of natural and cultural resources. She works with the Gwaii Haanas. Her Haida name is Kii7iljuus (Key-IL-juice), which means "Born Big." Barb is the Matriarch of the Saw-Whet Owl clan of Cumshewa.

Q. You work as a cultural liaison specialist in Gwaii Haanas. What does that mean?

A. My role is conservation. I look after our village sites and other elements of the Haida culture. For example, I work with Elders to gather information about Haida place names and traditional ecological knowledge, and I make sure that what is written about this place is correct and in its proper context. I speak to people from all over the world about who we are and what is important to us. I want to give people the right stories.

When Europeans first came here, the Haida had to adopt new ways. There was a lot of illness, and we had to move to one or two centres in order to survive. This changed the way we used our land. Now we are re-establishing and broadening our connection to our history and our lands once more.

Q. How do scientific views about the classification of living things fit with traditional Haida views?

A. For the Haida to live on the land, everyone had to learn about all things in the natural world: what to eat, where to find it, how to gather it, and so on. You had to know your "fellow travelers"—the plants and animals as well as soils, rocks, and water. There were specialists in different aspects of Haida lifestyle, such as medicinal plants, but everybody had to know a little about everything. This knowledge ensured survival on the land and ocean every day.

Before you can understand individual parts of the land—such as certain plants or animals—you first have to understand them as parts of an ecosystem. It is all connected. To us, everything is alive. Everything has energy and can be communicated with. Everything—plants, animals, rocks, soil—is part of the whole, part of the web of life.

In my work, I take from traditional Haida knowledge and from Western science to show they complement each other. We're all in the same place, looking at the same things. We've just learned about it in different ways.

Q. How does Haida tradition see the relationship between people and place?

A. What we eat, how we live—everything is sacred. We are part of the land and it is part of us. We are so interconnected that we define each other. The ruggedness of the land defines us. We are at the mercy of nature, so we learn to respect it. We looked after the land in a good manner so the land looked after us. Respect goes back and forth

Q. How does the National Park Reserve bring scientists and Haida together?

A. When scientists and Haida are studying organisms, we work together. If a scientist is going out to do some research on the land, then a Haida goes along, too. We both learn something. The Haida learn what we need to from the scientific world, and we also teach the scientists from the Haida perspective. It shows we can look after the land together.

It is important to build a learning portion into our work. Stories, knowledge of the people who live here, and the spirituality of the place must be recognized along with the physical land. For example, we study how to preserve our totem poles. Scientists can teach us physical ways to preserve wood. But for us it's never just a piece of wood. Our ancestors' prayers and discussion went into the creation of every pole: what face would be carved on it, where it would stand, who came to the pole-raising, and who ate what at the potlatch. All this is in the energy of the place. It is not just those of us who are here standing on the land who are standing on the land.

Q. What do you see as the role of young people in looking after the land?

A. Young people need to go out on the land and to the ocean. They need to learn about the natural world in the place they live. The more you know, the better you can look after your place. Learning about the land means learning respect for the land and for its past. Not to get so wrapped up in the past that you can't see now—but to recognize that how we came here makes us who we are. Scientific knowledge and stories together give a bigger appreciation of where we are and our share of the responsibility for looking after the land.

EXPLORING Further

Knowing your place

Over the next few days, find three times when you can spend 10 minutes outdoors in front of your home. One 10-minute period should be in the morning, one in the middle of the day, and one in the evening or at night. During each period, observe as much as you can about the area around you. For example, what plants can you see? Do you see or hear the same animals in the day and at night? Where does the moon rise and set? Which way does the wind blow? Draw a picture or write a story to describe the environment around you.

What animals live in your neighbourhood?

Ask a Marine Biologist

Jane Watson

Professor Jane Watson is a marine biologist. In winter, she teaches at Malaspina University College in Nanaimo. In summer, she studies the world of sea otters, sea urchins, and kelp forests off Vancouver Island.

Q. How did you become interested in marine biology?

A. I've wanted to be a marine biologist since I was old enough to say the words. I grew up in Vancouver, loving anything to do with the ocean. I saved up to buy a mask and snorkel. The first look through that mask had me hooked—I was going to be a marine biologist.

Q. Tell us about your sea otter project.

A. At one time, sea otters were common along the British Columbia coast. By the early 1800s, they had vanished, hunted out for their luxurious fur. When that happened, their prey—such as sea urchins—became much more abundant. Sea urchins eat kelp. As sea urchin populations grew, kelp forests disappeared.

In the 1970s, sea otters from Alaska were released near Nootka Sound. It was an experiment to see if sea otters could survive where they had once been so common. And it worked!

A research project often begins with a specific question. In this case, I asked myself: how has the return of sea otters on the west coast of Vancouver Island affected other living things?

Q. And what's the answer?

A. There are fewer sea urchins, and kelp forests grow back. But the interaction of living things means there's more to the answer: kelp forests provide food and shelter for many marine plants and animals. Free-swimming larvae of snails, clams, and other invertebrates are less likely to be swept out to sea, and many more members of those species survive. They grow faster, too, because there's more to eat. Fish become more abundant as kelp forests recover, up to forty times more abundant. Fish attract seabirds, seals, and sea lions. Seals and sea lions attract killer whales.

On the other hand, some of the species that sea otters prey on, like sea urchins, become less abundant. That can make life difficult for fishers or organisms harvesting the same species.

Q. Who benefits from the work you do?

A. I try to build a better understanding of how the members of marine ecosystems work. That knowledge should help us work in nature more wisely.

An important part of my work is communicating what I've learned through scientific papers, presentations to the general public, and meetings with other experts. Science is a cooperative process. Scientists exchange and discuss their ideas, test them, and discard any that prove incorrect. All scientists must contribute to that process.

Q. What do you enjoy most about your job?

A. I love working outdoors. Every day I learn something new. I've watched a kelp crab shed its old shell and run for cover. (The new shell is very soft!) I've watched coho salmon and halibut feed on herring. I've watched a leatherback turtle eat a large jellyfish. I've had killer whales check me out—thinking, perhaps, that I was something to eat. I've had seals blow bubbles at me and sea lions pull at my flippers.

Q. What advice would you give to students who might be interested in biology as a career?

A. Biology will always have a future. To live in this world, we need to understand it. Plants and animals are important parts of human survival: plants make oxygen and capture carbon dioxide from the atmosphere; bacteria clean water; plants and animals provide us with food. For our own good, we need to understand how different organisms interact with each other and with their environments.

EXPLORING Further

Sea otters have big appetites. Every day they eat an amount of food equal to almost 30% of their body weight. We eat just 2% of our body weight. Sea otters spend all of their time in bitterly cold water. They do not have the blubber for insulation that other marine mammals have. Thick fur and lots of body heat, fuelled by all that food, are what keep them warm.

Because they eat so much, sea otters have a big effect on their physical environment. When the sea otter population goes up or down, their ecosystem changes.

If you're interested in learning more, use your favourite Internet search engine to query "sea otter" and "kelp forest." The Vancouver Aquarium Marine Sciences Centre [www.vanaqua.org], the Bamfield Marine Science Centre [www.bms.bc.ca], and the Monterey Bay Aquarium [www.mbayaq.org] are all good sources of information.

UNIT 1

Project

What Is It?

SKILLCHECK
- Questioning
- Interpreting Data
- Problem Solving
- Communicating

Field guides, books, and web sites about nature contain information about different types of organisms in one place. They give people who want to learn more about different organisms one source for information. In this project, you will create a presentation that people can use to learn more about animals in a particular phylum.

Challenge

With other students, create a booklet, mural, electronic presentation, or other format that provides an interesting way to learn more about four organisms in a phylum.

Materials

art supplies
other supplies depending on student project

Design Criteria

A. Your project must include information on four different animals that all belong to one phylum.

B. Your project must include the following for *each* animal:
- the common name(s) and scientific name
- the classification (Kingdom, Phylum, Class)
- a physical description
- a photograph or illustration of the animal
- a description of the biome in which the animal lives
- other information about its home and how the animal has adapted to its environment
- information on what it eats, how it catches food, and how it interacts with other animals (for example, which animals are its enemies?)
- a classification key that would help someone identify all four animals
- other information of your choice

C. You may use a variety of research tools, including books, magazine articles, textbooks, films, and the Internet.

D. You must list the sources that you used to find information for your project.

Plan and Construct Group Work

1. Choose one animal phylum from those discussed in this unit. (For example, Phylum Arthropoda includes animals with jointed legs, such as insects, crabs, and spiders. Phylum Chordata includes mammals, birds, and fish.) Choose four animals that belong to this phylum.
2. Choose the method that you will use to present your data. You should choose a method where you can use both words and pictures. For example, you could use a brochure, a computer presentation, or even a mural with labels and small blocks of text. Your teacher may also approve a play, a song, or even a classroom "lecture."
3. Make of list of the sources you will use to conduct your research.
4. Create a list of jobs that must be done to complete the project. Decide which group member will be responsible for each job.
5. Create a plan that includes resources, jobs assigned, an outline of what your project will look like, and a timeline for getting the work done. Have your teacher approve your plan. Adjust your plan if necessary.
6. Carry out your group's plan and complete your project.
7. Ask your classmates what they learned from your group's presentation.

Evaluate

1. **(a)** When you were doing your research did you notice that the information you found from different sources did not always agree? Give some examples.
(b) Explain why this might be so.
(c) What did your group do when the facts didn't agree?

2. **(a)** How did your presentation help people learn about animals in an interesting way?
(b) How would you improve your group's presentation?

UNIT 2

Electricity

No matter where you live, transmission lines like the ones in this photograph are probably a familiar sight. They connect homes and communities across British Columbia into huge electrical systems. But what is happening inside these lines? What is electricity, and how does it travel from place to place? How does electricity get into these wires in the first place? These are some of the questions you will answer in this unit on electricity.

You might think electricity is a modern invention, but in fact electricity has always been with us. However, it is just within the last few hundred years that we have learned to harness it for our use. In this unit you will learn how this powerful natural force can be safely captured and controlled. You will examine different kinds of electricity, and see how each can be put to work in objects and systems that are used today.

Over the course of the unit you will find out what electricity can do for you, and what you can do to respect the electricity you use. You will learn how to stay safe around electricity at school, at home, and outdoors. You will also find out how the choices you make will help your household and community generate and use electricity responsibly, now and in the future.

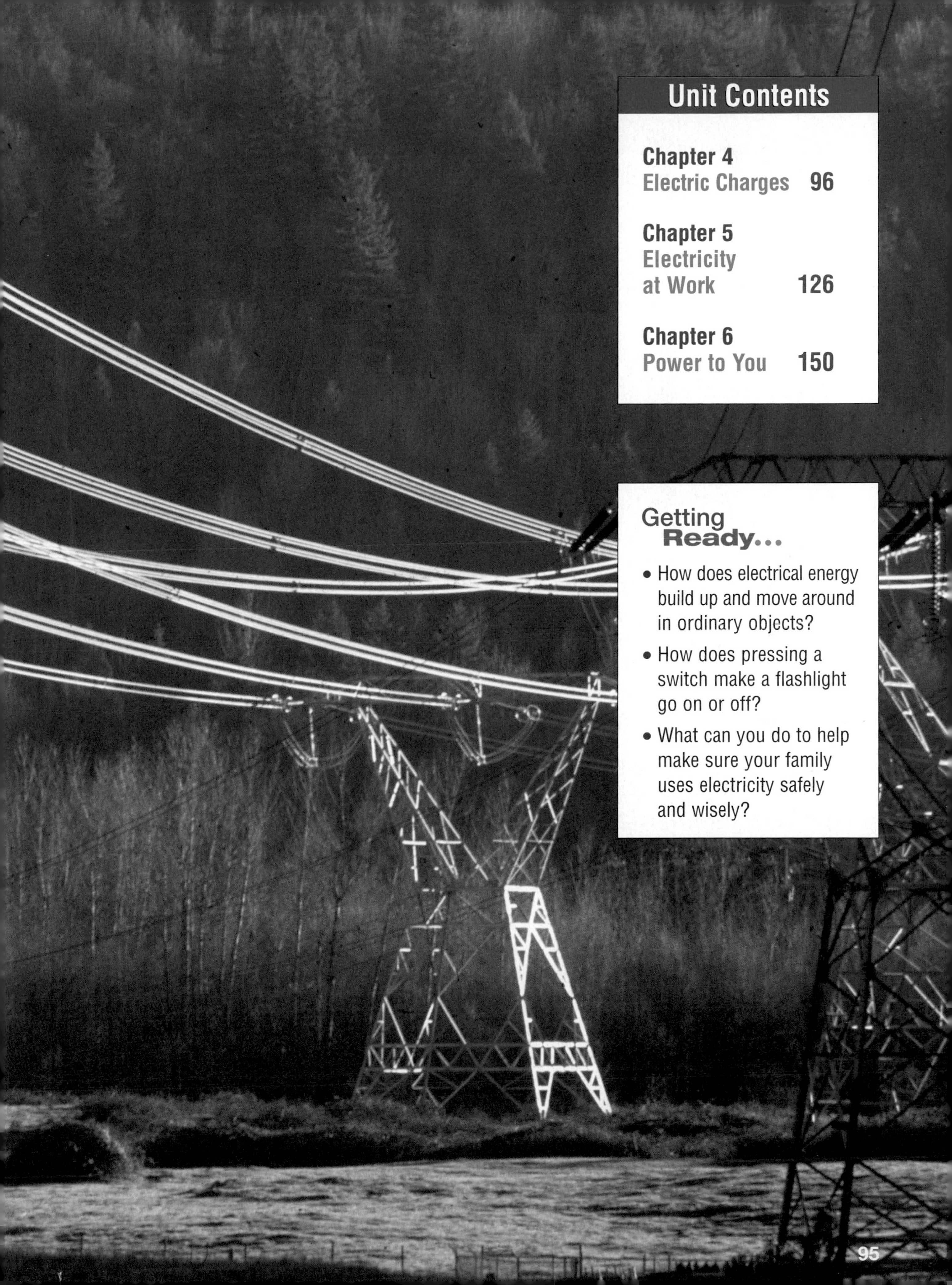

Unit Contents

Chapter 4
Electric Charges 96

Chapter 5
Electricity at Work 126

Chapter 6
Power to You 150

Getting Ready...

- How does electrical energy build up and move around in ordinary objects?
- How does pressing a switch make a flashlight go on or off?
- What can you do to help make sure your family uses electricity safely and wisely?

CHAPTER

4 Electric Charges

Getting Ready...

- Where does electricity come from?
- Why does the dryer make your cotton socks stick to your silk shirt, but not to each other?
- Is there a difference between the electricity in a storm cloud and the electricity in a flashlight?

The slow-moving Pacific electric ray can catch fast-moving fish by stunning them with an electric shock.

Imagine that you are scuba diving in the ocean near Vancouver Island. In the glow of your flashlight you see a Pacific electric ray gliding through the water. Watch out! The electric ray can produce an electric shock strong enough to knock a full-grown person unconscious. It might seem strange that an animal is able to produce electricity. But did you know that your own body possesses electric charges? In fact, so does every person and object around you. You probably know that you are using electricity when you turn on your flashlight. However, you don't usually see the effects of electric charges in your body and the ordinary objects in your classroom or home.

In this chapter, you will find out more about what we mean by "electricity" in ordinary objects that you don't think of as being electric. You will use objects in your classroom to study how electric charges behave. You will learn about two types of electricity, static and current, and will see how each can be used to perform useful tasks.

What You Will Learn

In this chapter, you will learn

- how to charge objects with electricity
- how electric charges behave
- why some materials conduct electricity while others do not
- the difference between static electricity and current electricity

Why It Is Important

- You use electricity every day.
- When you understand what electricity is, you can learn to control and work with it.
- Understanding how electricity works will help you stay safe around it.

Skills You Will Use

In this chapter, you will

- build a model to show the electric charges in an atom
- observe what happens when you charge an object with static electricity
- classify electric charges
- classify objects as insulators and conductors
- observe what happens when electric charges flow through a conductor

This lightning over Vancouver is a dramatic example of electricity in action.

Starting Point ACTIVITY 4–A

Balloon Buddies

What to Do

1. Inflate a balloon and tie it off. With a marker, gently draw a face on it so that the face will be right side up when you hang the balloon from its knot.
2. Using tape and string, attach the balloon from the ceiling so that it hangs at the same level as your head. This is your balloon buddy.
3. Walk slowly past the balloon without touching it. **Record** what happens in your science notebook.
4. Rub the balloon's face with a wool cloth. Walk slowly past the balloon again without touching it. **Record** what happens.
5. Touch the face of the balloon several times with your hand. Make sure you touch the whole face. Now walk slowly past it and **record** your findings.
6. Make a second balloon into a balloon buddy. Hang it close to the first one. Make sure they are facing each other. Rub the face of both balloons with the wool cloth. **Record** how the two balloons behave.

What Did You Find Out?

1. Compare your group's observations with those of the rest of your class.
2. Think about other situations in which objects stick together without glue or tape. What do those situations have in common with this activity?

Section 4.1 Explaining Electricity

Key Terms

- atoms
- nucleus
- positive charge
- electrons
- negative charge
- uncharged
- imbalance
- charged

Figure 4.1 Which of the objects in this picture contain electric charges? How do you know?

In Starting Point Activity 4-A, did you know that you were actually charging a balloon with electricity? You couldn't see the electricity, but you could see how the balloon behaved after you rubbed it. Where did that electricity come from?

Sometimes electricity seems to come out of nowhere. Actually the opposite is true: electricity is everywhere, all around you. Would you be surprised to know that in Figure 4.1, everything you see—including the video games, the people, and even the air—contains electric charges? Before you began your Starting Point Activity, electric charges were already in the rubber balloon, in the wool cloth, and in you. By rubbing the balloon with the wool, you transferred those charges from one place to another. You created electrical energy.

Sometimes the electricity around you is not noticeable at all. At other times electricity shows up in many different ways. It can flow steadily to turn on a light, or it can "jump" in sharp bursts. Examples of a sharp burst are lightning or the shock you get when you touch a doorknob after walking across a carpet. In an oven, electricity creates heat. In a video game, electricity is used to create images on a screen. Some electric appliances need to be plugged in, while others work from batteries. A lamp has to be turned on, but lightning happens naturally.

How can the effects of electricity be so different at different times? What do all these effects have in common? To understand electricity, you need to study some of the smallest particles that make up all objects.

Figure 4.2 You are full of electric charges, but do you feel electric?

Everything Is Made of Atoms

What do you have in common with a towering skyscraper, a babbling brook, and the air you breathe? Everything—whether it is alive or not, and whether it is solid, liquid, or gas—is made up of tiny particles called **atoms**. Atoms are so small that we cannot see them even with the most powerful microscope. A single cell in your body is made up of millions of atoms! Scientists cannot see atoms, but they have done experiments to figure out what atoms are made of and how they behave. From these experiments, scientists have built models to describe atoms. Figure 4.4 shows a simple model of an atom.

The **nucleus** is the name given to the centre of the atom. The nucleus contains particles with a **positive charge**. Around the nucleus are even smaller particles called **electrons**. Electrons swirl around the nucleus in special orbits. Each electron has a **negative charge**.

Figure 4.3 The mug and marshmallows, the hot chocolate, and even the steam are all made of different combinations of atoms.

Positive and Negative Charges Balance Each Other

Normally, the number of positive charges in the nucleus is the same as the number of negatively charged electrons orbiting the nucleus. An atom that contains an equal number of positive and negative charges is said to be electrically neutral or **uncharged**. It contains both positive and negative charges, but these opposite charges balance each other. The atom shown in Figure 4.4 is uncharged: its negative charges balance its positive charges.

An atom is made up mostly of empty space. If the nucleus of an atom were the size of a grain of salt in the middle of a football field, the electrons would be at the very edges of the field. That's a lot of empty space!

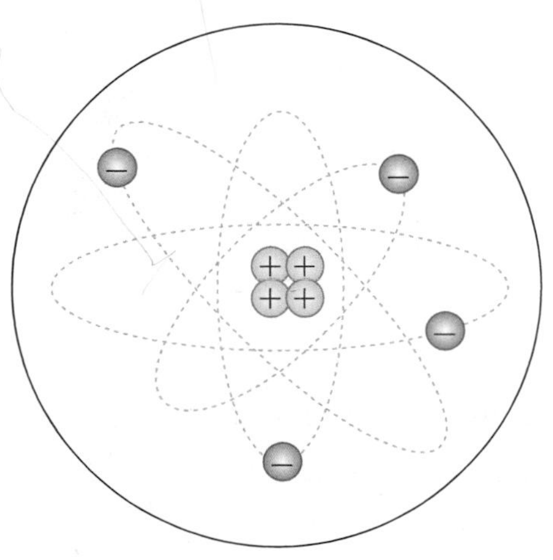

Figure 4.4 This model of an atom shows a positively charged nucleus and negatively charged electrons. Positive charges are shown with a plus sign (+). Negative charges are shown with a minus sign (−).

Electricity Comes from Unbalanced Positive and Negative Charges

One kind of atom has only a single electron, while others have two or more, and some more than 100. However, having more electrons does not necessarily mean having more total electric charge. As long as the number of negative charges from the electrons is equal to the number of positive charges in the nucleus, the atom as a whole will be uncharged.

What happens if the positive and negative charges are not equal? In this case, we say the charges in the atom are imbalanced. An **imbalance** of charges happens when the number of negative charges that orbit the nucleus of an atom is not equal to the number of positive charges in the nucleus of that atom. The result of an imbalance of charges is an overall electric charge.

Figure 4.5 shows two atoms that have imbalanced charges. An atom that has more positive charges than negative charges has an overall positive charge. An atom that has more negative charges than positive charges has an overall negative charge. An atom with an imbalance of positive or negative charges is said to be electrically **charged**.

DidYouKnow?

The word "electricity" comes from the Greek word, *elektron*, meaning "amber." Amber is a hard brownish-yellow fossil resin from trees. Amber is used for making jewellery and other ornamental objects. More than 2500 years ago, in 600 B.C., a Greek mathematician noticed that if he polished amber with fur, the amber would attract small objects. For the next 2200 years, scientists believed that electricity was found only in amber.

Amber was once more valuable than gold.

READING Check

If an atom has three electrons and three positive charges, what is the electric charge on the atom? If an atom has three electrons and five positive charges, what is the electric charge on the atom?

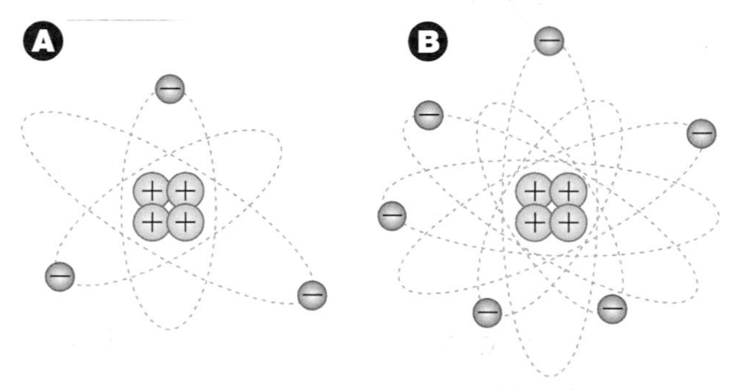

Figure 4.5 If the number of negative electrons is not equal to the number of positive charges, the atom is electrically charged. (A) This atom has *fewer* negative charges than positive charges. It has an overall positive charge. (B) This atom has *more* negative charges than positive charges. It has an overall negative charge.

In other words, the electric charge carried by an atom doesn't depend on the total *number* of positive or negative charges it has, but on the *balance* between the two. In Find Out Activity 4-B, you will model your own neutral (uncharged) and charged atoms.

Find Out **ACTIVITY 4-B**

Make an Atomic Model

What to Do

1. Using a coloured marker, draw two positive (+) signs on a piece of masking tape and attach it to one of the large Styrofoam™ balls. This represents the nucleus of an atom.
2. Using a different colour, draw a negative (−) sign on a piece of masking tape and attach it to one of the small Styrofoam™ balls. This represents an electron.
3. Stick one end of a short toothpick into the small ball. Stick the other end of the toothpick into the large ball. **Record** the overall charge of your atom as positive, negative, or neutral (no charge).
4. Repeat steps 2 and 3 to make another electron and stick it onto the atom using another short toothpick. **Record** the overall charge of the atom as positive, negative, or neutral (no charge).
5. Make a third electron and add it to the atom using the longer toothpick. **Record** the overall charge of the atom as positive, negative, or neutral (no charge).

What Did You Find Out?

1. When did the atom have a neutral charge?
2. When did the atom have a positive charge?
3. When did the atom have a negative charge?
4. What would you do to turn a negatively charged atom into an uncharged atom:
   (a) Take away one or more electrons?
   (b) Add one or more electrons?

Atoms Can Gain or Lose Electrons

When an atom has imbalanced positive and negative charges, the atom is electrically charged. But how do the charges become imbalanced? They become imbalanced when different kinds of atoms come close to each other, and one picks up electrons from the other.

Figure 4.6 shows what happens when uncharged atoms (represented by clowns) pick up or lose an electron. The atom that loses an electron becomes positively charged. The atom that gains an electron becomes negatively charged.

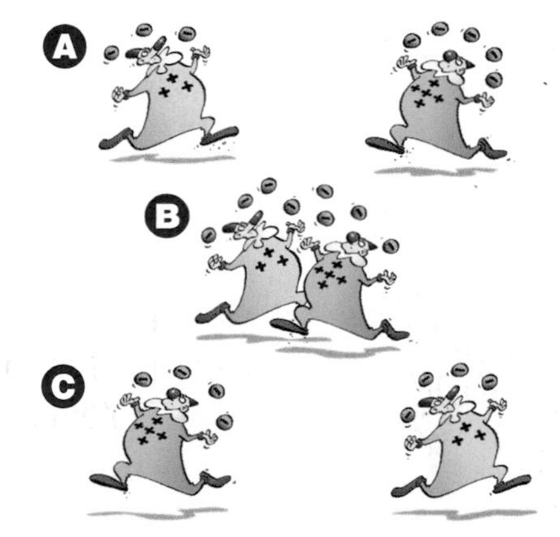

Figure 4.6 (A) These two "uncharged" clowns are juggling negative balls. (B) When the clowns bump into each other, some of their balls are transferred. (C) How have the clowns' charges changed? What happened to the total number of negative charges (balls)?

The Total Number of Electrons Does Not Change

New electrons are not created when an atom gains electrons. Look again at Figure 4.6, and notice what happens to the balls the clowns are juggling. One clown passes a ball to the other, and both clowns become charged. The total number of balls, however, stays the same.

The same is true of atoms and electrons. An electric charge results when one atom loses an electron to another. Both atoms become charged, but the total number of electrons does not change. People sometimes talk about "creating" electricity as though it comes out of nowhere, but that is not correct. In fact electrons do not appear or disappear—they just move from one place to another.

Is electricity caused by creating new electrons?

Electricity is a result of the movement of electrons, whether they are being transferred from one surface to another or flowing within an object. When scientists study electricity, they cannot see individual electrons moving from atom to atom. Instead, scientists explore the overall effects of the movement of electrons from one object to another or within an object. In the rest of this unit, you will explore the effects of the movement of electrons. You will see what happens when electric charges stay in one place on the surface of an object, and what happens when they move through an object.

Figure 4.7 You cannot see individual atoms moving—even with the most powerful magnifying glass—but you *can* observe the overall effect of their movement.

Section 4.1 Summary

In this section, you learned the following:

- Everything is made up of atoms.
- Every atom contains positive and negative charges.
- The nucleus of the atom contains positive charges.
- Electrons are negatively charged particles that circle the nucleus.

An atom that has an equal number of positive charges and negative charges is electrically neutral or uncharged. Sometimes an imbalance occurs between positive and negative charges in the following ways:

- If an uncharged neutral atom gains electrons, it becomes negatively charged.
- If an uncharged neutral atom loses electrons, it becomes positively charged.

Electricity is the effect of the movement of electrons from one atom to another.

Check Your Understanding

1. **(a)** What is the centre of an atom called?
   (b) What type of charge is in the centre of an atom?
2. When is an atom electrically charged?
3. An atom has four positive charges in the nucleus and six electrons that move around the nucleus. What is the overall charge of the atom? Show how you calculated the charge.
4. Draw a picture of the atom described in question 3. Label the parts of the atom.
5. **Apply** Josie is decorating her house for a party. She rubs a rubber balloon on her hair. The balloon develops a negative charge. What kind of charge does Josie's hair develop? Draw a diagram to show the transfer of electrons between the balloon and Josie's hair in this example.
6. **Thinking Critically** Name three situations that show electricity in action.

Key Terms

atoms
nucleus
positive charge
electrons
negative charge
uncharged
imbalance
charged

Section 4.2 Static Electricity

Key Terms
- static electricity
- attract
- repel
- discharge

Figure 4.8 Do your socks go with your shirt a little too well? Static electricity can make different fabrics stick together.

You pull a shirt out of the clean laundry basket and some other clothing is stuck to it. You drag your feet across the carpet and feel a shock when you reach for the doorknob. You brush your hair on a winter day and it sticks out from your head. You pull out an extra blanket on a winter night and sparks crackle. All of these are examples of **static electricity**. Static electricity is a buildup of electric charges on two objects that have become separated from each other. The word "static" means "not moving." Static electricity is an electric charge that does not move from the place on the surface of an object where it built up.

But what makes these charges build up? And why is static electricity noticeable sometimes, but not always? In this section you will explore the properties of static electricity.

Explaining Static Electricity

In Section 4.1 you learned that atoms can pick up electrons from other atoms. When this happens, both atoms become electrically charged. This picking up and losing—or *transfer*—of electrons takes place between the atoms of different objects or materials. If you rub two pieces of wool cloth together, they won't pick up electrons from each other. But if you rub a balloon and a wool cloth together, one will pick up electrons from the other. When you separate the objects, you leave a static charge on each object.

Rubbing Builds a Charge

In Starting Point Activity 4-A, you charged a balloon with static electricity by rubbing it with a piece of wool. As you can see in Figure 4.9, the surface of the balloon picked up electrons from the wool and became negatively charged. The surface of the wool lost electrons and became positively charged.

If you keep rubbing the balloon and the wool together, electrons will continue to be transferred from the wool to the balloon. The longer or harder you rub, the larger a static charge you build.

Pause & Reflect

What would happen if you rubbed the balloon only *once* with the wool? Would you still see the effects of static electricity? Why do you think the effect is different when you rub the balloon *several times* with the wool?

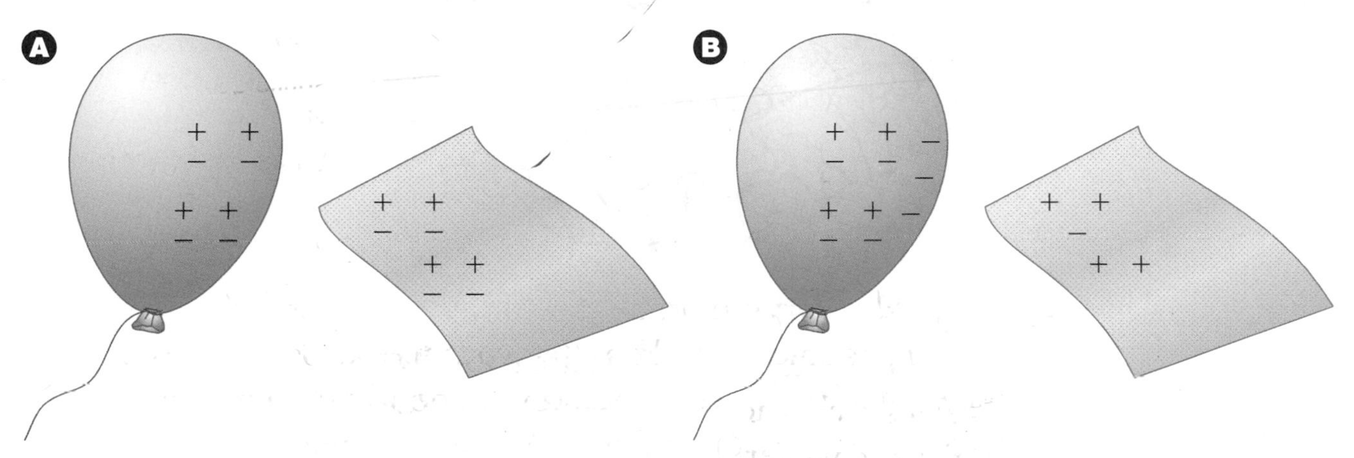

Figure 4.9 (A) Before the balloon is rubbed with the wool cloth, the charges on both objects are balanced (equal). (B) After the balloon is rubbed with the wool cloth and the objects are separated, the wool is left with a positive charge. The balloon is left with a negative charge. Electrons have transferred from the wool to the balloon on the spot that was rubbed.

After the balloon in Activity 4-A became charged with electricity, it behaved in a specific way. As you moved toward the charged balloon, the balloon moved toward you. Electricity can "pull together" or **attract** other objects. When the charged balloon was placed near another charged balloon, the two charged balloons moved away from each other. Electricity can also "push away" or **repel** other objects. In Investigation 4-C, you will charge different objects to find out more about how electric charges behave.

When two negatively charged objects are brought close together, do they attract or repel each other?

CONDUCT AN

INVESTIGATION 4-C

SKILLCHECK

- Predicting
- Observing
- Controlling Variables
- Interpreting Observations

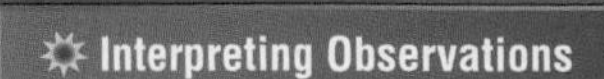

Get Ready, Get Set, Charge!

In the 1600s, scientists began to study carefully the behaviour of electric charges to find out what was causing static electricity. In this investigation, you will play the role of a scientist trying to discover how static electricity works. You will find out which types of objects can be charged with electricity, and you will observe what happens when charged objects are brought near other objects. You will also change the strength of an electric charge.

Question

Which objects can be charged with static electricity, and how do charged objects behave?

Safety Precautions

- Small static shocks may occur.

Apparatus

2 plastic spoons
2 glass rods
piece of wool cloth
piece of silk

Materials

paper punches or confetti

Procedure

Part 1

1. Copy Table 1 into your science notebook. Give it a title.
2. Before you do each trial, **predict** how the objects will behave. **Record** your predictions in Table 1.
3. Put a small pile of paper confetti on your desk.
4. Rub the two plastic spoons together. Place them near the paper confetti. **Record** your observations as "Trial 1" in your table.
5. Rub the two glass rods together. Place them near the confetti. **Record** your observations as "Trial 2" in your table.
6. Rub one of the spoons with a piece of silk. Place the spoon near the confetti. **Record** your observations as "Trial 3" in your table.
7. Rub one of the glass rods with a piece of silk. Place the rod near the confetti. **Record** your observations as "Trial 4" in your table.
8. Rub the bowl of each spoon with the silk. Do not touch the bowl of the spoon after you have rubbed it. Put one spoon down on your desk. Hold the other spoon by the handle and bring it close to the spoon on the desk. **Record** your observations as "Trial 5" in your table.
9. Rub the ends of both glass rods with the silk. Do not touch the ends after you have rubbed the rods. Put one rod down on your desk, making sure it cannot roll off. Bring the second glass rod close to the first. **Record** your observations as "Trial 6" in your table.
10. Rub one glass rod and the bowl of one spoon with the silk. Put the spoon down on your desk and bring the glass rod close to it. **Record** your observations as "Trial 7" in your table.

Skill POWER

For tips on making a data table, turn to SkillPower 4.

Table 1

Trial	Predictions	Observations
Trial 1: Plastic rubbed with plastic and put near paper		
Trial 2: Glass rubbed with glass and put near paper		
Trial 3: Plastic rubbed with silk and put near paper		
Trial 4: Glass rubbed with silk and put near paper		
Trial 5: Two plastic spoons rubbed with silk and put near each other		
Trial 6: Two glass rods rubbed with silk and put near each other		
Trial 7: Plastic spoon rubbed with silk and put near glass rod rubbed with silk		

Part 2

1. Copy Table 2 into your science notebook. Give it a descriptive title.
2. Rub one plastic spoon with a piece of wool 10 times and put it near the paper confetti. **Record** your observations under "Trial 1" in your table.
3. Rub the spoon with the wool 60 times and put it near the confetti. **Record** your observations under "Trial 2" in your table.

Table 2

Trial	Observations
Trial 1: 10 rubs	
Trial 2: 60 rubs	

4. Clean up your work area and return all your objects to your teacher.

Analyze

1. How many different types of effects did you observe in Part 1? Describe each type of effect.
2. Which trials in Part 1 did not show the effect of an electric charge? Why was there no electric charge even though the objects were rubbed together?
3. What was the effect of rubbing more times in Part 2? Explain what happened to the electric charges.

Conclude and Apply

4. Write a few sentences and draw and label a diagram that explains your observations when
   (a) you rubbed silk on glass and held the glass near the paper.
   (b) you rubbed glass on glass and held the glass near the paper.
5. If you wash cotton socks and put them in the dryer, will they cling to one another? What would happen if you put a silk shirt into the dryer with the socks? Give a reason for your answer.
6. When you brush your hair, your hair sometimes stands on end and is attracted to your brush. Will brushing your hair for a longer time help to settle it? Explain.
7. In Part 2, what do you think happens to the wool in each trial? Explain.

Extend Your Skills

8. Try this experiment using other materials around the classroom: Styrofoam™, metal, wood, rubber.

Pause& Reflect

"Nothing could exist without electric charges." Do you agree with this sentence? Think about the structure of the atom. What do you think holds an atom together?

DidYouKnow?

Scientists can place materials in order of how easily they give up electrons. If you rub together two materials from the following list, the one higher on the list will give up electrons. It will become positively charged. The one lower on the list will gain electrons. It will become negatively charged. Does this help to explain your observations in Investigation 4-C?

positive (+)
human skin
glass
hair
wool
silk
paper
cotton
rubber
plastic
negative (−)

Three Rules of Electricity

When scientists began to study static electricity, they did experiments as you did in Investigation 4-C. Through this research, scientists concluded that the behaviour of electric charges follows three rules.

Rule 1. Opposite Charges Attract

You might have heard this expression: "opposites attract." In the case of static electricity, it is definitely true. Objects with a positive charge will attract objects with a negative charge. If you dry a silk shirt and cotton socks together in a dryer, they rub together while the dryer runs. Atoms in the shirt will give up electrons to atoms in the socks. The shirt becomes positively charged and the socks become negatively charged. When the laundry is done, the socks and shirt cling together.

Figure 4.10 A bad hair day? When you pull off your sweater, static electricity may make your hair stand on end.

Rule 2. Like Charges Repel

Objects that have the same charge will push away from one another. When you rub two balloons with wool, both balloons become negatively charged. Then when you hang them side by side, they move apart.

Two positively charged objects will also repel each other. You may have experienced this yourself if you have ever pulled off a woollen hat or sweater on a dry winter day. As the wool rubs your hair, it picks up electrons. Each strand of hair becomes positively charged. Each strand of hair will try to move away from every other strand of hair because they have the same charge. The result: a positively hair-raising experience!

Rule 3. Charged Objects Attract Uncharged Objects

In Investigation 4-C you found that pieces of paper will move toward a charged object. The paper itself was uncharged—it was electrically neutral. Why did it move?

Remember that an uncharged object still has both positive and negative charges. These charges can move around within an object. Figure 4.11 shows what happens when a negatively charged balloon is brought close to an uncharged wall. The negative charges in the balloon repel the charges in the wall. The result is that the part of the wall nearest the balloon becomes positively charged, even though no electrons have actually been transferred from the balloon to the wall or from the wall to the balloon.

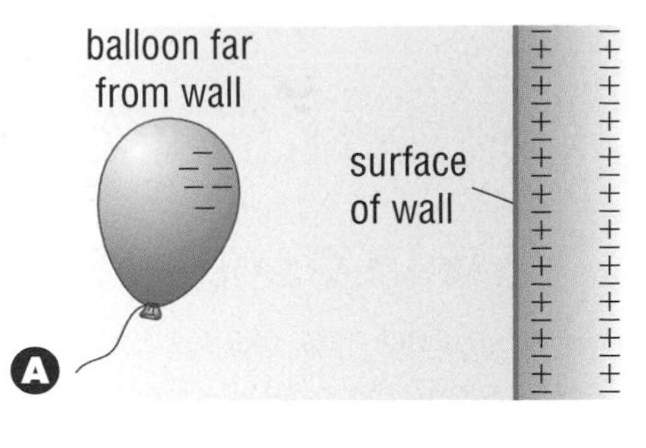

Figure 4.11 (A) When a negatively charged balloon is held far from an uncharged wall, the charges in the wall are evenly distributed. (B) When the charged balloon is held close to the wall, the electrons in the part of the wall next to the balloon move as far away from the balloon as possible. That part of the wall becomes positively charged. The negatively charged balloon and the positively charged part of the wall are attracted to each other. The balloon sticks to the wall.

Using Static Electricity

Making your hair stand on end or sticking balloons to the wall can be fun, but is it useful? Some technologies use the properties of charged objects to do important work.

For example, some pulp mills and factories use static electricity to clean the smoke from their smokestacks. To clean the smoke, charged plates are placed in the smokestack. Oppositely charged particles in the smoke are attracted to the plates and stick to them. The particles can then be collected and removed from the smokestack, allowing the cleaner gas to be released into the air. In the next activity you will build a simple version of another common machine that uses static electricity.

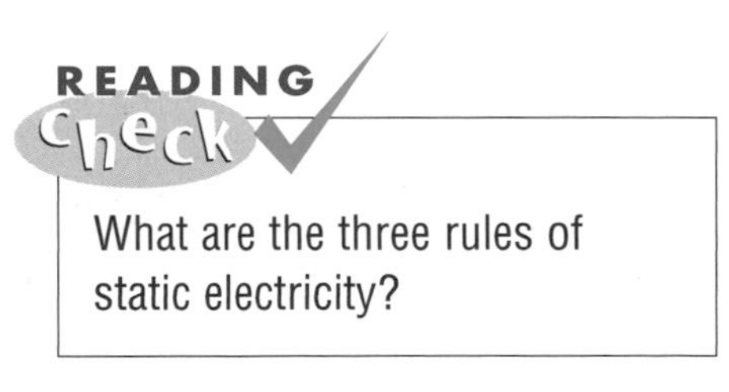

What are the three rules of static electricity?

Figure 4.12 The Teck Cominco smelter in Trail, British Columbia, uses static electricity to reduce pollution from its smokestacks.

Find Out ACTIVITY 4–D

Make a Pepper Copier

Your friend has drawn the funniest cartoon you have ever seen. If you wanted to keep the cartoon, you would use a photocopier to make an exact copy for yourself. In this activity, you will model what goes on inside a photocopier.

What You Need

a plastic petri dish with a lid
ground pepper
clear adhesive tape
paper
wool cloth
scissors

What to Do

1. Make a stencil by tracing the outline of the petri dish lid onto a piece of paper. Cut out the circle of paper. Then cut the shape of your first initial out of that circle.
2. Put a small amount of pepper into the petri dish.
3. Shake the dish to spread the pepper evenly so that it covers the entire bottom of the petri dish.
4. Place the lid on the petri dish.
5. Trial 1: Turn the petri dish upside down and then right side up again. Remove the lid and **observe** the pepper.
6. Attach the stencil to the top of the petri dish lid with small pieces of tape at the edges.
7. Trial 2: Turn the petri dish upside down and then right side up again. Remove the lid and **observe** the pepper.
8. Trial 3: Replace the lid on the petri dish. Place the Petri dish on a flat surface, holding the dish at the edges. Using a wool cloth, rub the lid only where it shows through your stencil. Rub hard and quickly with two fingers for 45 to 60 seconds.
9. Remove the stencil from the lid. Be careful to handle the lid only at the edges.
10. With the lid still on, turn the petri dish upside down, and then right side up again. Remove the lid and **observe** the pepper.

What Did You Find Out?

1. What happened to the pepper in your first and second trial? What happened in the third trial? Explain why the results were different.
2. What do you think happens to electrons in the pepper when they come close to plastic that has a negative charge? Explain.

Extension

3. Why was it important to do the three different trials in this activity? Explain.
4. Use your school library or the Internet to research how a photocopier uses static electricity. Make a diagram or write a short report to communicate what you learned.

Static Shocker: Electricity Can Jump

You have seen that objects charged with static electricity can attract and repel other objects. For example, a negatively charged balloon will attract a positively charged piece of wool. When this happens, the charges stay on the surface of each object. The electrons themselves do not transfer between the wool and the balloon unless they touch.

Sometimes, however, electrons do jump from one object to another. You may feel the burst of electrons as a shock, and you may see it as a spark. You can even hear it: think about the crackle when you separate clothes that have stuck together in the dryer. The crackle is the burst of electrons transferring from one object to another. This transfer of electrons is called a **discharge** of static electricity. That is, the transfer puts the electric charges back in balance on each object. Figure 4.13 shows what happens when you get a static electric shock from a doorknob.

DidYouKnow?

Lightning strikes are millions of times more powerful than the static shocks you get from a doorknob.

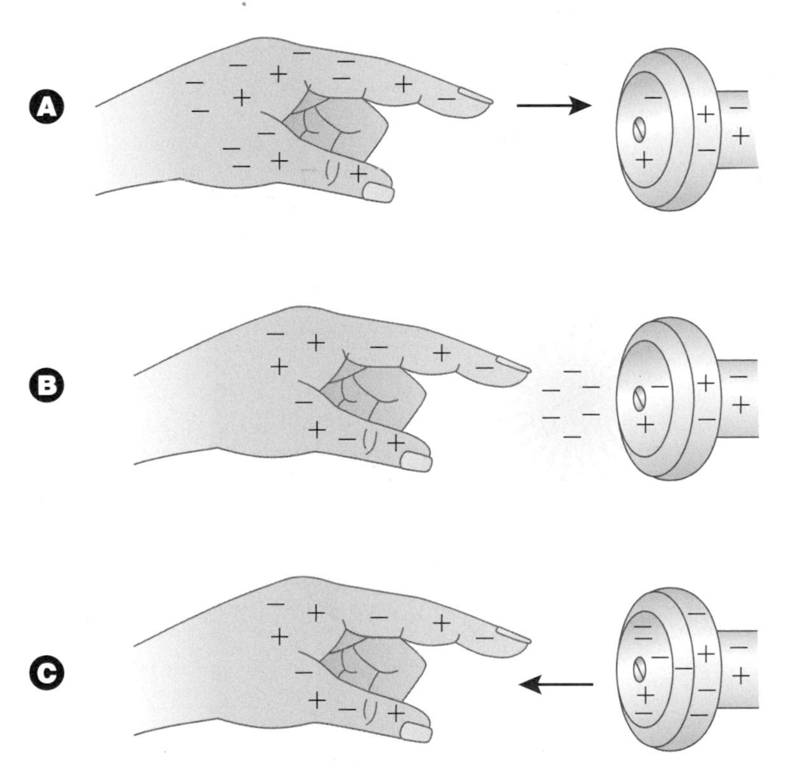

Figure 4.13 (A) If you shuffle your feet across a carpet, you may pick up electrons from the carpet. Your body may build a negative charge. (B) As you reach for the doorknob, the extra electrons are discharged in a sudden burst, which we often call a "shock." The shock happens when your finger and the knob are close together, but not yet touching. Electrons leap between them, making the air light up as they pass. (C) Your body now has a balanced charge again, and the extra electrons have moved into the doorknob.

You sometimes feel a static shock when you step out of a car and touch the metal around the door. During the drive your body has been rubbing on the seat and building a negative charge. In most cars, this shock will be several times stronger if you are wearing nylon than if you are wearing cotton.

What causes a static electric shock?

Figure 4.14 Throughout history, people have often thought of lightning in connection with supernatural beings. In the traditional stories of the Kwakwaka'wakw people of northern Vancouver Island, lightning is made by a powerful being called Thunderbird.

Lightning Is a Gigantic Static Shock

In the time it takes you to say the word "lightning," lightning strikes about 100 times around the world. A lightning strike is a very powerful natural event. A bolt of lightning can explode a tree, shatter rock, and melt sand into glass. And yet lightning is just a more powerful version of the shock you get from a doorknob.

During a thunderstorm, air currents cause water droplets and ice particles to collide and rub together inside a thundercloud. This action builds an enormous imbalance of charges within the cloud. Figure 4.15 shows how this static electricity is released, or discharged, in the form of lightning.

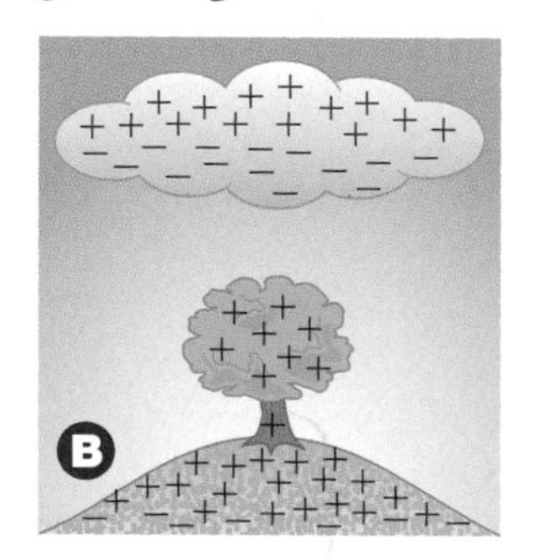

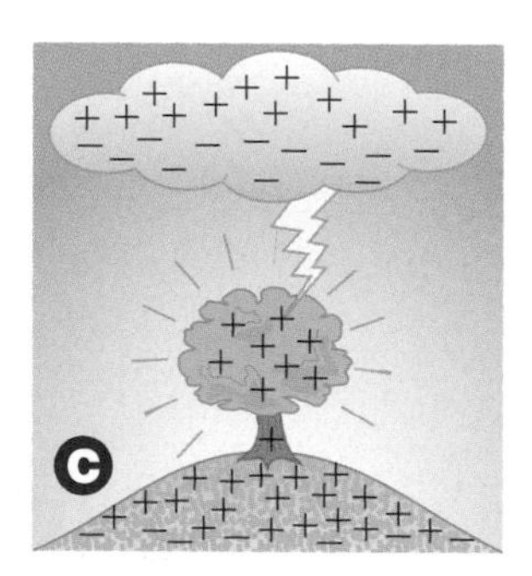

Figure 4.15 Charges that build up during a thunderstorm are released as lightning.
(A) Both the cloud and the ground are neutral (uncharged) to begin with.
(B) Within the cloud, particles of ice and water rub together. Some particles lose electrons and other particles gain electrons. Negative charges build up on the bottom of the cloud. The strong negative charge from the cloud repels negative charges on Earth's surface. The ground under the cloud as well as objects on the ground, such as trees or buildings, become positively charged.
(C) A gigantic spark leaps from the cloud to the ground. What you see as a single flash of lightning is actually many sparks that travel between the cloud and the ground.

Figure 4.16 An average bolt of lightning is about 10 km long. A lightning bolt can reach a temperature several times hotter than the surface of the Sun.

INTERNET CONNECT

www.mcgrawhill.ca/links/BCscience6

Lightning happens suddenly and is over in an instant. The speed of this event makes it hard to study. Benjamin Franklin was an American scientist in the 1700s who studied lightning. He was lucky to survive his own experiment. He flew a kite in a storm to show that storm clouds carry electricity. Go to the web site above and click on **Web Links** to find out where to go next. What did Franklin do?

A lightning strike will usually take the shortest route between the negatively charged cloud and the positively charged ground. This is why lightning tends to strike tall buildings and trees. In the following activity, you will find out facts and fiction about lightning. You will also learn how to protect yourself from lightning.

Why does lightning usually strike tall objects?

Figure 4.17 Some people try to protect themselves during a rain or lightning storm by taking cover under tall trees. Is this a good idea?

Pause & Reflect

If lightning strikes tall buildings, can you think of how lightning rods might work to prevent lightning damage? In Section 4.3, learn how lightning rods work.

At Home ACTIVITY 4-E

How Shocking!

You may have heard the saying "lightning never strikes twice in the same place." Or maybe someone has told you that you can tell whether a thunderstorm is moving toward you or away from you just by listening to it. But are these ideas really true? In this Activity you will collect information about lightning and then sort fact from fiction.

What to Do

1. Talk with friends and members of your family to find out what they know and believe about lightning. Make a list of all the "facts" you hear from them.
2. Use reference books in your library or the Internet to investigate which of the statements on your list are true, and which are not.
3. As you conduct your research, add more statements to your list. Try to collect statements that are true and statements that are false. For every false statement you collect, write a true statement to correct it.

What Did You Find Out?

1. How much of what you heard in the past about lightning was true?
2. Will any of the things you learned about lightning change how you and your family behave during a thunderstorm? Explain why.
3. Using the information you have collected, prepare a poster or a presentation that will **communicate** accurate information about lightning. Include safety tips that will help people protect themselves from lightning both indoors and outdoors.
4. Share this information with your family and friends at home.

Section 4.2 Summary

In this section, you learned that you can build a static electric charge by rubbing together objects made of different materials and then separating them. One object transfers electrons to the other with the following results:

- The object that gives up electrons becomes positively charged.
- The object that picks up electrons becomes negatively charged.

The imbalance of charges is called static electricity. This electric charge stays in one place on the surface of each object. The longer you rub the two objects together, the stronger the charge becomes.

Static electric charges follow three basic rules:

- Opposite charges attract each other.
- Like charges repel each other.
- Charged objects attract uncharged objects.

When two oppositely charged objects are close together, electrons may jump through the air between them to balance the charge. This discharge of static electricity causes a static electric shock. Lightning is an example of a large static electric shock.

Key Terms

static electricity
attract
repel
discharge

Check Your Understanding

1. The diagram illustrates the charge on three different objects, A, B, and C.
   (a) Is object A positively charged, negatively charged, or uncharged?
   (b) What will happen to object B if object C is brought close?
   (c) What would have to be done to object C to make it uncharged?

A
+ + − + − −
− − + − + +

B
+ + + + − +
− − + + + +

C
− − + − − +
− − − + − +

2. You rub two identical wool cloths together, and then hold them close together (but not touching). What would you expect to observe? Explain.

3. What is the term used to describe electrons leaving an object to balance a charge?

4. **Apply** A Van de Graaff generator is a device that charges objects with electricity. When you touch a Van de Graaff generator, your hair stands on end. Why does this happen?

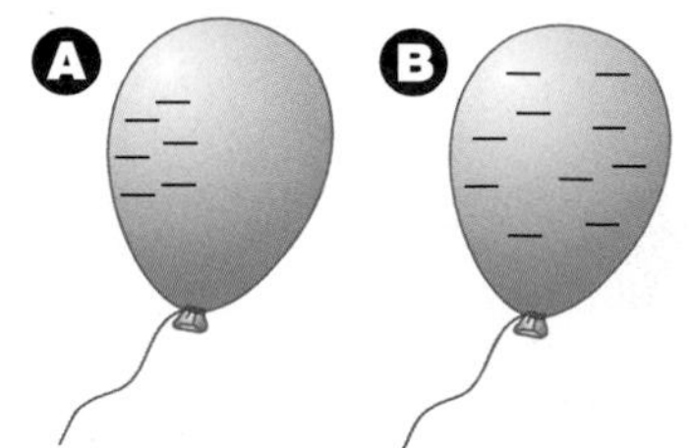

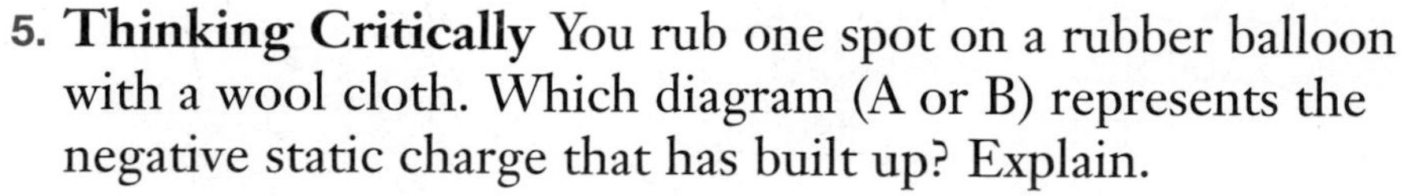

5. **Thinking Critically** You rub one spot on a rubber balloon with a wool cloth. Which diagram (A or B) represents the negative static charge that has built up? Explain.

Section 4.3 Current Electricity

Electricity on the Move

Key Terms
- current electricity
- conductors
- insulators
- ground
- battery

In Section 4.2 you learned about static electricity. Static electricity results from the transfer of electrons *between* objects (from one object to another). The imbalance in electric charges then stays in one place on the surface of the object. Another kind of electricity is **current electricity**. Current electricity results when the electrons from an imbalanced charge move *through* an object.

The movement of electrons through an object is called an *electric current*, or a flow of electricity. When electrons move through an object, the atoms themselves do not move. Instead, electrons are "passed" from one atom to the next. Current electricity is like a line of people passing buckets of water from one person to the next. The people do not move, but the water does. But when does electricity stay in one place, and when does it flow in an electric current? In the next investigation, you will study different materials to see which ones allow the flow of electricity.

Figure 4.18 A lightning rod turns static electricity into current electricity, and provides a path so the lightning discharge doesn't flow through the building.

READING Check

What is the main difference between static electricity and current electricity?

Figure 4.19 If the people are atoms and the buckets are electrons, which scene represents an electric current?

CONDUCT AN

INVESTIGATION 4-F

SKILLCHECK
- Controlling Variables
- Observing
- Interpreting Observations
- Communicating

Lighten Up!

An electric current is the flow of electricity through an object. Electricity can pass through some types of items and not others. In this investigation you will study different types of items to see which ones allow the flow of electricity.

Question

How can you tell when electricity is flowing?

Safety Precautions

- Do not connect more than one battery
- Do not touch the metal part of the alligator clips.

Apparatus

1 1.5 V D-cell battery
1 battery holder
3 copper wires with an alligator clip on each end
1 1.5 V light bulb
1 light holder
4 beakers
spoon

Materials

Solid test items:
glass rod
piece of silk
piece of wood
a penny coin
a nickel coin

Liquid test items:
tap water
lemon juice
salt-water solution
baking soda solution

Procedure

1. Copy the table below into your science notebook. Give it a title. **Predict** what you will see if a material allows the flow of electricity.

Solid Test Items							Liquid Test Items			
Results	No item	Glass	Silk	Wood	Penny	Nickel	Tap water	Lemon juice	Salt-water solution	Baking soda solution

2. Put the battery into the battery holder. Attach the end of one wire to one terminal (or end) on the light holder, and the other end to one terminal on the battery holder. Attach one end of the second wire to the other terminal on the light holder. Attach one end of the third wire to the other terminal on the battery holder. You should now have two alligator clips available. Attach the two clips together, making sure that your fingers touch only the plastic on the clips, not the metal. **Record** your observations in your table under "No item."

3. Attach one of the solid test items between the two available alligator clips. Make sure the metal part of each clip touches the test item. **Record** your observations in your table.

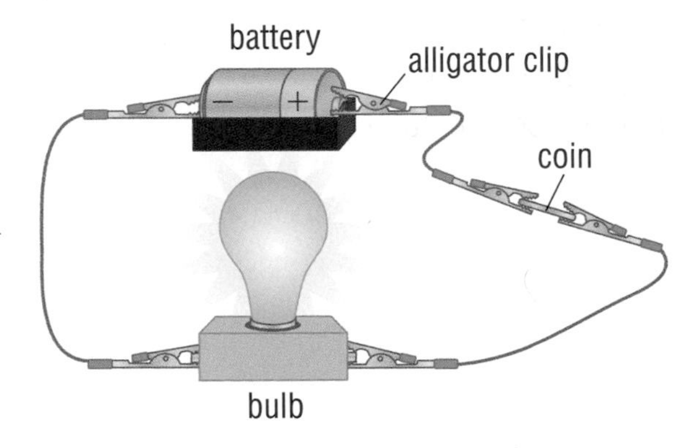

4. Repeat for each of the solid test items.

5. Test each of the liquid items by placing the alligator clips into a beaker containing the liquid. Make sure the clips do not touch each other because that will make a connection. Wipe the clips clean and dry between each test. **Record** your observations in your table.

6. Clean up your work area and return all materials to your teacher.

Analyze

1. What is the purpose of doing step 2 with no items?
2. Which items allowed the electricity to flow through them? How could you tell?
3. Which items did not allow electricity to flow through them? How could you tell?
4. Why was it important to clean the wires between tests of the liquid items?

Conclude and Apply

5. Write a paragraph or draw and label a diagram that shows how electrons flowed through the wires
   (a) when you put a penny between the two alligator clips
   (b) when you put wood between the two alligator clips
6. Do you think you can create a static charge (that is, one that stays in one place) on a type of object that allows electricity to flow through it? Explain why or why not.

Extend Your Skills

7. Do you think your body allows electricity to flow through it? How do you know?

What is a conductor? What is an insulator?

Pause & Reflect

In Investigation 4-F, which test items were conductors? Which ones were insulators?

Conductors and Insulators

Some materials allow electrons to flow through them easily. These materials are called **conductors**. Metals such as copper are good conductors. An electric charge at one end of a metal object will spread over the whole object. Other materials block the flow of electrons. These materials are called **insulators**. Rubber and wool are both insulators. An electric charge on one part of an insulator is static—that is, it will stay in one place. You can build static electricity only on objects that are insulators.

A

B

Figure 4.20 Which cartoon road represents a conductor? Which cartoon road represents an insulator? If more electrons travelled along each road, what would happen to the charge on the road in (A)? in (B)?

Figure 4.21 Humid air conducts electricity. That means that in humid areas, a static charge will flow away into the air instead of building up on the surface of an object. It is therefore much easier to create static electricity on an object in drier parts of the province like the Chilcotin region.

What makes one material a conductor and another an insulator? In a conductor, atoms hold their electrons loosely. Electrons in a conductor can move easily from one atom to the next. In an insulator, atoms hold their electrons more tightly than in a conductor. Experiments such as Investigation 4-F allow you to find out whether an object is a conductor or insulator.

Some materials allow electrons to flow through them, but not very well. These materials are called *fair conductors*. Table 4.1 shows some examples of good conductors, fair conductors, and insulators.

Table 4.1 Conductors and Insulators

Good Conductors	Fair Conductors	Insulators
aluminum	carbon	cotton
copper	earth	rubber
gold	human body	glass
iron	humid air	pure water
nickel	salt water	wool

Grounding an Electric Charge

Figure 4.22 Earth can absorb even big electric charges, the same way the ocean absorbs a cup of water. The overall effect is so small it cannot be measured.

"You're grounded!" Those words are not usually ones you want to hear. But when you are using electricity, being grounded can be very important. It can even save your life. To **ground** an object means to connect it through a conductor to the ground, or Earth. Grounding is a way to prevent an electric charge from building on an object, or to get rid of an electric charge. Because Earth is so large, it can accept or give up electrons without any noticeable change in its charge. Extra electrons can flow into the ground from an object with a negative charge so that the object is uncharged again. We say that the charge has been *neutralized*. Electrons can also flow from the ground into an object to neutralize a positive charge on that object. If an uncharged conductor is grounded, it will not build a static charge.

Grounding works for big charges like lightning, but it also works for smaller charges. Remember that you can charge an insulator such as a balloon by rubbing it with wool. If you touch the balloon with your hand, you ground the static charge.

How does grounding discharge electricity?

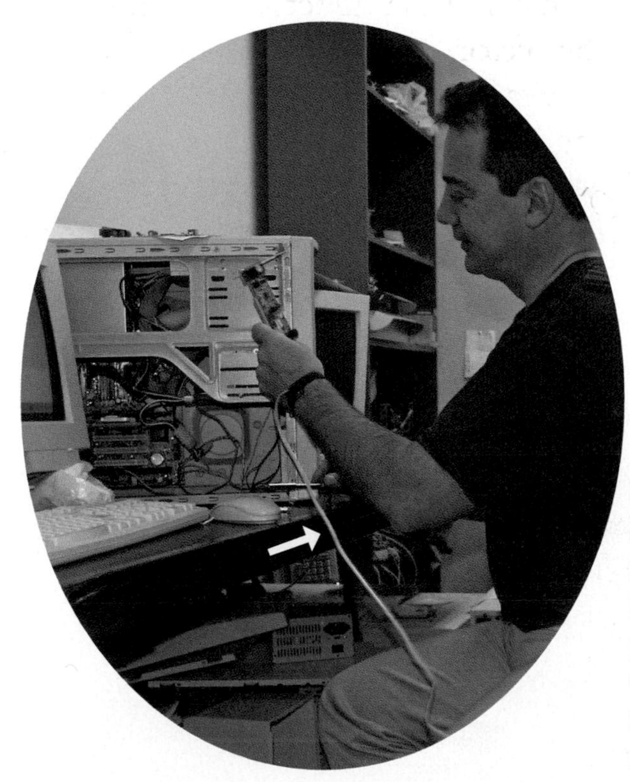

Figure 4.23 Static electric charges can damage sensitive electric equipment like computers. The arrow points to a special ground wire often wore by computer technicians to get rid of static electricity.

Pause & Reflect

When you take your clothes out of the clothes dryer, some of them stick together. By the time you've finished folding the laundry, the static cling has gone. What has changed?

Figure 4.24 What if you always had to wait for a static charge in order to start an electric current?

Creating an Electric Current

In Investigation 4-C, you charged different insulators with a static charge. If you touch a conductor such as a metal wire with that charged object, the static electricity can be turned into an electric current. But static electricity, when discharged, produces only a one-time burst of electric current. You need a steady electric current for most objects that use electricity. In the next activity, you will use a lemon to make a steady electric current.

Find Out ACTIVITY 4–G

Electric Lemon

You can use a static charge to start an electric current, but that is not always very useful. Chemical reactions can also create electric currents. In this activity you will find out how to make electrons flow *without* first building a static charge.

What You Need 

1 pair of headphones
1 galvanized nail
1 piece of copper wire (heavy gauge)
1 lemon

What to Do

1. Stick the copper wire down into one end of the lemon. Push it almost all the way through.
2. Stick the nail down into the other end of the lemon. Again, push it almost all the way through.
3. Bend the top of the wire so that it is almost touching the nail.
4. Put on the headphones. Hold the plug end of the headphones on the top of the nail. **Observe** what you hear.
5. With your other hand, touch the copper wire to the plug as shown in the illustration. Make sure that the plug is touching both the nail and the wire.
6. Wait one to two minutes. **Observe** what you hear.

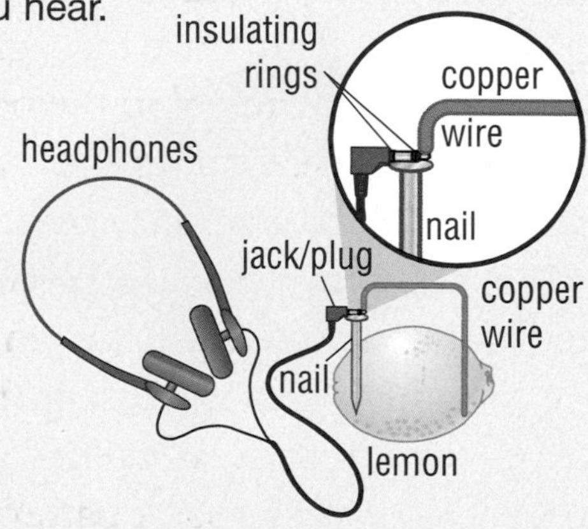

What Did You Find Out?

1. Did you hear anything in step 4? How did it differ from what you heard in step 6?
2. How do you know that electricity is flowing?

Extend Your Skills

3. Trace the path of electrons to explain what you hear.

Battery Power

In Activity 4-G, you made electrons move in a steady current from one metal to another, through a liquid conductor and then a wire conductor. You made a simple battery. A **battery** is a device that turns chemical energy into electricity. It is basically a can containing two different metals in an acidic solution. A chemical reaction between these materials causes electrons to gather at the negative terminal. Unlike the one-time burst of electrons that happens when static electricity is discharged, a battery provides a steady flow of electrons. This makes it very useful as a source of electric current.

How many times have you used a battery today? Whenever you ride in a car, use a watch, or talk on a cellphone, you have used battery power. To operate these electric devices, you need something less awkward—and often a lot more powerful—than a lemon. Figure 4.26 shows you the inside of one type of battery. When the positive and negative terminals are connected on the outside by a conductor, a chemical reaction starts inside the battery. Electrons then move through the conductor from the negative terminal to the positive terminal.

The chemical reaction in a battery happens only when the two terminals are connected by a conductor. You can start the electric current by connecting the conductor. You can also stop the electric current by disconnecting the conductor. In most devices, we use a switch to connect and disconnect the conductor. A battery can also be stored between uses without using up its energy. In the next chapter, you will use batteries to put electricity to work.

INTERNET CONNECT

ww.mcgrawhill.ca/links/BCscience6

The lemon battery you made was a "wet cell" battery. Another kind of battery is called a "dry cell." Using the Internet, research the difference between these two kinds of batteries. Go to the web site above and click on **Web Links** to find out where to go next. Which kind will you find

(a) in a car?

(b) in a laptop computer?

(c) in a flashlight?

In what way is a battery more useful than static electricity as a source of electric current?

Figure 4.25 You can make a battery using a lemon, but do you really want to?

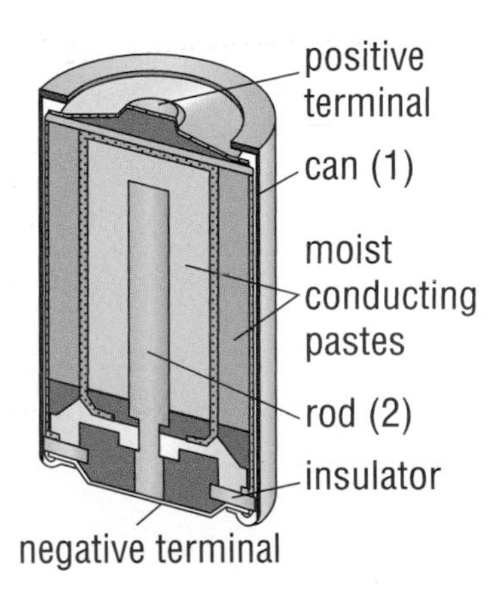

Figure 4.26 A simple battery contains two different metals (labelled 1 and 2) that react to cause the movement of electrons through the conducting pastes to the negative terminal. Insulators prevent the end of the battery from reacting with the can.

DidYouKnow?

Lewis Urry was a Canadian inventor who developed the first practical long-life battery in the 1950s. These are the small, long-lasting alkaline batteries that power a Walkman™, Gameboy™, and other portable devices. About 80 percent of the dry cell batteries in the world today are based on the work of Urry. He died in October 2004.

Key Terms

current electricity
conductors
insulators
ground
battery

Section 4.3 Summary

In this section, you learned the following:
- Current electricity is caused by the movement of electrons from one atom to the next through a conductor.

Electrons can move easily through some materials and not others:
- A conductor is a material that allows the flow of electrons.
- An insulator is a material that blocks the flow of electrons.
- A fair conductor is a material that allows the flow of small amounts of electrons (for example, your body or salt water).

One way to get rid of an electric charge is to ground the charged object. Grounding an object means connecting it through a conductor to Earth. Earth is so big that it can give up or accept enough electrons to neutralize even a very big electric charge.

To use electricity effectively, you should know the following:
- The electric devices we use every day require a steady electric current instead of a sudden discharge of electrons.
- A battery uses a chemical reaction to produce a steady electric current.

Check Your Understanding

1. If an object is to be classified as a good conductor, what characteristics must it have?
2. List three examples of fair conductors. What do these three examples have in common?
3. Explain how grounding a charged object neutralizes the charge on that object.
4. **Apply** Think of lightning and the electric shocks you may get when you rub your feet across a carpet. What makes the "flash" of a lightning strike larger than that of a carpet shock?
5. **Thinking Critically** When the headphones are connected to the lemon in this arrangement, you can hear the electricity flow.
   (a) What does this tell you about the properties of lemon juice?
   (b) What would happen if you injected pure water into the lemon while you were listening to the headphones? Explain.

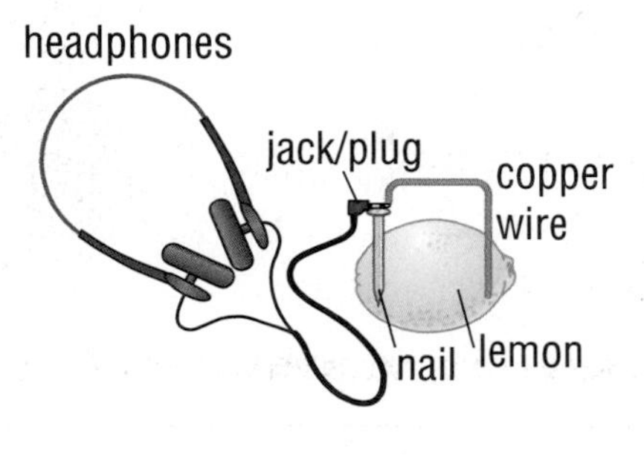

CHAPTER at a glance

Now that you have completed this chapter, try to do the following. If you cannot, go back to the sections indicated.

(a) Define the following: nucleus and electron. (4.1)

(b) Draw a diagram of an atom that has 4 electrons and a nucleus with 3 positive charges. Label the features. (4.1)

(c) Explain how an atom can have an imbalanced charge. (4.1)

(d) Explain what causes electricity. (4.1)

(e) Explain what causes static electricity. (4.2)

(f) Draw and label a diagram that shows what happens when a positively charged and a negatively charged object are placed beside each other. (4.2)

(g) Describe what happens when a rubber balloon is rubbed with a piece of wool. (4.2)

(h) List the three rules of electricity. (4.2)

(i) Compare static electricity and current electricity. (4.3)

(j) Explain the main difference between conductors and insulators. (4.3)

(k) Describe what the term "grounding" means. (4.3)

(l) Explain how a battery creates an electric current. Draw a diagram of a battery. Label the parts. (4.3)

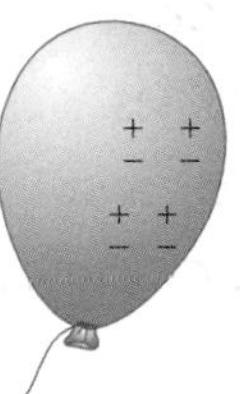

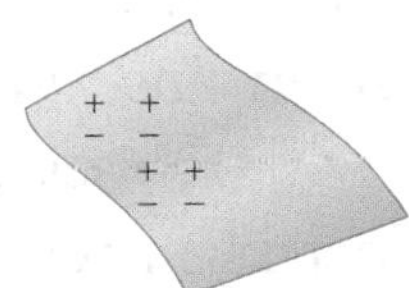

Prepare Your Own Summary

Summarize this chapter by doing one of the following. Use a graphic organizer (such as a concept map), produce a poster, or write a summary to include the key chapter ideas. Here are a few ideas to use as a guide:

- Use the structure of atoms to explain how a positive or negative charge can be formed on an object.
- Make a Venn diagram (see SkillPower 1) to explain the similarities and differences between static electricity and current electricity.
- Design a poster with pictures of items that will conduct electricity and others that will not. Label each one as a conductor or insulator. Provide a definition of conductors and insulators that could be understood by younger children.
- Write a set of instructions for charging a balloon with static electricity. Include labelled diagrams to show what is happening to the electrons.
- Write a flow chart to describe the discharge of electricity from your hand to a doorknob, starting with what happens when you are walking across a carpet and ending with balanced electric charges in your hand again.

CHAPTER

4 Review

Key Terms

atoms
nucleus
positive charge
electrons
negative charge
uncharged
imbalance
charged
static electricity
attract
repel
discharge
current electricity
conductors
insulators
ground
battery

Reviewing Key Terms

If you need to review, the section numbers show you where these terms were introduced.

1. In your science notebook, match each description in column A with the correct term in column B. Use each description only once.

A	B
(a) a negatively charged particle in an atom	• electron (4.1)
(b) a word to describe an atom that contains an unequal number of positive and negative charges	• nucleus (4.1)
(c) a buildup of separated positive and negative charges	• grounding (4.3)
(d) a device that uses a chemical reaction to produce a steady electric current	• battery (4.3)
(e) connecting an object through a conductor to Earth	• charged (4.1)
(f) the centre of an atom	• static electricity (4.2)

2. Describe the difference between:
   (a) positive charge and negative charge (4.1)
   (b) charged and uncharged (4.1)
   (c) attract and repel (4.2)
   (d) static electricity and current electricity (4.3)
   (e) conductor and insulator (4.3)

Understanding Key Ideas

3. What are the two kinds of electric charges? (4.1)

4. An uncharged object picks up some electrons. What type of charge does the object have now? (4.1)

5. Draw a diagram of an atom that has one particle orbiting the nucleus and two particles inside the nucleus. Label the names of and charges on each of these three particles that make up the atom. What charge does this atom have? (4.1)

6. Consider the charges on the spheres shown. Which spheres would be repelled by each other? Explain why this happens. (4.2)

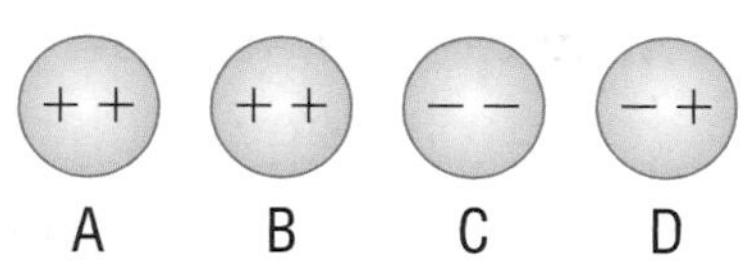

7. List four objects that are conductors and four that are insulators. What properties of each type of object make it a good conductor or insulator? (4.3)

8. Why are wires usually made of metals? (4.3)

9. How does grounding a charged object neutralize the charge on that object? (4.3)

Developing Skills

10. A cotton T-shirt is placed in a clothes dryer with another garment. If you don't want static cling, what fabric would the second garment have to be made from?

11. Write a step for each of the trials in Part 1 of Investigation 4-C to provide a control.
12. List two devices that use the properties of static electricity.
13. If you build a log cabin like the one shown in the diagram, where would you place a lightning rod? Why?

Problem Solving

14. Explain why a student holding a metal rod is unable to charge the rod with static electricity by rubbing it with a piece of wool.
15. Explain how you could prevent clothes from sticking to each other when they are dried in the dryer.
16. You rub one side of a rubber balloon with wool, and try to attach the opposite side to the wall.
    (a) Will the balloon stick?
    (b) What will happen if you try the same thing with a metallic balloon?

Critical Thinking

17. Electric charge passing through a material that is not a good conductor causes the material to heat up greatly. Use this information to explain how lightning can cause forest fires.
18. Approximately 100 bolts of lightning strike Earth every second. If all of these lightning bolts deliver electrons to Earth's surface from the clouds, speculate why Earth does not become so negatively charged that lightning stops altogether.

Pause& Reflect

Go back to the beginning of this chapter on page 96, and check your original answers to the Getting Ready questions. How has your thinking changed? How would you answer those questions now that you have investigated the topics in this chapter?

CHAPTER

5 Electricity at Work

Getting Ready...

- How does flicking a switch on your wall make the light go on and off?
- What would you do to an ordinary flashlight to make the batteries last longer?
- How could you use electricity to turn an ordinary iron bar into a magnet?

Electric devices such as flashlights put electricity to work—so you can have fun!

British Columbia has some of Canada's deepest and longest caves. Exploring in these caves can be exciting if you and your friends have good flashlights. But would you be able to explore safely without a flashlight?

By using electricity, you can go places and do things that would otherwise be difficult or even impossible. You probably use many different kinds of electric devices every day. Look around the classroom or think about your home. How many electric devices can you name? Did you know that every one of these devices is built around an electric circuit? Only two kinds of electric circuits serve as the basic building blocks for almost every electric system in the world.

In Chapter 5, you will explore these two kinds of electric circuits. You will build your own circuits to do different tasks. You will also see how circuits convert electricity into light, sound, heat, and motion that can make your day—and night—fun and safe.

What You Will Learn

- how to control the flow of electricity in a circuit
- how to design and build different electric circuits to perform different tasks
- how circuits convert electrical energy into other kinds of useful energy

Why It Is Important

- Every electric device you use, whether it is plugged in or works from a battery, is part of an electric circuit.
- When you understand how electric circuits work, you can build circuits that are safe and that meet different needs.
- Some people think that the link between electricity and magnetism is one of the most useful scientific discoveries of all time.

Skills You Will Use

- classify electric circuits according to their design
- solve problems by designing, building, and testing electric circuits
- observe the relationships between electricity and magnetism
- make inferences about the flow of electricity in a circuit by controlling variables

Even the most complicated electric circuits are built using the same few simple rules.

Starting Point **ACTIVITY 5-A**

Light Your Way!

Can you make use of the flow of current electricity?

What You Need 

1 1.5 V D-cell battery in a holder
2 aluminum foil strips
(10 cm long × 1 cm wide)
1 1.5 V light bulb

What to Do

1. Using only the items you have been given, try to arrange all the items so that the bulb lights up.
2. Draw a picture of every arrangement you try. Label each part and note whether or not the arrangement worked.
3. When your teacher tells you to stop, clean up your work area and return all items to your teacher.
4. Compare your findings with the rest of the class.

What Did You Find Out?

1. Which arrangements worked to light the bulb?
2. Which arrangements did not work? Why do you think they did not work?
3. Using what you learned about current electricity in Chapter 4, draw a diagram to show the flow of electricity through each of the parts of a successful arrangement (e.g., the battery, the foil, and the light bulb).

Section 5.1 Series Circuits

Key Terms

circuit
source
load
closed circuit
open circuit
switch
hazard
short circuit
series circuit
voltage
current
resistance

Figure 5.1 At night, the Legislature buildings in Victoria are lit up with 3,330 lights. You will learn whether it would make sense to wire these in a series circuit.

Electricity Flows in a Circuit

When you did Activity 5-A, were you able to make the bulb light up? Congratulations! Your system allowed the steady movement of electrons around a specific pathway. You built a working electric **circuit**. A circuit is a complete pathway or loop for the flow of electricity.

Every electric circuit has three basic components:

- A **source** of electrical energy: The source provides the "push" that causes electrons to move through the circuit. In the electric circuit you built, the source was the battery.
- A conductor to carry electricity: In an electric circuit, a conductor is usually the wire or wires that carry electricity. In Activity 5-A, the conductor was the foil strip. In most of the circuits you will be working with, the conducting wire is wrapped in an insulating material.
- A **load**: The load is any component along the circuit that uses the electricity. In the circuit you built, the bulb was the load.

Closed and Open Circuits

What is the difference between a closed circuit and an open circuit?

In Activity 5-A, you found that a circuit works only when it is complete. That is, the bulb lights up only when there is a complete and unbroken pathway to carry the electricity from the battery to the light and back to the battery. This is called a **closed circuit**. If there is a gap or break anywhere along the path, electricity does not flow. A circuit with a gap in it is called an **open circuit**. Figure 5.2 shows the difference between an open and closed circuit.

A
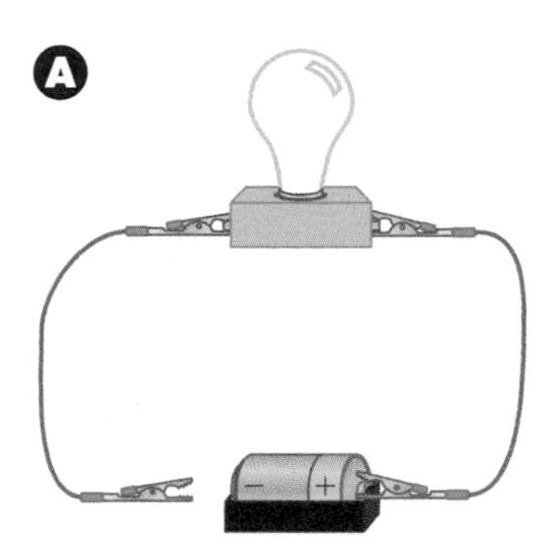

B
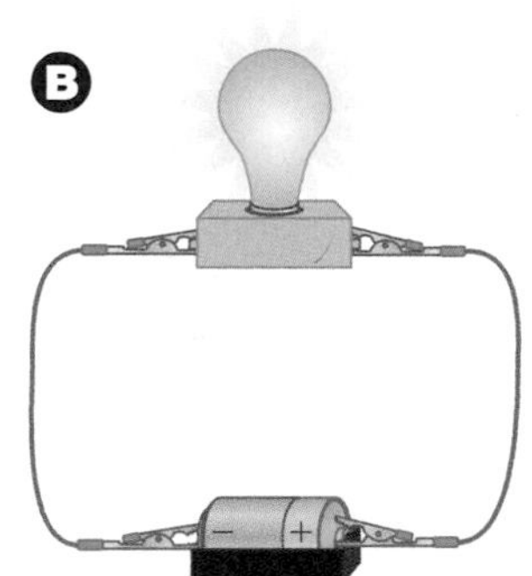

Figure 5.2 (A) A circuit with a gap in it is an open circuit. Electricity cannot flow through an open circuit. (B) When the circuit is fully connected, it is called a closed circuit. Electricity can flow through a closed circuit.

Opening and closing an electric circuit is a way to control the flow of the current. Most useful electric circuits have a fourth component: a switch. A **switch** is a device that closes or opens the circuit to start or stop the flow of electricity. Figure 5.3 shows how an ordinary light switch works to turn a light on or off.

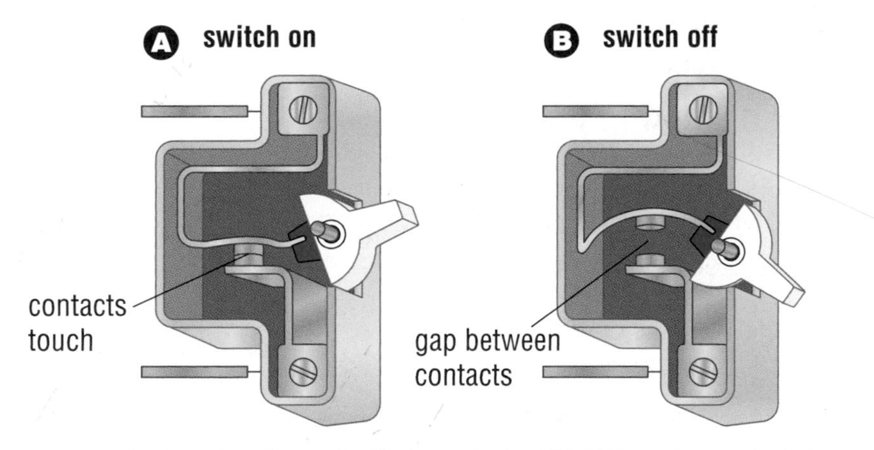

Figure 5.3 An inside view of a light switch. (A) When the switch is on, two metal parts (called contacts) touch. When the contacts touch, the electric circuit is closed and the light turns on. (B) When the switch is off, the contacts separate and the circuit is open. The light does not work.

Pause & Reflect

An electric current is the movement of electrons through a conductor. What do you think must happen within the circuit to allow a light to go on as soon as you turn on the switch? Discuss your ideas with your class.

Sometimes electricity can flow through a closed circuit in a way that is not controlled or safe. We call this danger a **hazard**. In Find Out Activity 5-B, you will study a possible electric hazard.

Find Out **ACTIVITY 5-B**

Out of Control

In this activity you will find out why a poorly designed circuit can be dangerous.

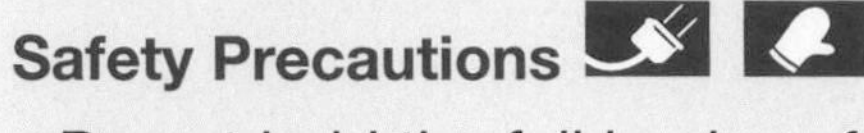

Safety Precautions

- Do not hold the foil in place for longer than 10 s.

What You Need

1 D-cell battery (1.5 V)
1 strip of aluminum foil (1 cm × 10 cm)

What to Do

1. Fold the aluminum foil so that it touches both ends of the battery.
2. Hold the foil in place for 2 s. **Observe** what happens.
3. Hold the foil in place for 10 s. **Do not hold the foil in place for longer than 10 s. Observe** what happens.

What Did You Find Out?

1. Did you create a closed circuit? How do you know?
2. This circuit did not include a light bulb or other device to use the electricity. What happens to the electrical energy when there is no useful load?
3. How could this kind of circuit be dangerous?

READING Check

What is a short circuit? Why do we want to avoid short circuits?

Short Circuits Are Dangerous

A closed circuit that does not have a useful load is called a **short circuit**. In Activity 5-B, you didn't include a useful load, so the conductor itself (the foil strip) became the load. The electrical energy flowing through the circuit was enough to heat up the conductor. In some short circuits, as shown in Figure 5.4, the current follows a different path instead of flowing through the intended load.

A hot or smoking conductor is one sign of a short circuit. Other short circuits can be much more dangerous than the one you just made—especially if *you* become the load! You will study this hazard in more detail in Chapter 6. Watch out for short circuits whenever you are working with electricity.

A

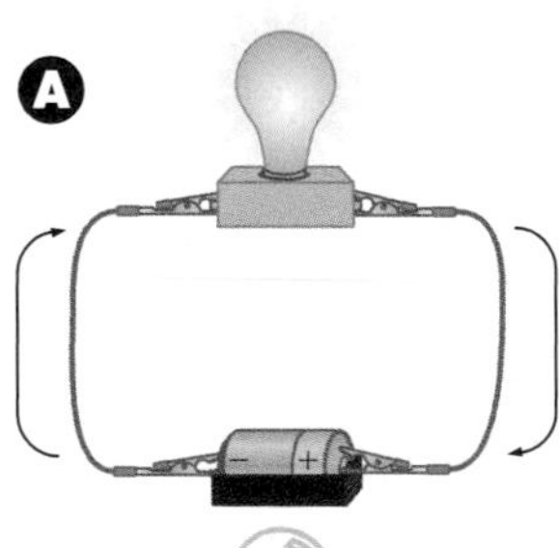

B

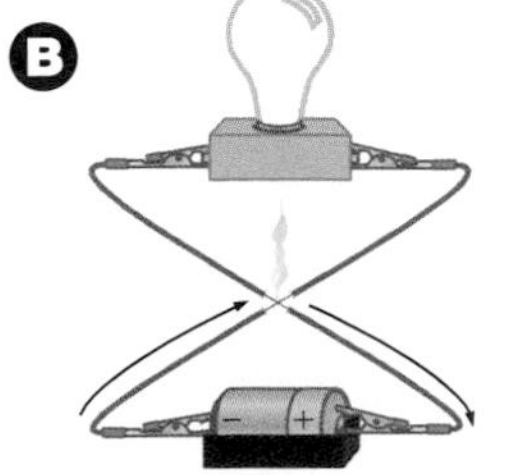

Figure 5.4 Arrows show the flow of electrons through each circuit. (A) This closed circuit is safe to work with. (B) In this circuit, the insulation around the wires is damaged where the wires cross. The electrons travel through the uninsulated wires back to the source without going through the light. This is a dangerous short circuit.

A Series Circuit Is a Single Pathway

In Activity 5-A you built a series circuit. A **series circuit** provides a single pathway for electrons to travel from the source(s) through the load(s). That single pathway then continues through the load(s) to the source(s). Figure 5.5 shows that a series circuit can have more than one load and more than one source, all connected in one continuous loop. Electrons flow from the negative terminal of the battery through the circuit components to the positive terminal of the battery. In the next pages you will take a closer look at series circuits.

Pause & Reflect

If the insulation on the wires in Figure 5.4B was not damaged, would electricity still flow from one wire to the other where they cross?

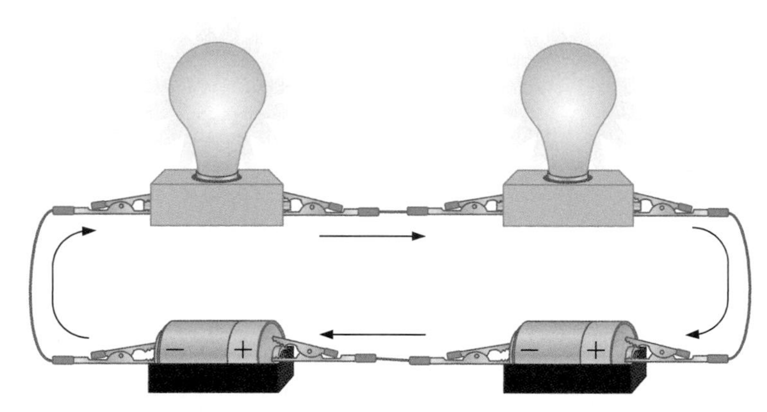

Figure 5.5 A series circuit can have many loads (lights) and sources (batteries). Electrons have only one path to follow (shown by arrows) through the entire circuit.

Find Out ACTIVITY 5–C

It's the Only Way to Go

A flashlight is an example of a series circuit.

What to Do Group Work

1. Carefully take apart a flashlight and draw a diagram to show how the parts fit together inside.
2. Using a coloured pencil or marker, draw arrows on your diagram to show the flow of electrons from the battery through the circuit.

What Did You Find Out?

1. How does your diagram compare to the diagrams drawn by others in your class?
2. Which parts of a flashlight do you think are needed for it to work?
3. What role do the other parts of the flashlight play?

CONDUCT AN

INVESTIGATION 5-D

SKILLCHECK
- Observing
- Controlling Variables
- Interpreting Observations
- Making Inferences

In Series

In a series circuit, electrons travel along a single pathway through all of the components. In this investigation you will build and test a series circuit. As you change some components, you will affect other parts of the circuit.

Question

How can you make a light bulb burn more dimly or more brightly?

Safety Precautions

- Connect the load first.
- Disconnect the battery first.

Apparatus

2 D-cell batteries in holders (1.5 V)
2 light bulbs in holders (3 V)
4 copper wires with an alligator clip on each end
1 switch

Procedure (Group Work)

Part 1: Sources in Series

1. You will use two wires for this step. Join one alligator clip from each wire to each terminal on the light bulb holder. Then join the other end of each wire to the terminals on the battery holder. **Record** your observations.

2. Add a second battery to the circuit. To do this, disconnect the wire from the negative terminal of the battery holder in the circuit. Use a new wire to connect the positive terminal of this battery holder to the negative terminal of the second battery holder. Your open circuit should look like the illustration below.

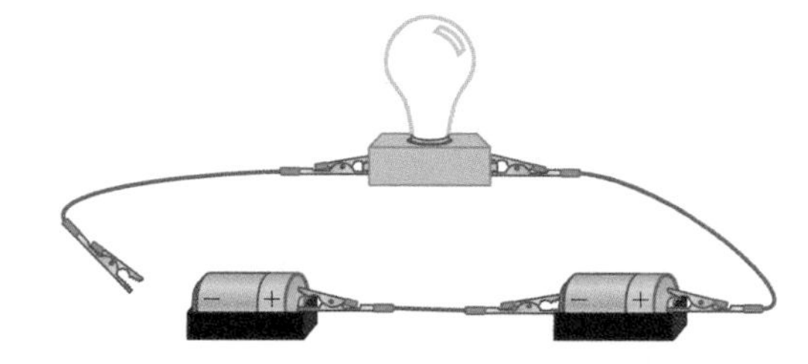

Working Safely with Electric Circuits

Electricity can be dangerous. Without special equipment, it is not always possible to tell when there is a current in a wire. Never conduct experiments using electric circuits without the supervision of an adult who is trained to work safely with electricity.

Always follow these guidelines when you work with electric circuits:

- Inspect your equipment before you begin. Watch out for and replace damaged wires, leaky batteries, broken bulbs, or damaged clips.
- Make sure your work area is clean, dry, and uncluttered.
- Remove metal jewellery that could conduct electricity.
- Handle wires by holding the plastic (insulated) coating or clips.
- When connecting a circuit, *connect the wires to the load first,* before connecting them to the battery.
- When disconnecting a circuit, *disconnect the wires from the battery first,* before disconnecting them from the load.
- Check to make sure that the battery is not too powerful for your circuit. If you connect a 6 V battery to a 1.5 V light, you may "blow" the bulb.

3. Connect the available alligator clip to the negative terminal of the second battery holder to close the circuit. **Record** your observations.

Part 2: Loads in Series

4. Build an electric circuit like the circuit in step 1. **Observe** what happens.

5. Add a second light to the circuit. To do this, first open the circuit by disconnecting one of the wires from the battery holder. Then disconnect the other wire from the light holder. Connect this wire to the second light bulb. Use a new wire to connect the two light bulbs. Your open circuit should look like the illustration below.

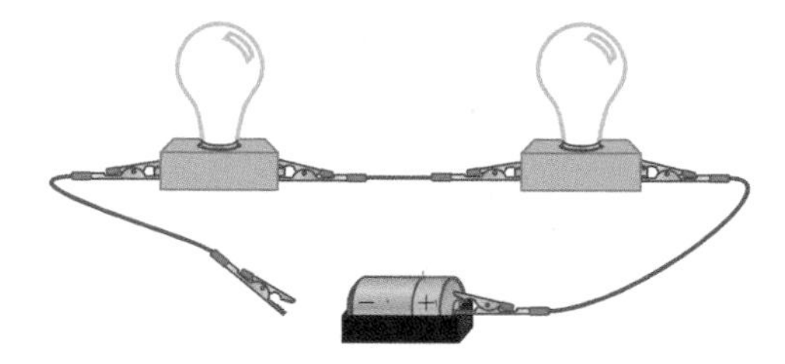

6. Close your circuit by connecting the available alligator clip to the available battery terminal, and **record** your observations.

7. Unscrew one light bulb in the series. Do not remove the wires from the bulb holder. What happens to the other lights? **Record** your observations.

Part 3: Circuit with a Switch

8. Screw the light bulb back in to the circuit you built in Part 2.

9. Using a switch and the fourth wire, try to connect the switch to your circuit so that you can turn the lights on and off. Remember to always disconnect the battery first when you change your circuit.

10. Test various locations for the switch.

11. Draw diagrams of any successful arrangements.

Analyze

1. What did you observe about the light bulb when you added a second battery in series? What can you **infer** about the flow of electrons?
2. What did you observe about the light bulbs when you added a second light bulb in series? What can you **infer** about the flow of electrons?
3. What did you **observe** about the light bulbs when you unscrewed one bulb? Can you explain your observation?
4. Were you able to turn the light bulbs on and off when you connected a switch to your circuit? Why do you think it worked or didn't work?

Conclude and Apply

5. Did the location of the switch affect the result in Part 3? Why or why not?
6. Draw a diagram to show the flow of electricity through a series circuit with two bulbs, one battery, and one switch. In what way is unscrewing a bulb like turning off the switch?

Extend Your Skills

7. You can also wire switches in a series. Can you think of any examples of common household devices that use more than one switch in a series? What is the advantage of this kind of circuit?

INTERNET CONNECT

www.mcgrawhill.ca/links/BCscience6

Ohm's Law is an equation that explains how voltage, current, and resistance are related in a circuit. Use the Internet to research this relationship. Go to the web site above and click on **Web Links** to find out where to go next. Can you use this equation to explain your observations of series circuits?

READING Check

What is the difference between current and voltage in an electric circuit?

Current and Voltage in a Series Circuit

The electricity in any circuit depends on three things:

- **Voltage** is a term used to describe the size of the "push" from the source that forces electrons through the circuit. This push is measured in units called "volts" (V).
- **Current** is a term used to describe electron flow through the circuit within a certain amount of time. Current is measured in units called "amperes" or "amps."
- **Resistance** is a term used to describe how difficult it is for electrons to pass through the load. Resistance is measured in units called "ohms" (Ω).

In Investigation 5-D, you found that adding more sources (batteries) in series to your circuit caused the attached light bulb to become brighter. Increasing the number of sources in series provides a larger "push" (voltage) through the circuit.

In Investigation 5-D, you found that adding more light bulbs (loads) in series to your circuit caused the attached light bulb to become dimmer. The current decreases because the battery cannot push as much current through the greater resistance in the loads.

You have observed some characteristics of series circuits. In the next section, you will study another kind of circuit.

DidYouKnow?

When you connect sources in series, the voltage of the sources add up. For example, if you connect two 1.5 V batteries in series, the voltage in your circuit is now 3 V. Inside a 9 V battery are six small batteries. They each generate 1.5 V, just as the large D-cell battery does. Because they're connected in series inside the casing, their voltages add up to 9 V.

Section 5.1 Summary

In this section, you learned that an electric circuit is a pathway for the flow of electricity. For the circuit to be useful, it needs

- a source of electric "push" or voltage.
- a conductor to carry electric current.
- a load that responds in some way to the current.
- a switch to control the electric current.

Circuits can be closed or open.

- A closed circuit allows the flow of electricity.
- An open circuit does not allow the flow of electricity.
- A short circuit is closed and allows the flow of electricity, but does not have a useful or intended load. It can be quite dangerous.

A series circuit has several characteristics.

- A series circuit connects all components in a single pathway.
- Removing any device from a series circuit (or turning the device off) opens the circuit.
- Adding more sources to a series circuit increases the voltage.
- Adding more loads to a series circuit increases the resistance and reduces the current in the circuit.

Check Your Understanding

1. How can you stop the flow of current around a circuit?
2. What are the four basic components that make up all circuits? Describe the role each component plays in the circuit.
3. Why are short circuits dangerous?
4. In the water system shown in the illustration, which one of the numbered elements is like the conductor in an electric circuit? Explain your answer.

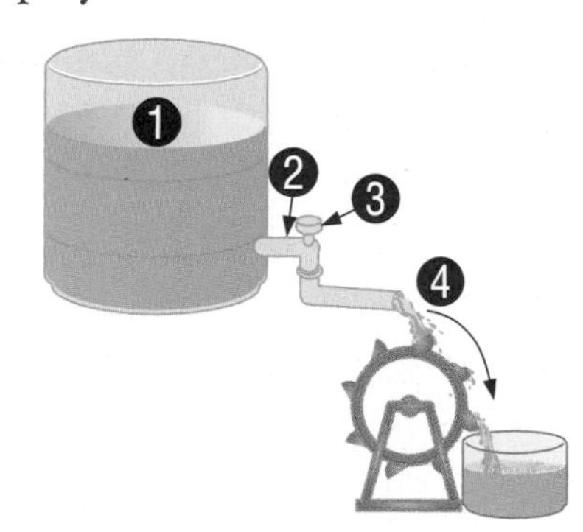

5. **Apply** Look at the photograph of the Legislature buildings at the beginning of Section 5.1. From what you know about series circuits, explain whether you think it would be a good idea to wire these lights in series.
6. **Thinking Critically** In question 4 we compare the illustrated water system to an electric circuit. Can you think of at least one characteristic that makes this water system *not* like an electrical circuit?

Key Terms

circuit
source
load
closed circuit
open circuit
switch
hazard
short circuit
series circuit
voltage
current
resistance

Section 5.2 Parallel Circuits

Key Terms

parallel circuit

READING Check

What is the difference between a parallel circuit and a series circuit?

What if every time you turned off the lights, the computer shut off too?

In your home, many electric devices are probably connected together in a circuit. However, you can turn each of them on or off without affecting the others. Within this circuit, electrons have more than one complete pathway to follow. This kind of circuit is called a **parallel circuit**. Figure 5.6 shows the difference between a single path and multiple pathways in circuits.

Figure 5.7 shows several loads connected in parallel. If you turn off the lamp, you stop the current from flowing to the lamp by opening that part of the circuit. However, there are other closed paths for electrons to follow. The circuits that connect the electric piano and the toaster to the source are still closed. Turning off the lamp does not affect the flow of current to these loads.

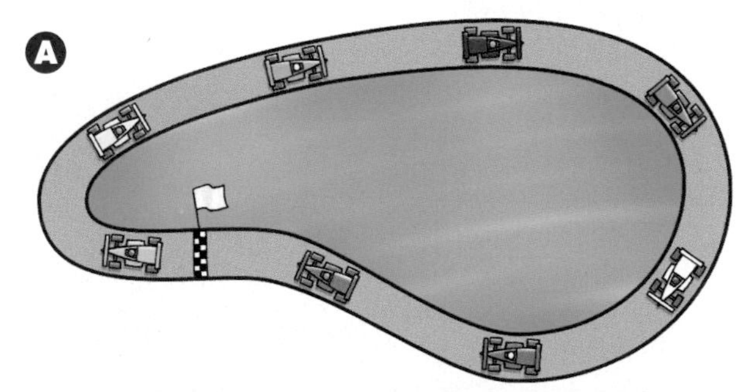

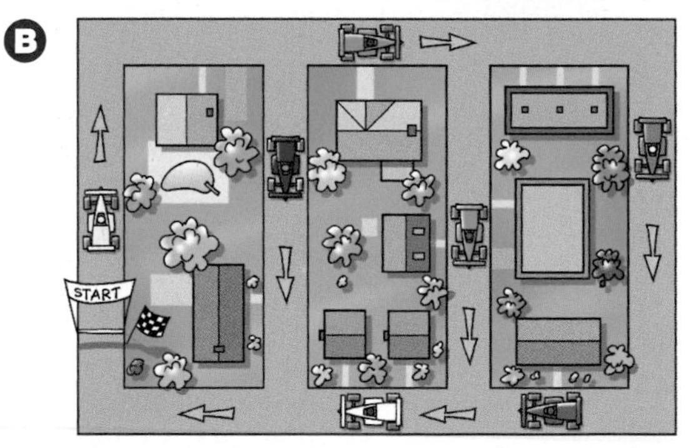

Figure 5.6 The difference between a series circuit and a parallel circuit is like the difference between a single-lane racetrack and a system of single-lane roads. (A) Cars on a racetrack all travel the same path to return to the starting point. (B) Cars on a road system can choose different routes to return to the starting point. In each case, what happens if one car stops and blocks the road?

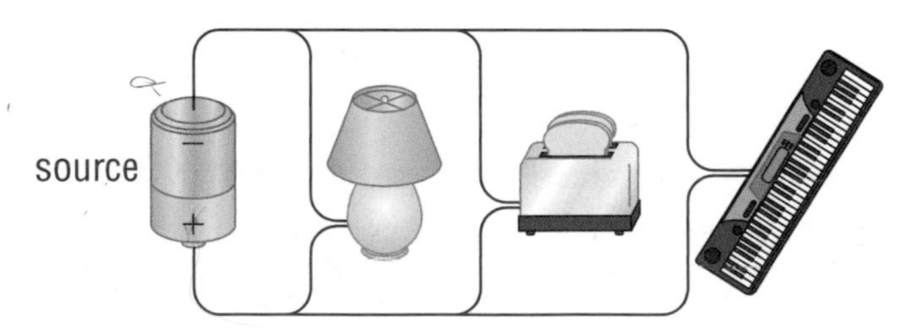

Figure 5.7 A parallel circuit. Each load can be turned on or off without affecting the others. Where could you place one switch to control all three loads?

At Home **ACTIVITY 5-E**

Current in Parallel and Series Circuits

You can use water as a model for the rate of movement of electrons (current) through a parallel and a series circuit.

What You Need

2 glasses (filled with water)
4 drinking straws
clear adhesive tape

What to Do

1. Put all four straws into one glass of water. Put the other end of all four straws into your mouth.
2. Draw the water up the straws with your mouth. **Observe** how much water enters your mouth.
3. Tape the four straws end to end and put one end into the second glass of water.
4. Draw the water up the four connected straws with your mouth. Try to draw with the same force and for the same amount of time as in step 2. **Observe** how much water enters your mouth.

What Did You Find Out?

1. **(a)** Which arrangement of straws (the first or the second) provided more than one path for water to flow?
   (b) Was this similar to a series circuit or a parallel circuit?
2. Which arrangement of straws allowed you to draw up more water into your mouth?
3. After drawing up the water once, what was the difference in the water level in the glass for each arrangement?
4. What does this model tell you about the flow of current in each type of circuit?

Current and Voltage in a Parallel Circuit

In Activity 5-E, you used straws to model the flow of electrons (current) through a parallel circuit and a series circuit. Remember that the current through a series circuit decreases as you add more loads to the circuit (the light bulbs got dimmer). In contrast, each time you add a new load to a parallel circuit, the total current through the system increases. If you had added even more straws in parallel, more water would have been pulled up, and the glass of water would have been drained further. If a battery is your source, the battery will be drained more quickly as more loads are added in parallel. In your household electric system, as you add more loads, the total current can increase to a dangerous level. You will learn more about safety features of household electric systems in Chapter 6.

READING Check

What happens to the total current flowing through a circuit as you add new loads in parallel?

CONDUCT AN

INVESTIGATION 5-F

SKILLCHECK
- Controlling Variables
- Observing
- Making Inferences
- Communicating (Reporting)

On Parallel Tracks

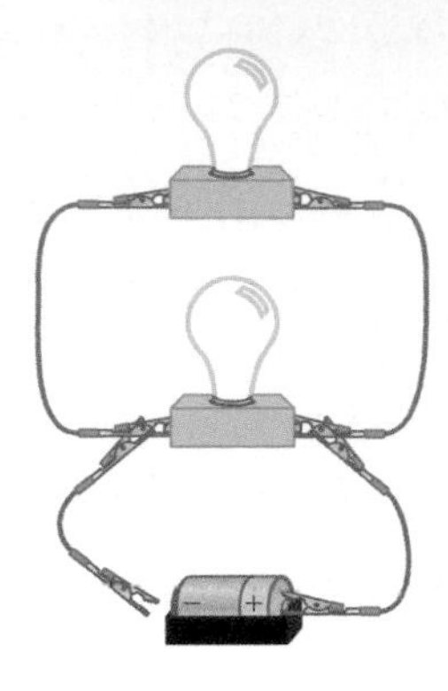

In a parallel circuit, electrons can travel along more than one pathway through all of the components. In this investigation you will build and test a parallel circuit. As you change some components, you will affect other parts of the circuit.

Question

How does electricity flow through a parallel circuit?

Safety Precautions 

Apparatus

1 D-cell battery (1.5 V) in a holder
6 copper wires with an alligator clip on each end
2 light bulbs (1.5 V) in holders
1 buzzer (1.5 V)

Procedure

1. Using the wires, connect the battery holder and light holder in a simple circuit so that the light goes on. **Observe** the brightness of the light.

2. Add a second light in parallel with the first. To do this, open your circuit by disconnecting one of the wires from the battery holder. Then without disconnecting the first light, connect a new wire to each of the terminals on the first light holder. Connect each available end of each of these wires to a terminal on the second light holder. Your circuit should look like the illustration.

3. Close your circuit by reconnecting the battery holder. **Observe** the brightness of both lights. **Record** what happens in your science notebook.

4. Using the same procedure as in step 2, add a buzzer in parallel with the two lights. Remember to disconnect the battery before you work with your circuit. Close your circuit and **observe** what happens.

5. **Record** your observations about what happens when you increase the loads in a parallel circuit. Then disconnect your circuit.

Analyze

1. What happened to the brightness of each light as more lights were added?
2. What happened to the lights when you added the sound buzzer?
3. From question 1 and 2, what can you **infer** about the current in the parallel circuit?

Conclude and Apply

4. How do the results of adding more lights in parallel compare with adding more lights in series? (Hint: See Investigation 5–D.)

Sources Can Be Connected in Parallel

In Investigation 5-F, you found that you can add *bulbs* in a parallel circuit without changing the brightness of each bulb. The current increases as you increase the loads, but the "push" (voltage) stays the same. In a parallel circuit, you can add or take away loads without affecting the voltage across other components of the circuit.

If you had added more *batteries* in parallel, you would have seen that the brightness of the light does not change in this situation either. The voltage in the circuit does not increase when you add batteries in parallel. Although the lights are not any brighter, the batteries will *last* longer because there are more of them. Figure 5.8 shows how the voltage is affected when batteries are wired in series and in parallel.

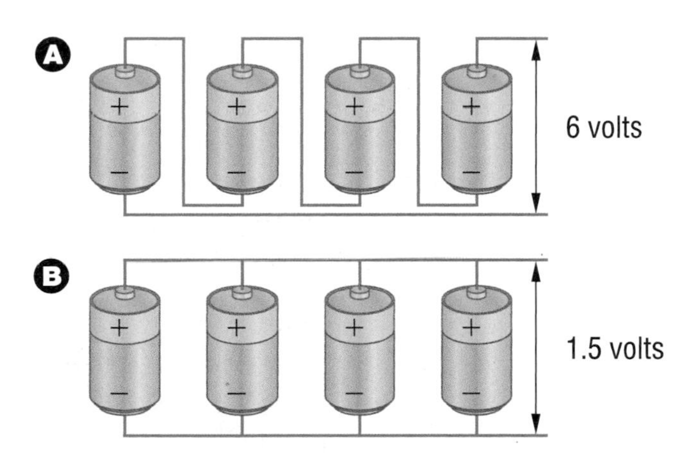

Figure 5.8 More than one battery is usually used in a device. Batteries can be grouped together either in series to produce a higher voltage, or in parallel to provide current for a longer time. (A) Four 1.5 V batteries in series. In a series circuit, electrons flow through one battery after another and the "push" (or voltage) of all the batteries adds up. (B) Four 1.5 V batteries in parallel. In a parallel circuit, electrons flow through only one battery, and receive a "push" from only one battery. The voltages do not add up.

So far in this chapter, you have learned how to build series and parallel circuits. You have also seen how to wire circuit components—including batteries, conductors, loads, and switches—to do different things. In the next Investigation, you will put your knowledge to the test.

Does the voltage in a circuit change as you add more identical batteries in parallel?

You may have heard that you should keep batteries in the refrigerator. This is not true. Batteries are best stored in a dry place at room temperature.

DESIGN YOUR OWN

INVESTIGATION 5-G

SKILLCHECK
- Communicating
- Interpreting Observations
- Problem Solving
- Testing and Evaluating

Build a Complex Electric Circuit

Challenge

Using the apparatus provided, design and build one electric circuit that will meet all the design criteria listed below.

Safety Precautions

Apparatus

batteries in holders
light bulbs in holders
buzzers
switches
copper wires with alligator clips

Design Criteria

1. Your circuit must include at least one light bulb and one buzzer.
2. Your circuit must be able to light the bulb without sounding the buzzer, and sound the buzzer without lighting the bulb.
3. Your circuit must also be able to light the bulb and sound the buzzer at the same time.
4. Your circuit must also be able to turn off both the light and the buzzer at the same time.

Skill POWER

For tips on technological problem solving, refer to SkillPower 6.

Plan and Construct 

1. With your group, discuss how you would set up this circuit.
2. Draw a labelled diagram to plan your circuit, and make a list of the apparatus you will need.
3. Discuss your plan with your teacher.
4. Build your circuit.
5. Improve your circuit until you are satisfied with the way it works.
6. Demonstrate your circuit to your class.

Evaluate

1. Did your circuit work as you expected?
2. If any part did not work,
   (a) What do you think happened?
   (b) How could you identify and fix the problem?
3. Compare your circuit to that of others in your class. What similarities and differences can you see?
4. After having looked at other circuits, what changes (if any) would you make to your own?
5. Think about your work as a group:
   (a) What helped you work together to make your circuit?
   (b) What would you like to do to improve on your group's performance next time?

Comparing Series and Parallel Circuits

Every electric system—from the simplest to the most complicated—is made up of either a series or parallel circuit, or combinations of series and parallel circuits connected together. You can use the same guidelines you have studied in these sections to understand the flow of current through any electric system.

Series and parallel circuits have different characteristics. Whether you use one or other depends on the needs of the system. Table 5.1 compares some of the main characteristics of series and parallel circuits.

Table 5.1 Comparison of Series and Parallel Circuits

Series Circuit	Parallel Circuit
Electricity follows only one path. Components are joined along a continuous closed loop.	Electricity can follow several different paths. Components are joined along more than one continuous closed loop.
If the circuit is opened at any point, no current flows through.	If the circuit is opened at one point, current will continue to flow through the other paths of the circuit.
As more lights (loads) are added in series, the lights get dimmer. The current decreases as loads are added.	As more lights (loads) are added in parallel, the lights will *not* get dimmer. The current increases as loads are added.
As more batteries (sources) are added in series, the lights get brighter. The voltage increases as sources are added.	As more batteries (sources) are added in parallel, the lights will *not* get brighter. The voltage does not increase, but the batteries will last longer.

DidYouKnow?

You have been working with small electric circuits. Some circuits are very large. Some are large enough to connect Canada and the United States. British Columbia uses electric circuits thousands of kilometres long to sell electricity to California.

In this chapter, you have also seen that the loads in a circuit can turn electrical energy into light, sound, and heat. In Section 5.3 you will find out more about the conversion of electrical energy to other forms of energy.

Pause& Reflect

What are two advantages of a parallel circuit? What are two advantages of a series circuit?

Figure 5.9 An electric problem has caused a blackout in part of the city. If another part of the city is *not* experiencing a blackout, what does that tell you about the kinds of circuits in the city's electric system?

DidYouKnow?

In 1873 two Canadian scientists, Henry Woodward and Matthew Evans, were the first to patent a light bulb. Thomas Edison bought the patent and improved the design. By the early 1900s, electric lights were common in cities across North America.

Key Terms

parallel circuit

Section 5.2 Summary

In this section, you learned that a parallel circuit is an electric circuit with more than one complete pathway for electrons to follow. If you remove one load from a parallel circuit, the electricity can still flow to other loads. Parallel circuits are a useful way to connect several appliances that must be able to work even if other appliances on the same circuit are not turned on.

- As you add more loads in parallel, the total current through the circuit increases.
- As you add more sources (batteries) in parallel, the total voltage remains the same. However, each battery will last longer.
- If the loads in the circuit are the same, batteries connected in parallel last longer than batteries connected in series.

Check Your Understanding

1. Name two characteristics of parallel circuits.
2. The diagram shows a circuit with three identical lamps connected in parallel.
   (a) Which light will be brightest when the switch closes the circuit?
   (b) If one light burns out, what will happen to the other lights?

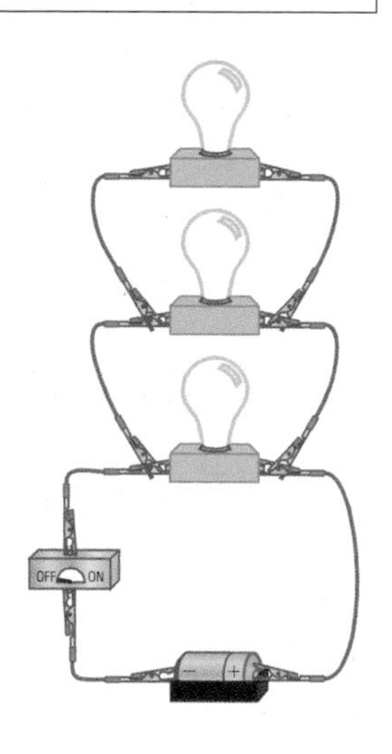

3. What happens to the current when additional loads are connected in parallel?
4. **Apply** Suggest a way to determine whether the bulbs in a string of decorative lights are connected in series or in parallel. Describe what you might **observe** and what you might **infer**.
5. **Thinking Critically** Examine the four circuit diagrams. Each diagram shows three light bulbs connected in parallel. Each circuit has a switch in a different location. In which circuit would the switch be able to turn on or turn off all of the light bulbs at the same time? Explain.

A
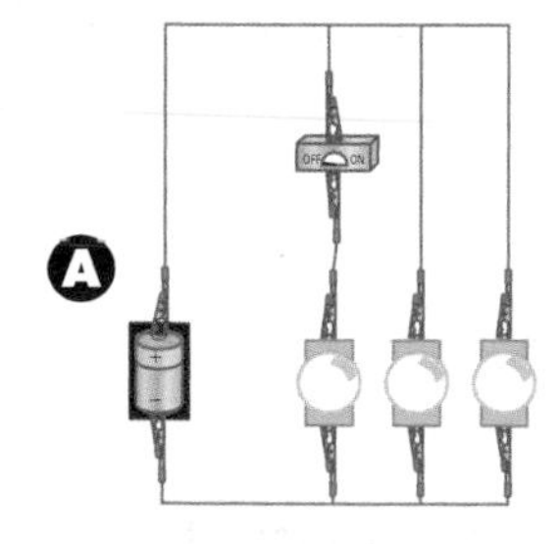

B
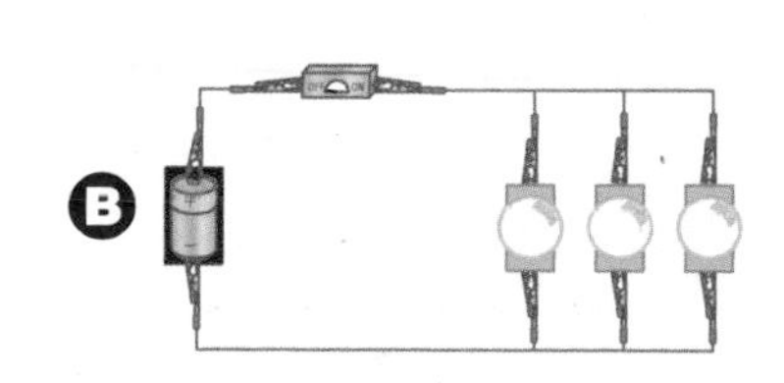

C
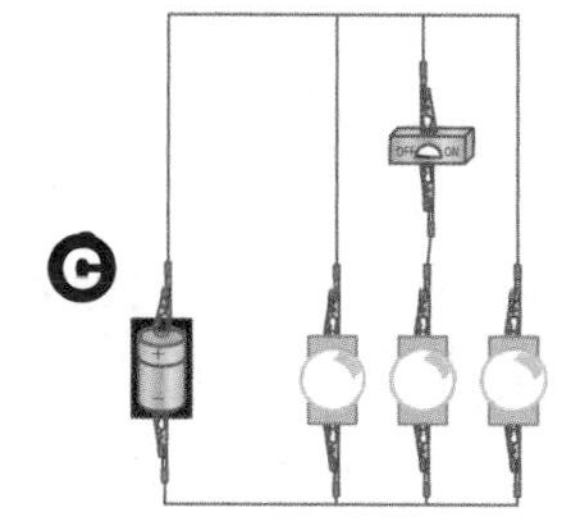

D
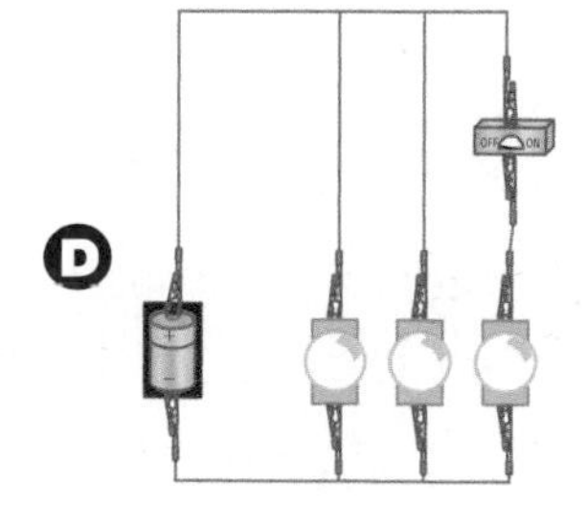

Section 5.3 Transforming Electrical Energy

Figure 5.10 An electric system turns electrical energy into both light and sound for this performance.

Key Terms

- light bulb
- filament
- electromagnetism
- electromagnet
- motor

For an electric circuit to be useful, it must change electrical energy into another form of energy such as light, sound, heat, motion, and magnetic effects. You learned in Sections 5.1 and 5.2 that electricity flows in a circuit and produces a response in connected loads. Every load in an electric system resists the flow of electrons (current). Remember that the resistance of a load is a term used to describe how hard it is for electrons to pass through the load. As electrons are forced through the load against this resistance, their electrical energy is converted into other forms of useful energy. But why does the electric current produce a different response in different kinds of loads? In the next few pages, you will learn how some loads convert electrical energy.

close-up of coiled filament
coiled filament
bulb filled with non-reactive gas such as argon
screw-type base (connects bulb to electrical source)
glass bulb
lead-in wires
glass rod (holds up the filament)

Figure 5.11 The filament of a light bulb is made of a metal that glows when it heats up. Over time, the filament becomes thinner and thinner until it breaks. When this happens, the circuit in the bulb opens and we say the bulb has burned out.

Electricity to Light

A **light bulb** is a device that turns electrical energy into heat and light energy. Take a close look at a clear light bulb that is not turned on. You will see a tiny coiled **filament** held between two wires. A filament is a thin wire with a very high resistance. (Imagine blowing through a thin straw versus a thick straw. It is harder to blow through the thin straw.) When you turn on the light, electrons are pushed through this thin filament, which resists the electrons. The resistance to the electric current causes the filament to heat up. This is similar to what happened to the foil when you experimented with a short circuit. In this case, as the filament heats up, it glows and produces light.

What happens to the filament in a light bulb when you turn on the light?

Electricity to Heat

Inside a light bulb, electricity is converted into heat energy. This causes the filament to glow and produce light energy. Many appliances produce heat alone, or with only a little light (think of the glow of the wires in a toaster). These loads work like the light bulb, but use different materials so that the right amount of heat or light is produced. The heating filament in a toaster or hair dryer is made of a metal that becomes hot when an electric current runs through it.

Figure 5.12 A light bulb and a hair dryer work in a similar way. Electricity is converted to heat in both devices. However, in a light bulb we use the light produced by the glowing filament, and in a hair dryer we use the heat.

Electricity to Magnetic Effects

Some of the most important scientific discoveries are made by accident. In 1819, a Danish physics professor named Hans Oersted was giving a lecture when he happened to put his directional compass on the desk near an electric circuit. A directional compass points to a magnetic pole, which is located near the north pole. However, when Oersted connected the circuit, the compass needle moved. In this way, Oersted discovered that an electric current creates a magnetic force. This was an important discovery. The magnetic force that surrounds a wire carrying a current is called a *magnetic field.*

Magnets are objects that can attract some metals. You can use a small magnet to pick up paper clips or pins. You can create the same kind of magnetism in some metal objects by using electricity. **Electromagnetism** is the name given to magnetism produced by electricity. An **electromagnet** is a temporary magnet created by an electric current. One of the most important uses of electromagnets is in an electric **motor**, where an electromagnet spinning within a permanent magnet causes other components, such as wheels and fans, to turn. In your next activity you will build your own electromagnet.

CONDUCT AN

INVESTIGATION 5-H

SKILLCHECK
- Observing
- Interpreting Observations
- Making Inferences
- Communicating (Reporting)

What's the Big Attraction?

What does your telephone have in common with your television set and your doorbell? Each of these devices contains an electromagnet. You can use a simple electric circuit to create your own electromagnet.

Question

How can you create magnetic effects with electricity?

Safety Precautions

- Avoid breathing in the iron filings.
- Dispose of iron filings carefully.
- Wear rubber gloves when handling iron filings.

Apparatus

1 battery (1.5 V) in holder
2 copper wires with alligator clips
1 iron rod (5 cm long)
1 piece of copper wire (15–20 cm)

Materials

iron filings
white paper
rubber gloves

Procedure

1. Place a white piece of paper under the circuit for easier cleanup. Build an electric circuit by connecting the wires to each end of the iron rod. Scatter the iron filings around the iron rod. Connect one of the wires to the battery. Do not close the circuit yet. Your circuit should look like the illustration.

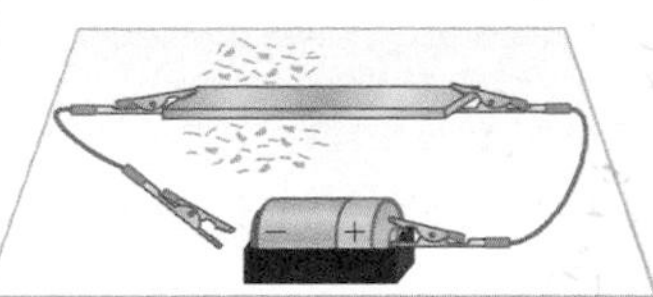

2. Close your circuit and **record** what happens.
3. Open your circuit and wrap the copper wire around the iron rod twice. Connect the circuit wires to either end of the copper wire. Place the iron rod back near the scattered iron filings. Your circuit should look like the illustration.

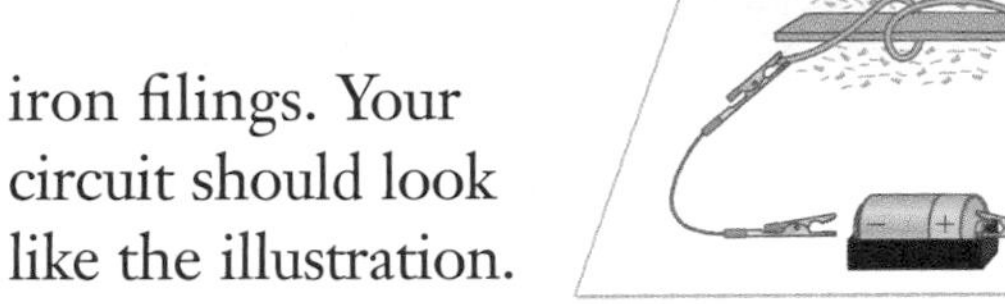

4. Close the circuit and **record** what happens.
5. Open the circuit and wrap the copper wire around the iron rod four more times. Place the iron rod back among the iron filings.
6. Close the circuit and **record** what happens.
7. Take apart your circuit, clean up your work area, and return all materials to your teacher.

Analyze

1. What happened to the iron filings
   (a) in step 2?
   (b) in step 4?
   (c) in step 6?
2. What happened as you wrapped the copper wire more times around the iron bar?

Conclude and Apply

3. What do you think caused the difference in your observations between steps 2, 4, and 6?
4. Based on these observations, how would you change your circuit to create a stronger electromagnet?

Extend Your Skills

5. Using the Internet or a library, find out how a television set uses an electromagnet. Write a brief report to **communicate** what you learned.

Section 5.3 Summary

In this section, you learned that the load in an electric circuit resists the flow of electrons. This resistance converts electrical energy into other forms of energy, such as light, heat, sound, motion, and magnetic effects.

You also learned about electricity and magnetism:

- An electric current has a magnetic field.
- Electromagnetism is the magnetism produced by electricity.
- An electric current will turn some metals into temporary magnets called electromagnets.
- We use electromagnets to convert electrical energy into motion.
- Electric motors use electromagnets.

Key Terms

light bulb
filament
electromagnetism
electromagnet
motor

Check Your Understanding

1. Which part of an electric circuit converts electricity into other forms of energy?
2. Sketch this diagram in your science notebook. Identify the three labelled parts of the light bulb.

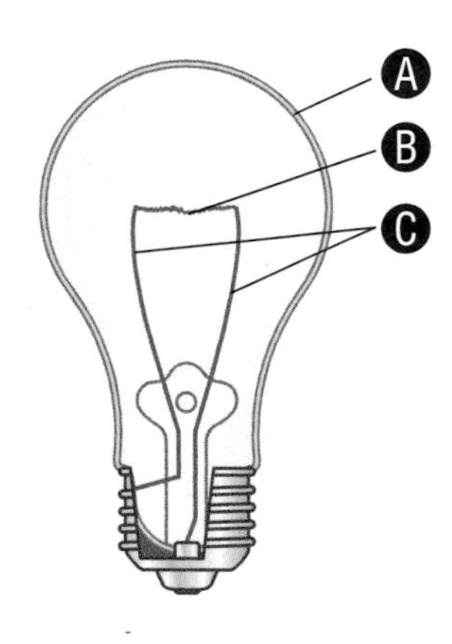

3. What does it mean when a material is described as having a low resistance? Use your knowledge of electron movement to explain your answer.
4. What materials can be used to create an electromagnet?
5. What happens to the strength of an electromagnetic field when you increase the current flowing through a coil of wire around an iron bar?
6. **Thinking Critically** The illustration shows a simple electromagnet. Which part is missing?

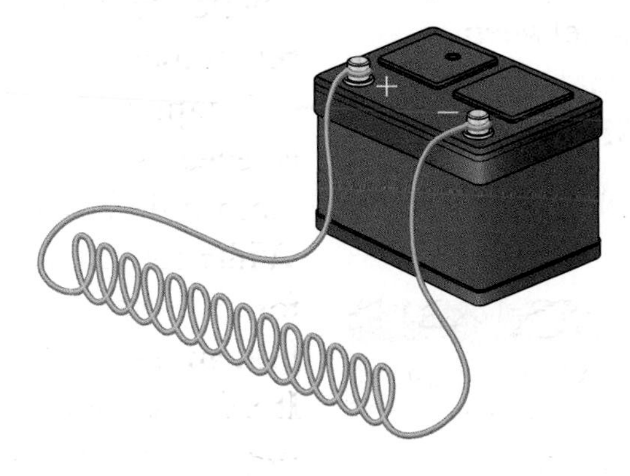

CHAPTER at a glance

Now that you have completed this chapter, try to do the following. If you cannot, go back to the sections indicated.

(a) Explain the difference between series and parallel circuits. (5.1, 5.2)

(b) Describe the four components of all circuits. (5.1)

(c) Explain how short circuits happen and give one reason they can be dangerous. (5.1)

(d) Define electromagnetism. (5.3)

(e) Describe how a light bulb transforms electrical energy into light energy. (5.3)

Prepare Your Own Summary

Summarize this chapter by doing one of the following. Use a graphic organizer (such as a concept map), produce a poster, or write a summary to include the key chapter ideas. Here are a few ideas to use as a guide:

- Describe how electric circuits work. Make sure to use each of the following terms at least once: load, source, switch, series circuit, parallel circuit, current, voltage, resistance.
- Describe how the current and voltage of a circuit will change when you
  (a) add more loads in series
  (b) add more batteries in series
  (c) add more loads in parallel
  (d) add more batteries in parallel.
- Make a Venn diagram to group the similarities and differences between series circuits and parallel circuits.
- Design a poster with pictures of devices that use electricity. Group them into devices that
  (a) use batteries
  (b) plug in
  (c) do both.

Skill POWER

For tips on Venn diagrams, refer to SkillPower 1.

- Prepare a chart titled "My Home." Along the top, write the following categories:
  (a) light **(c)** motion
  (b) heat **(d)** sound.
  On the left, identify at least one device in each room of your home that uses electricity. For each device, write "YES" under the kind of energy it converts electricity to. One example is shown here for you.

Device	Light	Heat	Motion	Sound
Toaster	*YES*	*YES*	*YES*	*YES (dings)*

- Create a Know-Wonder-Learn chart about using electricity safely (see SkillPower 1). In the first column, labelled "Know," write things that you remember learning about the subject in Chapter 5. In the second column, labelled "Wonder," write the things that you would like to know about the topic. While you are reviewing Chapter 5, look for answers to those questions. As you find answers, fill in the "Learn" column. If there are answers you can't find, use the Internet to find more information.

CHAPTER

5 Review

Key Terms

circuit
source
load
closed circuit
open circuit
switch
hazard
short circuit
series circuit
voltage
current
resistance
parallel circuit
light bulb
filament
electromagnetism
electromagnet
motor

Reviewing Key Terms

If you need to review, the section numbers show you where these terms were introduced.

1. Describe the difference between:
   (a) current and voltage (5.1)
   (b) switch and load (5.1)
   (c) series and parallel circuits (5.1, 5.2)

2. In your science notebook, match each description in column A with the correct term in column B. Use each description only once.

A	B
(a) converts electrical energy into light energy	• electron (4.1)
(b) provides a single pathway for electrons to travel from the source to the load and back again	• resistance (5.3)
(c) movement of electrons past a point in a circuit in a given period of time	• current (5.1)
(d) provides a means of opening the circuit	• switch (5.1)
(e) tendency of a device or substance to resist the flow of current	• light bulb (5.3)
	• series circuit (5.1)

Understanding Key Ideas

3. Give two reasons why you would not want your reading lamp and computer connected in series. (4.1)

4. You learned that in a series circuit, the lights dim as you add more lights. Which of the following explains why: (4.1)
   (a) The voltage decreases.
   (b) The current decreases.
   (c) The circuit is open.
   (d) There is a short circuit.

5. Imagine one circuit has a 1.5 V battery and two lights connected in *parallel*. A second circuit has a 1.5 V battery and two lights connected in *series*. If both circuits are connected at the same time, which battery will burn out faster? Explain why. (Hint: Look back at the results of Activity 5-E.) (4.2)

Developing Skills

6. An ordinary electrical cord contains copper wires wrapped in a rubber sheath. Copper is a kind of metal. These wires carry electric current to a light bulb, where the current flows through a filament made of a different kind of metal. Explain how each of these metals is suited to its purpose.

7. As part of an experiment, Joel set up the circuit illustrated below. The bulbs did not light. What is the problem with his arrangement?

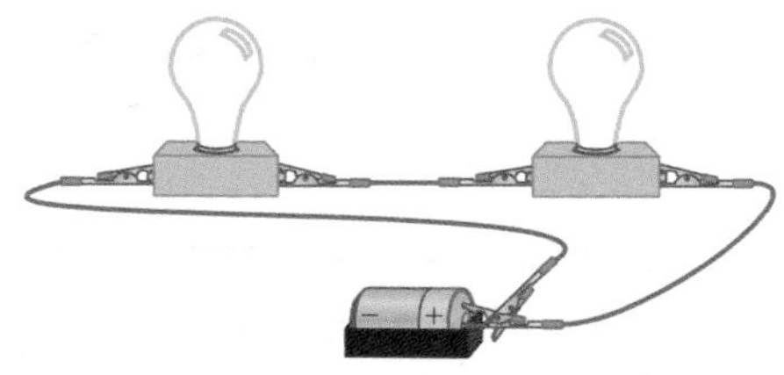

8. Draw a diagram of a circuit in which three light bulbs are arranged in parallel and are each controlled by their own switch. In the diagram, include a fourth switch that must be closed before any of the other bulbs can be turned on.

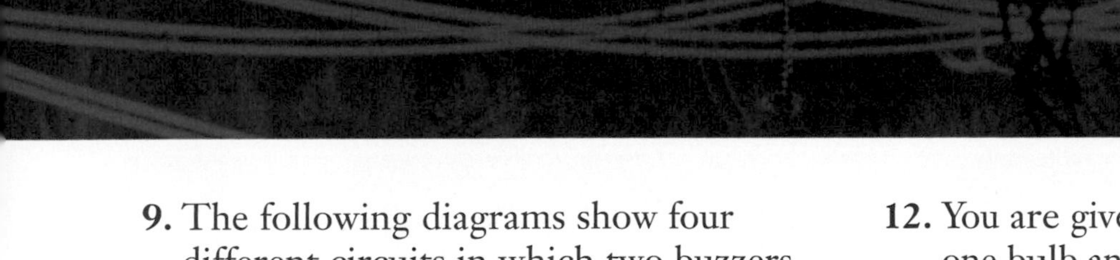

9. The following diagrams show four different circuits in which two buzzers, a lamp, and a switch are connected to a battery.
   (a) Which illustration shows a series circuit?
   (b) In which of the circuits shown will the buzzers work *only* when the switch closes the circuit?

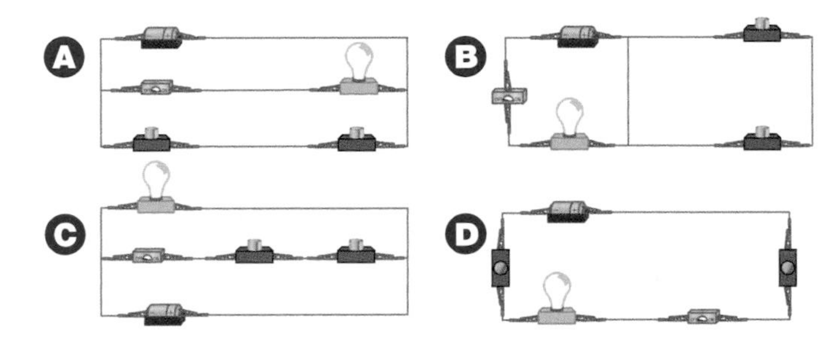

Problem Solving

10. Suggest a method of determining whether the bulbs in a string of decorative lights are connected in series or in parallel. Describe what you might observe and what you might infer. Identify one advantage of each type of circuit for this purpose.

11. In Stephanie's office, a computer, a printer, and a desk lamp are all plugged into a single power bar (a bar that contains several places to plug in different devices). The plug sockets (outlets) in the power bar are arranged in one row. Is it likely that this power bar is wired so that all the attached components are in series? How could you test your prediction? (**Note:** You should be able to test without unplugging any of the devices.)

12. You are given an ordinary flashlight with one bulb and one battery, wired in series. Draw and label this circuit. You are also given one more battery. Decide whether you would add this battery in parallel or series to do each of the following:
    (a) make the light bulb burn more brightly
    (b) make the batteries last longer

13. In a series ciruit, what could you do to prevent attached lights from becoming dimmer as you add more lights?

Critical Thinking

14. Draw a simple electromagnet. Then describe two modifications you could make to this electromagnet to increase the strength of the electromagnet.

15. Rennie has just been hired as the manager of an auto-wrecking company. He is touring the grounds and sees an employee using a large electromagnet to move scrap metal from the lot to the crushing pit. "I wonder if we should switch over to using permanent magnets for that work," thinks Rennie. Should he? Explain.

Pause& Reflect

Go back to the beginning of this chapter on page 126, and check your original answers to the Getting Ready questions. How has your thinking changed? How would you answer those questions now that you have investigated the topics in this chapter?

CHAPTER

6 Power to You

Getting Ready...

- How far is your TV from its source of electricity?
- What does a nuclear power station have in common with a wind turbine?
- How can turning off your bedroom light help salmon stocks in faraway rivers?

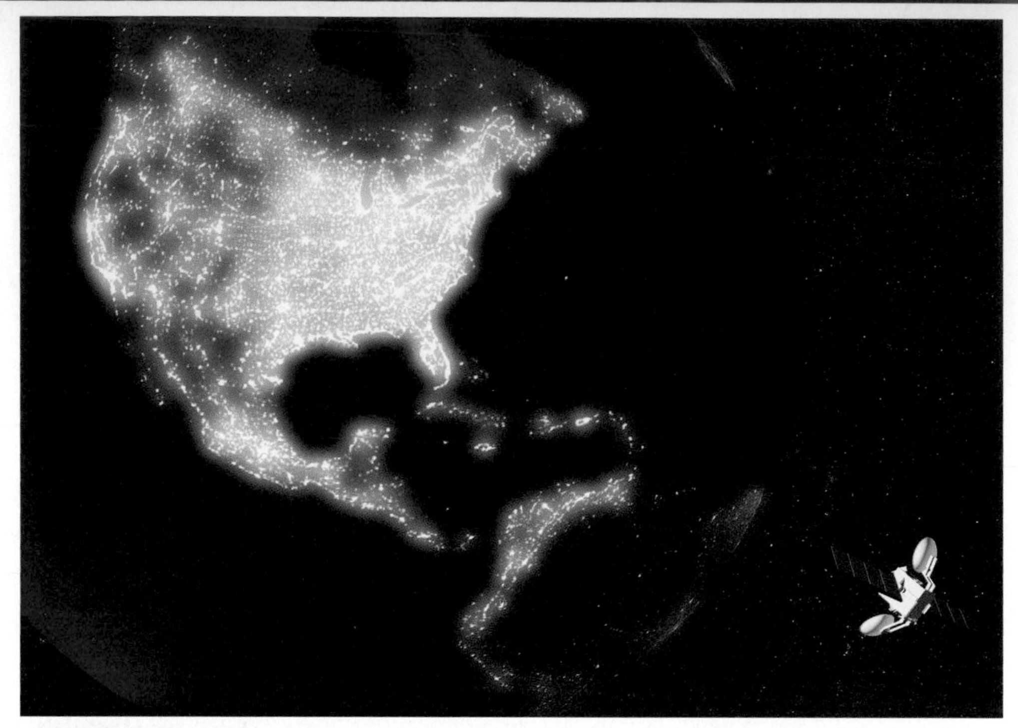

Bright spots on this satellite image show cities and towns that use large amounts of electricity. Can you find British Columbia in this image?

Nighttime images make it easy to see which parts of the country use the most electricity. Did you know that many people in British Columbia live hundreds or even thousands of kilometres away from the power station that produces the electricity they use every day? This can make it easy to forget that producing electricity is more than just flicking a switch.

Electricity seems clean and easy to use. When you turn on a lamp, you don't see any polluted air, damage to the environment, or waste products. All electricity comes from the natural environment. All methods of producing electricity affect communities and ecosystems locally and around the world. In this chapter, you will study how we generate electricity in Canada, and how new technologies are changing electricity production. You will also learn how you can use electricity responsibly.

What You Will Learn

In this chapter, you will learn

- how electricity is produced in British Columbia and in Canada
- what the advantages and disadvantages are of different ways of producing electricity
- how to find out the efficiency of electric devices in your home

Why It Is Important

- Canadians use an enormous amount of electricity. The choices we make about producing and using electricity affect our communities and environment.
- Wise energy use can help to protect the environment and can save money.
- Storms, earthquakes, and other events can cause power blackouts anywhere in the province. It is important to be ready to live without electricity during a blackout.

Skills You Will Use

In this chapter, you will

- model the operation of a hydro-electric dam
- classify different sources of electricity
- communicate advantages and disadvantages of different methods of producing electricity
- develop strategies for household and community electricity conservation

Some people in British Columbia use the Sun to produce electricity.

Starting Point ACTIVITY 6–A

Electric Lunch

What to Do

1. In your science notebook, make a list of every item that you will have (or had) for lunch today. Include the food as well as any wrappers or packaging.
2. Share your list with a classmate. Together, brainstorm as many steps as you can think of that are involved in making each item, transporting it, and storing it. Be as specific as you can. Draw a flow chart to **record** the steps.
3. Try also to include steps that are not obvious uses of electricity. For example, the kitchen lights may be on while you cut bread for a sandwich, or you may use an electric can opener to open your can of tuna. You could say that these uses of electricity are *indirectly* involved in making your lunch.
4. Put a star beside any steps that might use electricity directly or indirectly.
5. Count the stars. The total is your "electric lunch score." **Record** your score in your science notebook.

What Did You Find Out?

1. How could you make the same lunch but reduce your electric lunch score?
2. As a class, discuss some changes you would be willing to make in your lifestyle to save electricity.

Section 6.1 Generating Electricity

Key Terms

- electric generator
- turbine
- hydro-electric power
- hydro
- hydro-electric dam
- fossil fuels
- coal
- oil
- natural gas
- nuclear energy
- renewable
- non-renewable

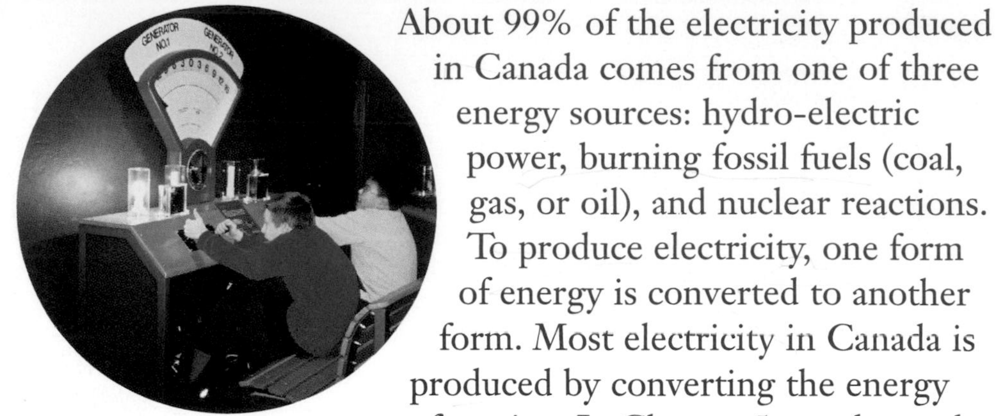

Figure 6.1 These people are using the movement and energy of their own bodies to generate electricity.

About 99% of the electricity produced in Canada comes from one of three energy sources: hydro-electric power, burning fossil fuels (coal, gas, or oil), and nuclear reactions. To produce electricity, one form of energy is converted to another form. Most electricity in Canada is produced by converting the energy of motion. In Chapter 5, you learned that an electromagnet can convert electrical energy into motion. An **electric generator** uses the energy of motion to produce electricity.

In Chapter 5, you created a magnet by passing an electric current through a wire conductor wrapped around an iron bar. It is also possible to create an electric current by moving a magnet across a wire conductor. This is the basic idea for an electric generator.

Figure 6.2 shows how an electric generator uses a spinning device called a **turbine** to create motion that produces electricity. But what makes the turbine spin? In the rest of this section you will learn what form of energy causes the turbine to spin for each of the three main sources of energy used to generate electricity in Canada.

How does an electric generator convert the energy of moving objects into electricity?

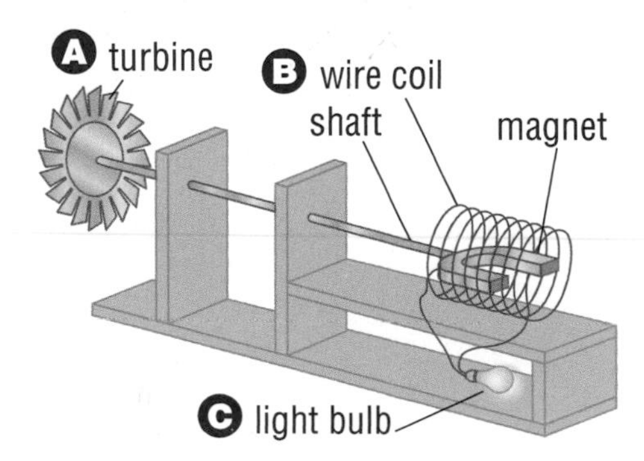

Figure 6.2 A simplified version of an electric generator.
(A) The generator contains a turbine with a set of blades. When moving air, water, or steam pushes on the blades, the turbine spins. As it spins, it rotates a shaft.
(B) The shaft spins a magnet within a coil of wire.
(C) The spinning magnet produces an electric current in the wire, which lights the bulb.

Find Out **ACTIVITY 6–B**

What's Your Source?

Most electricity produced in Canada is either from hydro-electric power, burning fossil fuels (coal, gas or oil), and nuclear reactions. Where do you fit in?

What to Do

1. Review the data in the table below.
2. Draw two bar graphs from this data. One bar graph should show the percentage of Canada's electricity that comes from each energy source. The other graph should show the percentage of British Columbia's electricity that comes from each energy source. Give each of your graphs a title.
3. Use the Internet, your school library, or community resources to locate the electrical generating station closest to your community. What source of energy does it use?

What Did You Find Out?

1. Find the kind of electrical generating station near your community on the bar graphs you made in step 2. Where does this kind of electricity generation fit in comparison to Canada and to British Columbia?
2. What factors might explain why the electrical generating station was built or placed in this particular location near your community?
3. British Columbia produces 85 percent of its electricity from hydro-electricity, and very little from burning coal. In Alberta, 82 percent of the electricity generated comes from coal, and only 4 percent comes from hydro-electricity. Why do you think the sources of energy are so different in these two provinces?

Skill POWER

For tips on creating bar graphs, refer to SkillPower 4.

Most of British Columbia's electricity is produced by hydro-electric power stations.

Energy Source:	Hydro	Nuclear	Fossil Fuels (coal, oil, gas)	Other
Percentage of Canada's electricity produced from this energy source	59	12	28	1
Percentage of British Columbia's electricity produced from this energy source	85	0	12	3

Pause& Reflect

Large construction projects affect the environment either directly or indirectly. Imagine that a hydro-electric dam blocks fish from swimming up a river to spawn. What are some of the possible indirect effects of this? Brainstorm your ideas with your class.

Electricity from Moving Water

As you saw in Activity 6-B, more than half of Canada's electricity comes from **hydro-electric power** (which many people call **hydro** power). Canada produces more hydro-electric power than any other country in the world. Hydro-electric power stations use falling water to spin turbines in an electric generator.

Figure 6.4 shows the main components of a hydro-electric station. Each large hydro-electric station in British Columbia uses a **hydro-electric dam** to store water. A dam is a large barricade that blocks the river. Instead of flowing, the water is stored in a lake or *water reservoir*. A *spillway* near the surface of the reservoir allows extra water to flow over the dam and into the river. Pipes control the flow of water through *waterways* from the base of the reservoir into the generator.

One of the main advantages of hydro-electric power is that it does not produce any pollution or harmful waste products. However, constructing a hydro-electric dam floods large areas of land and affects the ecosystems of rivers, lakes, and the surrounding areas. These changes affect plants, fish and other animals, as well as the people who use them. In the past, entire communities have had to move when dams were built in their area.

READING Check

What is one advantage of hydro-electric power? What is one disadvantage?

Figure 6.3 It is possible to make use of the energy of flowing water without a hydro-electric dam or large power station. For people with a source of running water near their homes, small water generators (called *micro-hydro turbines*) are a very reliable source of renewable energy. This small turbine measures less than 50 cm across. It will produce electricity non-stop, as long as running water is available, no matter what the weather. These are called "*run-of-river*" generators because they use the energy of running water without a dam.

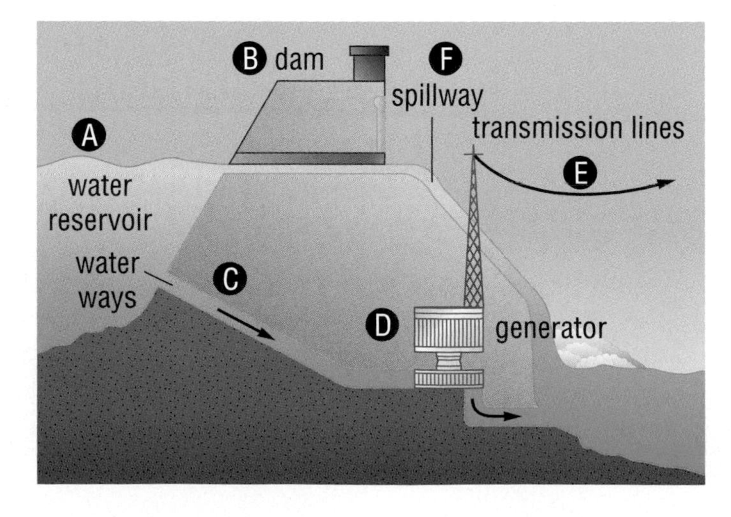

Figure 6.4 A hydro-electric power station.
(A) The reservoir stores water.
(B) The dam keeps the water in the reservoir.
(C) Pipes control the flow of water from the reservoir through the generator.
(D) The fast-flowing water spins the turbines to turn the generator.
(E) Transmission lines carry electric current away from the power station.
(F) A spillway near the top of the reservoir lets water flow over the dam into the river.

CONDUCT AN

INVESTIGATION 6-C

SKILLCHECK
- Predicting
- Observing
- Making Inferences
- Developing Models

Model Hydro-Electric Power Station

The waterfall that flows over the top of a hydro-electric dam has a lot of energy. However, as you saw in Figure 6.4, this waterfall is not used for the hydro-electric generator. You can find out why when you build your own model of a hydro-electric power station.

Question

Why is the waterway for a hydro-electric generator located near the bottom of the water reservoir, rather than near the surface?

Safety Precautions 

Materials	Apparatus
water	scissors
masking tape	1 23 cm × 33 cm baking pan
nail	
small pinwheel	
1 2 L plastic pop bottle	

Procedure

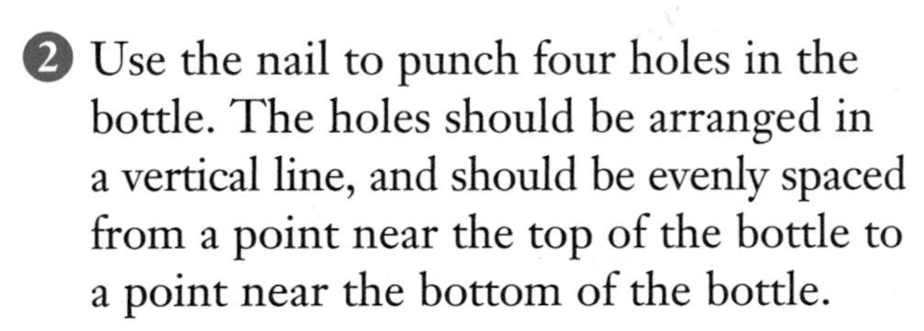

1. Using the scissors, carefully cut off the top of the plastic bottle.
2. Use the nail to punch four holes in the bottle. The holes should be arranged in a vertical line, and should be evenly spaced from a point near the top of the bottle to a point near the bottom of the bottle.
3. Cover all the holes with one strip of masking tape. Make sure the masking tape is pressed firmly onto the plastic bottle.
4. Stand the bottle at one end of the baking pan with the holes facing toward the centre of the pan.
5. Fill the bottle with water. **Predict** what will happen when you remove the strip of masking tape.
6. Remove the strip of masking tape. **Observe** what happens. Then empty the bottle and the pan.
7. Repeat steps 3–5. Remove the tape then hold the pinwheel under the stream of water from each hole in the bottle. Hold the pinwheel so that the stream of water spins the pinwheel. **Observe** what happens at each hole. **Record** your observations.
8. Clean up your work area and return all materials to your teacher.

Analyze

1. How did the streams of water from each hole differ?
2. What effect did this difference have on the pinwheel?

Conclude and Apply

3. Based on your results, explain why the waterways are located as you see them in Figure 6.4.

Extend Your Skills

4. Some hydro-electric power stations do not use dams, but instead capture the energy of water as it flows in a river. This is called a "run-of-river" generator (see Figure 6.3). Use your findings to explain why "run-of-river" hydro-electric generators are used to produce only small amounts of electricity.

Figure 6.5 Mining and drilling for fossil fuels can produce waste and also damage the environment, as seen in this coal mining operation.

Electricity from Burning Fossil Fuels

What would happen if you buried a pile of dead plants and animals, then waited millions of years? Under the right conditions, with enough heat and pressure, you could make fossil fuels. **Fossil fuels** are sources of energy that are created in nature from the rotted body parts of ancient plants and animals. The three types of fossil fuel are

- **coal** (a solid fossil fuel)
- **oil** (a liquid fossil fuel)
- **natural gas** (a fossil fuel that is usually found as a mixture of gases).

Fossil fuels burn easily and at high temperatures. When coal, oil, or natural gas is burned in a power station, the heat that is produced is used to boil water, turning it to steam. The steam spins the turbine of an electric generator. Figure 6.6 shows the main components of a fossil fuel-burning power station.

Fossil fuels are an important source of electricity in Canada. Several Canadian provinces, including British Columbia and Alberta, have large deposits of fossil fuels. However, when we burn fossil fuels, we release air pollution that causes health problems and damages the environment. This pollution may even be changing the world's climate. New technologies help fossil fuel-burning power stations operate in a more environmentally friendly way.

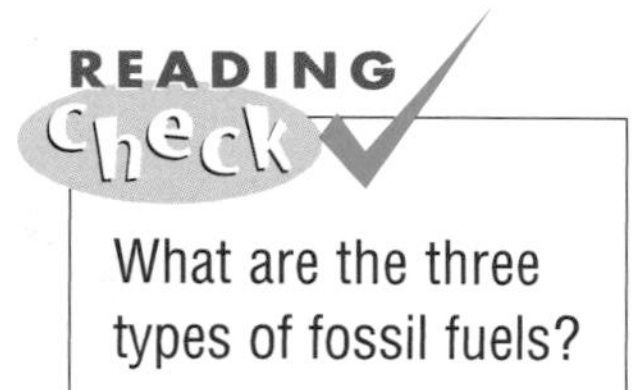

What are the three types of fossil fuels?

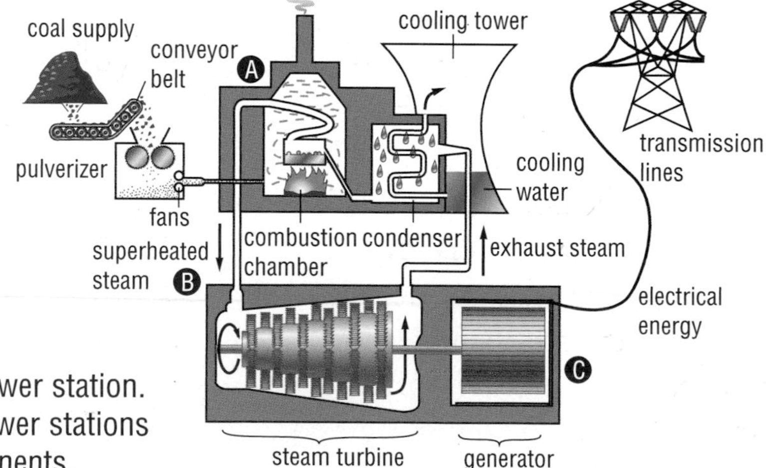

Figure 6.6 A coal-fired power station. Other fossil fuel-burning power stations have the same basic components.

(A) In a combustion chamber, the burning fuel boils water. (The smoke passes through filters before it is sent into the air.)

(B) Steam from the boiling water flows through a pipe to a turbine. Pressure from the steam causes the turbine to spin.

(C) The spinning turbine drives an electric generator. Transmission lines carry electricity away from the power station.

Electricity on the Go

Figure 6.7 Portable generators can supply electricity far from any power lines.

Figure 6.8 A car battery can also be used as a portable generator. It can power only small electric devices for a short period of time.

Fossil fuels can also be used in portable gasoline or diesel generators. In these machines, a small engine spins a small electric generator. A car engine works the same way to charge the car battery whenever the engine runs. Portable generators are useful for working in remote areas and for emergency backup systems.

Electricity from Splitting Atoms

In Chapter 4, you built a simple model of an atom. In that model, the nucleus was one solid piece. A more complicated model shows that the nucleus of an atom is made up of different kinds of small particles held together. When the nucleus of an atom is split apart in what is called a nuclear reaction, it releases a large amount of **nuclear energy**. This process is called *nuclear fission*.

The generator in a nuclear power station works in the same way as other turbine generators. The heat energy from the nuclear reaction is used to boil water. The steam produced from the boiling water spins a turbine to run an electric generator. Nuclear power can produce a great deal of electricity from very little fuel. The reaction also produces wastes that release harmful particles for many years. The handling and storage of these wastes is a serious challenge that has not been solved yet.

Figure 6.9 The Canadian-designed CANDU reactor produces electricity from nuclear energy. CANDU reactors are used in countries around the world. Unfortunately, nobody has yet found a lasting solution to the problem of storing nuclear wastes.

Renewable and Non-renewable Energy Sources

Canadians are among the top users (and wasters) of electricity. Almost all of this electricity comes from hydro-electric, fossil fuel-burning, and nuclear power stations. If we continue to rely on these sources, what do you think will happen over time?

Some sources of electricity are **renewable**. A renewable source of energy is one that can renew or replace itself. Hydro-electricity is an example of a renewable electricity source, because the water that flows through a hydro-electric power station is not changed or used up in the process.

Why are fossil fuels a non-renewable resource? What do you think this means for the future of fossil fuel as a source of electricity?

Other sources of electricity are **non-renewable**. A non-renewable energy source is one that cannot be replaced within a human lifetime. Fossil fuels and nuclear energy are examples of non-renewable energy sources because the materials for fossil fuel stations and nuclear energy can be used only once. Eventually, the supply of these materials will be gone.

Our demand for electricity in Canada will continue to grow. We will need to build new power stations and find sources of energy that are renewable. In the next section you will study some new technologies that may play a larger role in electricity production in the future.

The world is using fossil fuels at a rate 100 000 times faster than they are being replaced.

Figure 6.10 Current sources of electricity affect water, land, and air as shown in (A) a hydro-electric power station, (B) a fossil fuel-burning power station, and (C) nuclear waste from a nuclear power station. How can we meet the demand for electricity without hurting the environment?

Section 6.1 Summary

In this section, you learned the following:

- Over half of Canada's electricity is hydro-electricity that is produced in large electric generators.
- An electric generator uses energy from motion to spin a turbine.
- The turbine spins a wire coil inside a magnet or a magnet inside a wire coil.
- The magnet produces an electric current in the coil.

Most of Canada's electricity comes from three energy sources:

- hydro-electric power
- burning fossil fuels
- nuclear reactions

Each of these energy sources has advantages and disadvantages. Sources of energy can be

- *renewable*: they can be renewed or replaced
- *non-renewable*: they cannot be renewed or replaced within a human lifetime

INTERNET CONNECT

www.mcgrawhill.ca/links/BCscience6

British Columbians disagree about whether or not to develop offshore oil and gas. Use the Internet to research the main points in the debate. Go to the web site above and click on **Web Links** to find out where to go next. What is your opinion?

Check Your Understanding

1. What source of energy supplies most of British Columbia's electricity?
2. Why is the waste produced from a nuclear reaction harmful?
3. What are some of the environmental problems related to the production of hydro-electricity?
4. Explain why fossil fuels are a non-renewable source of energy.
5. **Apply** The diagram illustrates the parts of a coal-fired power station. In your science notebook, match the component of the power station to the correct letter in the diagram.
   (a) Steam turbine
   (b) Cooling water
   (c) Combustion chamber
   (d) Generator
   (e) Superheated steam
   (f) Transmission lines

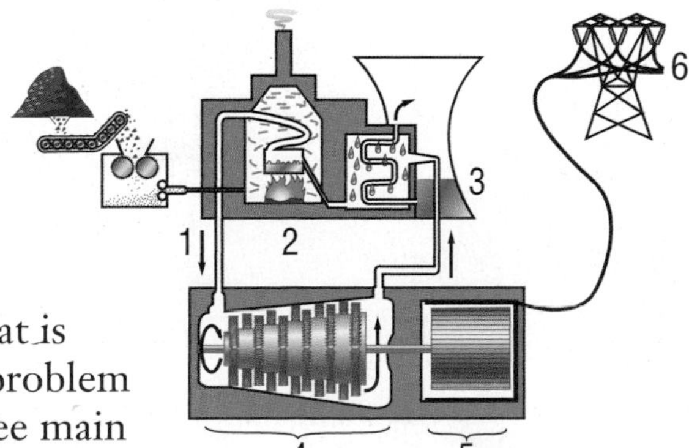

6. **Thinking Critically** What is the main environmental problem caused by each of the three main sources of energy used to produce electricity in British Columbia?

Key Terms

electric generator
turbine
hydro-electric power
hydro
hydro-electric dam
fossil fuels
coal
oil
natural gas
nuclear energy
renewable
non-renewable

Section 6.2 Electricity Ahead

Key Terms
- solar energy
- wind energy
- tidal energy
- geothermal energy
- biomass energy

In Section 6.1 you learned about some of the social and environmental costs of producing electricity in Canada. These costs have caused researchers to find new ways to generate electricity from renewable sources. In the next few pages, you will study some of the technologies that might play a more important role in the future.

Each of these new technologies has advantages and disadvantages. For example, one disadvantage is that some technologies work only in specific locations. To decide which technologies to use, we must consider the following:

- How expensive is the technology to develop?
- How much electricity can the technology produce in this location?
- How might the technology affect the environment?
- How might the technology affect people and communities?

The answers to these questions help describe the advantages and disadvantages of using different technologies. Communities can use these answers to decide on the best technology to meet their future electricity needs.

What are the questions to ask before deciding on a suitable technology for generating electricity?

Figure 6.11 Each region of British Columbia has special natural features, as shown in these photos of (A) Terrace, (B) Fort St. John, and (C) Prince Rupert. Some technologies for producing electricity will work best in particular regions. What are the features of your region?

Electricity from the Sun

For thousands of years, Aboriginal peoples have used the Sun's energy, or **solar energy**, for drying and preserving food and clothing, and for other purposes.

The Sun's energy can be converted into electrical energy in a *photoelectric cell*. A photoelectric cell is a device coated with a material that conducts electricity. When sunlight strikes the surface of the cell, some atoms in the coating release electrons. These electrons flow through a conductor and start an electric current in a wire.

Figure 6.12 The Nisga'a people use solar energy to dry fish.

Figure 6.13 Solar panels can be gathered together in a "solar farm." A solar panel is made up of many photoelectric cells connected together.

DidYouKnow?

Scientists have discovered that solar panels can be "painted" on a surface. It is five times more efficient than current panels, and can collect energy even on a cloudy day. Perhaps in a few years this technology will be available to us.

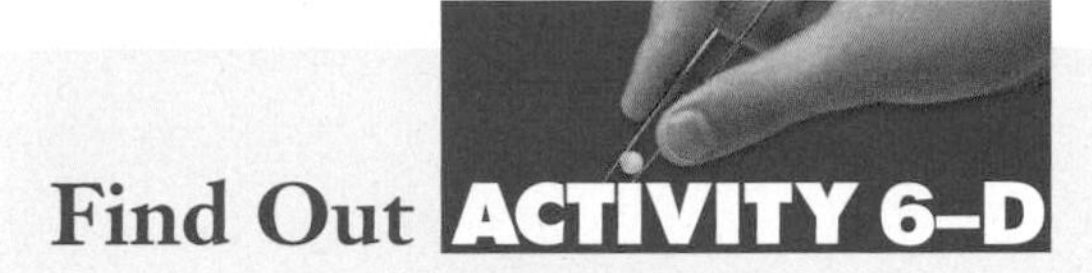

Find Out **ACTIVITY 6–D**

Catch Some Rays

Solar cells convert solar energy into electricity. In this activity, you will observe the effect of solar energy.

What You Need

2 glass jars
2 thermometers
water

What to Do

1. Copy the table into your notebook. Give your table a descriptive title.

Sample	Starting Temperature (°C)	Temperature after 5 min (°C)
Jar A		
Jar B		

2. Label one jar "A." Label the other jar "B." Half-fill each jar with water.
3. Put one thermometer into each of the glass jars with water. Wait 2 min. Then **record** the starting temperature in each jar.
4. Place jar A in a sunny spot. Place jar B in a shady spot. Wait 5 min. Then **record** the temperature of each jar again.
5. Empty your jars and return your materials to your teacher.

What Did You Find Out?

1. How did the temperature in each jar differ after 5 min? Explain your findings.
2. What energy source was used?
3. What form of energy was produced?

Electricity from Wind

Wind energy is the energy of moving air. Advances in the technology of wind turbines have made wind energy the fastest-growing source of electricity in the world.

A wind turbine works in much the same way as other turbine generators. When moving air pushes on its blades, the spinning turbine drives an electric generator. Most wind turbines need winds of at least 15 km/h to operate, and work best in areas where the wind stays fairly constant.

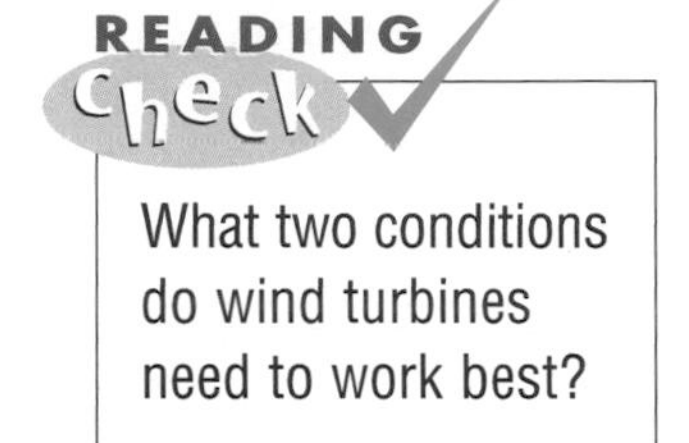

What two conditions do wind turbines need to work best?

Wind power is renewable, and wind turbines do not create pollution or wastes. However, some people feel that the appearance and sound of wind farms spoil the beauty of the natural environment. Birds are sometimes killed when they fly into the turbines, but new ways of designing the turbines are creating a safer environment for birds.

Modern wind turbines make hardly any noise.

This wind turbine, installed in Toronto, Ontario, generates enough electricity for 250 homes.

Figure 6.14 In 2005, work began on the construction of British Columbia's first wind farm near the community of Holberg on the coast of Vancouver Island. This photo of the proposed locations for the wind turbines shows that the turbines can become part of the environment without spoiling the natural beauty.

Electricity from the Ocean

All along the British Columbia coast, ocean tides cause the water level to rise and fall twice a day. When the tide rises, water flows into bays and inlets, and flows out again when the tide falls. The energy of the moving water in the tides is called **tidal energy**. Just as in hydro-electric power stations, generators in tidal generating stations can convert the energy of moving water into electricity.

Tidal generating stations can work only in certain locations. A tidal generating station must be in an area where the difference between high and low tides is very large. This means that the tidal waters will have a great deal of energy. The area must also have a natural bay or inlet that can be easily blocked by a dam to trap tidal waters, as shown in Figure 6.16.

Tidal generating stations have many of the same advantages and disadvantages as hydro-electric power stations. They do not produce pollution or harmful emissions, but building a tidal power station can damage the ecosystem of tidal areas. New technologies are being developed to capture the energy of flowing tidal waters without building barricades or large power stations.

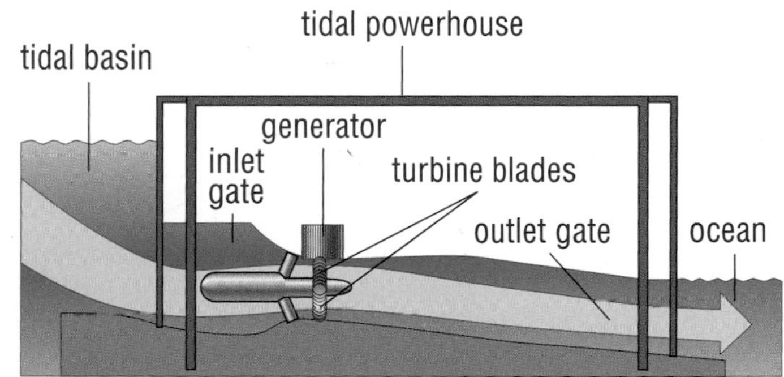

Figure 6.16 A tidal bay or inlet with a narrow opening is a suitable place for a tidal generating station. A barricade across the mouth of the opening traps tidal waters. When the tide drops, water flows through a waterway to turn a turbine and generator. Some tidal generating stations are designed to capture the energy of water flowing into and out of the basin.

Figure 6.15 The Annapolis tidal generating station in Nova Scotia was the first in North America.

Figure 6.17 Ocean waves can also serve as a source of electrical energy. Near Ucluelet, on the west coast of Vancouver Island, ocean waters have the strong and steady wave energy that can be used to produce electricity. British Columbia does not yet have a tidal generating station.

Figure 6.18 The red-hot lava flowing from this volcano is a dramatic example of the energy contained in Earth's crust.

Electricity from Earth

Earth's crust contains a great deal of heat energy known as **geothermal energy**. This term comes from *geo* meaning Earth and *thermal* meaning heat. Although most of this energy is deep within Earth's crust, pockets of hot rocks or water are close to Earth's surface. Have you ever seen hot springs or steam vents? In these places, hot water or steam rises all the way to the surface from far below the surface. In other locations, hot water collects in underground reservoirs. Each of these reservoirs could be used to generate geothermal electricity. Of all the Canadian provinces, British Columbia has the best potential for using geothermal energy.

Geothermal energy is renewable, and some systems are nearly pollution-free. However, some geothermal systems create exhaust which releases harmful substances into the air or nearby rivers.

DidYouKnow?

Geothermal energy can be used to control the temperature in buildings. *Geothermal heat-exchange systems* draw heat from Earth's crust to warm buildings in the winter, and also put heat back into Earth's crust to cool buildings in the summer. Heat-exchange systems do not produce electricity, but they are an alternative to using electricity or fossil fuels to heat buildings.

Figure 6.19 The Geysers is the largest geothermal electricity station in North America. It produces enough electricity for the city of San Francisco.

Figure 6.20 Meager Creek Hot Springs near Pemberton. The Lil'wat people call these springs *Teiq'* and consider them a sacred place where people can train to seek visions and knowledge. The Meager Creek area is the largest non-commercial hot springs in southern British Columbia, and is also the site of the province's first tests on geothermal electricity.

Electricity from Plants and Animals

In Section 6.2, you learned that dead plants and animals can create fossil fuels, an important but non-renewable source of energy. Another energy source also comes from plant and animal tissues, but without having to wait for millions of years! **Biomass energy**, or *bioenergy*, is the energy stored in all plant and animal tissues. We can release biomass energy in several ways.

Burning plant fibres such as wood and straw releases heat energy. This energy can be used, just like that of fossil fuels, to drive steam turbines to generate electricity. These fibre fuels are renewable, as long as we carefully manage the forests or agricultural lands they come from.

An alternative to burning fibre fuels to release bioenergy is to feed plant and animal tissues to micro-organisms inside special compartments. As the micro-organisms digest the tissues, they release a form of natural gas called *biogas*. We can collect the biogas and use it as fuel. Both fibre fuels and biogas are renewable. However, they both release pollutants when they are burned.

Figure 6.21 Wood fuel probably supplied the earliest human uses of biomass energy.

Figure 6.22 All organisms, including animals and people, release biogas as they digest their food.

What is one advantage of using bioenergy to produce electricity? What is one disadvantage?

Figure 6.23 Bioenergy is available wherever there are living organisms. In this small biogas generator in the Cook Islands, micro-organisms inside the metal box produce methane gas from pig manure. The methane gas is collected to heat small cooking stoves. This particular biogas system is not used to generate electricity.

Did You Know?

In Section 6.1, you learned that nuclear energy can be used to create electricity. Today's nuclear power stations rely on *nuclear fission*, which is the splitting of atoms into smaller pieces. *Nuclear fusion* is the opposite. A fusion reaction combines two small atoms to create a single, larger atom and releases energy at the same time. Nuclear fusion has the potential to generate much more energy than nuclear fission. Fusion reactions do not create nuclear waste products. However, these reactions normally require temperatures in the millions of degrees Celsius. Scientists around the world are exploring nuclear reactions to find ways to make fusion energy possible.

The immense energy of stars comes from nuclear fusion reactions.

INTERNET CONNECT

www.mcgrawhill.ca/links/BCscience6

One of British Columbia's important contributions to the field of renewable energy has come from the work of a Vancouver-based company, Ballard Power Systems, to develop a new kind of *hydrogen fuel cell*. Hydrogen fuel cells use the energy of a chemical reaction between hydrogen and oxygen to create electricity. The only waste product from this reaction is water. Hydrogen fuel cells can be used to run vehicles as well as devices such as laptop computers. Some people refer to fuel cells as a type of battery. Is this accurate? Use the Internet to find out. Go to the web site above and click on **Web Links** to find out where to go next. In what ways is a hydrogen fuel cell similar to a battery? In what ways is it different?

This bus is powered by a Ballard hydrogen fuel cell. The only gas coming out of its exhaust pipe is water vapour. Over time, electric cars powered by hydrogen fuel cells could replace gasoline-powered cars.

Figure 6.24 This home on Lasqueti Island uses electricity from solar panels in the summer time. In the winter, its electricity comes from a small run-of-river hydro-electric generator in a nearby stream. A portable gasoline generator provides emergency backup.

Combining Electricity Sources

You have learned that there are many ways of using the energy stored in nature to generate electricity. Not all of these ways can supply steady electricity. For example, solar cells work only when the sun is shining, and wind turbines turn only when the wind blows. For this reason, a community's renewable energy strategies often combine more than one source of electricity. Another option is to use rechargeable batteries to store electrical energy.

PROBLEM-SOLVING

INVESTIGATION 6-E

SKILLCHECK
- Identify the Problem
- Decide on Design Criteria
- Plan and Construct
- Evaluate and Communicate

Production Teams

Imagine that your community needs to build a new power station. In this investigation, you will research and present to your class the advantages and disadvantages of one way of producing electricity.

Challenge

Group Work

Working in a group, research and design a presentation on a possible source of energy for your community. Decide whether you would recommend that source for the proposed new power station.

Apparatus

textbook, library books, and the Internet

Materials

pens or pencils
paper

Design Criteria

A. Your presentation must consider whether the energy source could be developed in your region of the province.

B. Your presentation must consider how developing this energy source will affect the environment and people.

C. Your presentation must consider the advantages and disadvantages of the energy source.

D. Your presentation should show whether the energy source would be suitable for your community. Make sure you justify your decision.

E. Your presentation could take a variety of forms. For example, you could make your presentation orally, prepare a play that you can film or act out, or create an interactive computer presentation.

Plan and Construct

1. With your group, discuss how you will begin your research. Assign tasks among group members.
2. After you have collected your information, work together to organize it. Create large data tables or drawings to show the class. Decide if any further research is needed.
3. As a group, decide whether you will recommend the energy source for your community.
4. Make your presentation to the class.

Evaluate

1. Discuss with your group what you could have done to make your presentation more convincing.
2. Did presentations by other members of your class talk about problems you had not considered? If so, name and describe them.
3. As a class, decide which energy source should be used in your community. Explain your reasons.

Skill POWER

For tips on technological problem solving, turn to SkillPower 6.

Section 6.2 Summary

In this section, you studied a number of ways that are being researched to generate electricity from renewable energy sources. These technologies involve capturing the energy in sunlight, wind, ocean waters, Earth, and plant and animal tissues. These sources could play a larger role in providing electricity for British Columbia and Canada in the future.

- Many renewable energy sources will work only in certain regions or under certain conditions.
- Some renewable energy sources cause pollution.
- There is no single "best" way to produce electricity. Future energy strategies are likely to combine different sources to provide a steady electricity supply.

Key Terms

solar energy
wind energy
tidal energy
geothermal energy
biomass energy

Check Your Understanding

1. What is the main source of energy used to run both hydro-electric turbines and wind turbines? (Hint: More than just water and wind.) Explain.
2. Researchers are trying to make solar cells work better to collect solar energy. Why would it be useful to make the collection of solar energy more effective? Explain your answer.
3. Identify three sources of renewable energy. For each source, identify one or more factors that could limit the use of this source for electricity production in Canada.
4. **Apply** Genevieve is the mayor of a small town. She and the other members of the town council have received a proposal from an energy company to build a geothermal power station on the outskirts of town. What are some of the things that Genevieve and the councillors should consider as they decide whether or not to approve the proposal?
5. **Thinking Critically** Many sources of electrical energy are less harmful to the environment than the burning of fossil fuels. However, these sources of energy are not in common use. Give two reasons why.

Section 6.3 Bringing It Home

Figure 6.25 Transmission lines carry electricity long distances across the province.

Key Terms
- fuses
- circuit breakers
- electrocution
- electric meter
- energy consumption
- energy efficiency
- energy conservation

Transmission and Distribution

You have learned that most of British Columbia's electricity is generated in huge power stations (see Figure 6.26). These power stations send electricity along high-voltage *transmission lines*, which carry electricity over long distances to distribution stations. These stations distribute the electricity to communities and neighbourhoods through smaller *distribution lines*. Service wires from distribution lines bring electricity through the meter into individual buildings.

DidYouKnow?

Electricity travels most effectively at very high voltages. The voltage in transmission lines can be more than 4000 times higher than the voltage in a household circuit. Special devices called *transformers* cause the voltage to drop, or "step down," as the electricity travels from the distribution line to a home. The high voltage in transmission lines makes them extremely dangerous. You should never climb near transmission lines.

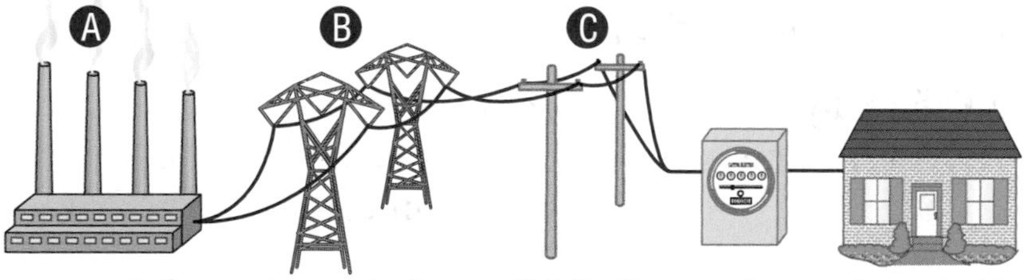

A B C
power station
transmission lines
distribution lines
meter
household

Figure 6.26 Power production and delivery include three steps: (A) generation, (B) transmission, and (C) distribution.

Figure 6.27 An electric meter. A technician can read the dials to find out how much electricity has been used since the last reading. Do you know where to find your home or building's electric meter?

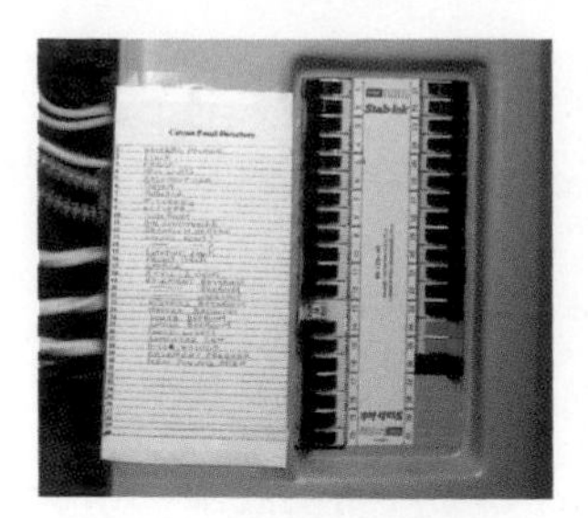

Figure 6.28 The service panel contains a circuit breaker for each household circuit. Every circuit is labelled.

Figure 6.29 A fuse box contains a fuse for each circuit.

How can a short circuit cause a fire?

Fuses and Circuit Breakers

Most homes have a *fuse box* or *service panel (circuit panel)* at the point where service wires deliver electricity to the many circuits in your home. The fuse box or service panel is an important electric safety device. In Chapter 5 you learned that a short circuit can cause wires to heat up because the current becomes too high. Short circuits in wiring can cause house fires. **Fuses** and **circuit breakers** are devices in fuse boxes and service panels that prevent too much current from flowing through a circuit. In Activity 6-F, you will see how a fuse works to protect a circuit.

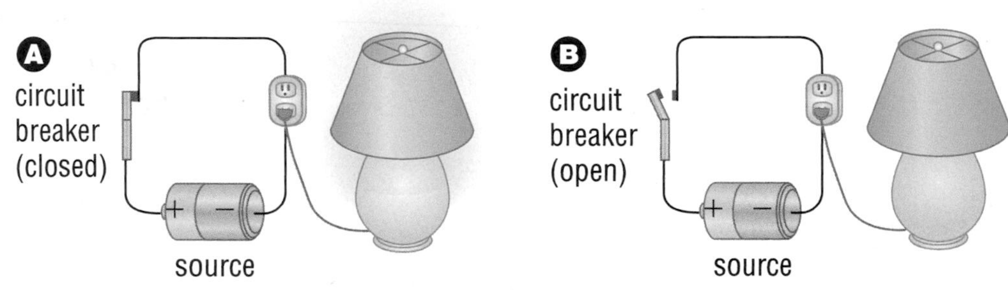

Figure 6.30 A circuit breaker.
(A) In a normal circuit, the breaker is straight, creating a closed circuit.
(B) If the conductor heats up too much, the circuit breaker bends and opens the circuit. After the problem is fixed, you can reset the breaker to close the circuit.

Find Out ACTIVITY 6–F

Blown Away

Have you ever plugged in a toaster, only to have the electricity go out in the kitchen? This is a signal that there is too much current in the circuit. You will use a simple electric circuit to see how a fuse could protect this circuit.

What You Need

2 1.5 V batteries in holders
5 copper wires with alligator clips
1 3 V light bulb
1 1 amp fuse

Safety Precautions 

What to Do

1. Examine the fuse. **Record** notes or draw a diagram to describe what you see.
2. Build a circuit that connects a single battery to the bulb and the fuse. Close your circuit and **observe** what happens.

continued

3. Open the circuit and add a second battery in series.

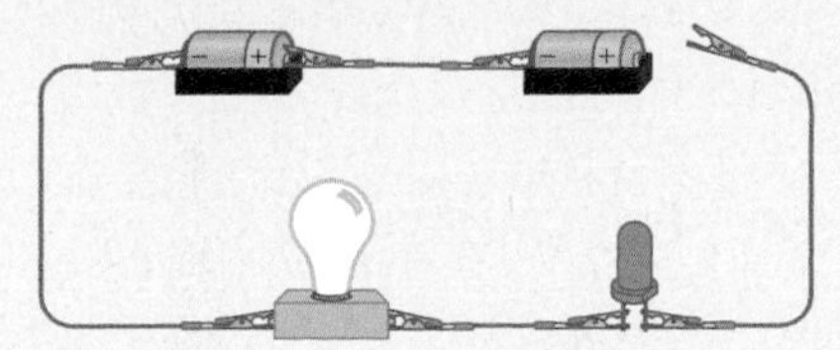

4. Close the circuit and **observe** what happens.
5. Clip each end of the fifth copper wire to each terminal on the light. **Observe** what happens.
6. Take apart your circuit and examine the fuse again. **Record** notes or draw a diagram to describe what you see now.

What Did You Find Out?

1. What happened as you added batteries to your circuit? Explain your observations.
2. What happened to the light when you attached the fifth copper wire to the light terminals? Explain your observations.
3. What happened to the fuse? Write a paragraph or draw a labelled diagram to explain how a fuse works as a safety device.
4. Would your results have been different if you arranged the batteries in parallel rather than in series? Explain.

Electric Safety

Fuses and circuit breakers can help protect you if there is a problem with your household electric circuits. They cannot protect you from every electric hazard. The circuits in your home carry enough electrical energy to cause **electrocution**. Electrocution is death caused by a strong electric current passing through a living organism.

Exposed wires and damaged electric cords are hazards. If you touch them, you could become part of a dangerous short circuit. For this reason, it is important to make sure you treat all electric devices, including appliances, cords, and outlets, with care. Plugging numerous devices into the same outlet can cause the circuit to become overloaded.

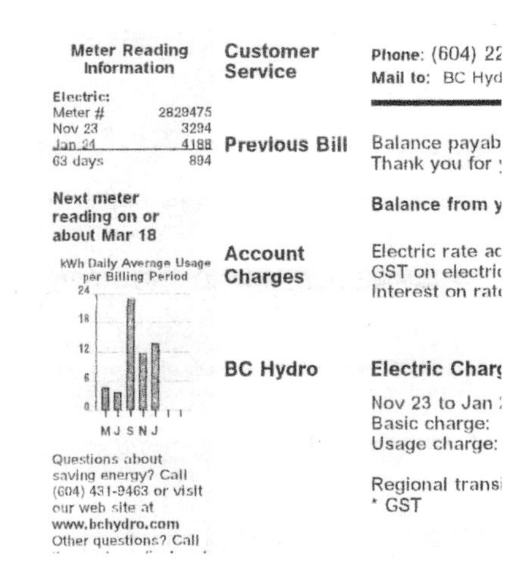

Figure 6.31 An electric bill. The bar graph shows how electricity use changes over the course of the year. Suggest several reasons to explain why this home uses more electricity in the winter than in the summer.

Measuring the Electricity You Use

Do you know how much electricity your home or school uses every day? Your **electric meter** does. The electric company uses information from a building's meter to calculate the cost of electricity.

You can also get an idea of how much electricity your home uses by calculating the electricity consumed by the appliances you use regularly. Every electric appliance should have a label like the one in Activity 6-G.

Watts Up?

A watt (W) is a unit of *power*. The rate at which a light bulb turns electrical energy into heat and light is the power of the light bulb. We call the watt usage of an appliance its *energy rating*. You will use the energy rating on appliance labels to estimate your home's electricity consumption.

What to Do

1. Copy the table below into your science notebook. Give your table a descriptive title.
2. Choose at least three electric devices that are often used in your home. Look for the wattage label on each device. If you cannot find a label on a certain device, you may substitute another device.

3. Use the table to calculate how much electricity those devices use.

What Did You Find Out?

1. Which device uses the most electricity each day?
2. Compare the energy ratings of similar devices with other members of your class. How different are they?
3. Suggest a reason why similar devices (for example, a microwave oven) can have different energy ratings.
4. Based on the results of this activity, what are some ways to reduce the amount of electricity your family uses in your home each day?

Column 1	Column 2	Column 3	Column 4
Device	Watts (W) (From label)	Number of hours device is used each day	Energy used by device each day in watt-hours (Column 2 × Column 3)

Conserving Electricity

Earlier in this chapter, you saw that building electricity-generating stations can affect the environment and surrounding communities. Using new sources of electricity can help reduce these effects.

Energy consumption is the amount of energy we use. Another strategy is to consume less electricity. You can do this by turning off lights and electric devices when you are not using them and by choosing devices that don't waste electricity.

Energy efficiency describes using less energy or electricity to accomplish the same task. **Energy conservation** describes doing without certain things to save electricity. In the next Investigation, you will design your own plan to conserve energy or to use energy more efficiently.

What are two ways to conserve electricity?

SKILLCHECK
- Identify the Problem
- Decide on Design Criteria
- Plan and Construct
- Evaluate and Communicate

From Consuming to Conserving

If your household is like most in British Columbia, you are using more electricity than you need to. You are going to look for ways to reduce the electricity you use at home.

Challenge

Group Work

Working in a group, prepare a presentation on the topic "Energy Conservation Is Everyone's Responsibility."

Apparatus
textbook, library books, and the Internet

Materials
pens and pencils, paper, other props

Design Criteria

A. Your presentation should show people how to conserve electricity at home. Include information on why energy conservation is important.

B. Your presentation should include information about energy conservation in British Columbia as well as in your community.

C. Your presentation should include some facts on energy use in Canada and the world.

D. Your presentation should focus on actions that people really can take to help conserve energy.

E. Your presentation should combine information from this textbook with information from at least three other sources.

F. Your presentation could be a proposal, a written or oral report, a poster, an electronic slide show, a musical performance (song/rap), or a drama (short skit).

Plan and Construct

1. With your group, brainstorm how you will gather information. Assign tasks to group members. You might produce a concept map of your ideas.

2. After you have collected your information, work together to organize it. Think about how you can make your presentation as convincing as possible.

3. Design and create any images, models, or props that you will use with your presentation.

4. Rehearse your presentation in front of another group. After your rehearsal, discuss your presentation with your audience. Decide whether you want to make any changes to your presentation.

5. Prepare your final presentation. Make your presentation to your class, to a school assembly, or to a group in your community.

Evaluate

1. Did you present enough data to convince your audience to conserve energy?

2. After seeing presentations by other groups in your class, would you make any changes to your own for a possible future presentation? Explain.

3. After working on this presentation
   (a) What is the most important step you could take to conserve energy?
   (b) Describe what effect you think your choices will have.

INTERNET CONNECT

www.mcgrawhill.ca/links/BCscience6

Do you use electricity safely? Make a list of things you can do to stay safe around power lines, appliances, plugs, and outlets. Use the Internet to check your list. Go to the web site above and click on **Web Links** to find out where to go next.

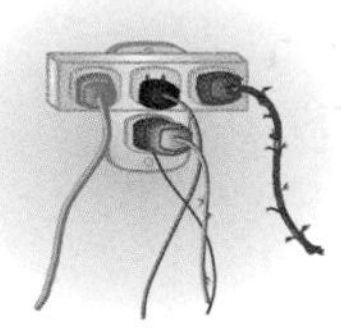

What safety issues can you see in this illustration?

Section 6.3 Summary

In this section, you learned the following:

- Electricity is transmitted from a power station to your community through high-voltage transmission lines.
- Electricity is distributed around your neighbourhood through distribution lines.
- Electricity enters your home through a meter and a service panel.
- Fuses and circuit breakers help keep circuits safe.
- You must always take steps to avoid causing a short circuit or electrical fire.
- You can reduce the amount of electricity you use by
  - turning off appliances when they are not in use, and
  - choosing more efficient devices.
- Over the long term, energy conservation will be good for people and the environment.

Key Terms

fuses
circuit breakers
electrocution
electric meter
energy consumption
energy efficiency
energy conservation

Check Your Understanding

1. What is the purpose of a circuit breaker?
2. How could you calculate the energy used by an electric device in the home?
3. **Apply** You have been hired by the government to help reduce the rate of electricity-related injuries and damage in British Columbia. As part of your strategy, you will create a flyer on electric safety. The flyer will be distributed to homes in the province. What three key safety suggestions will you include in the brochure?
4. **Apply** A new type of light bulb rated at 25 W gives off approximately the same amount of light as a standard 60 W bulb. How much energy (in watts) can be saved per day if you switch from a 60 W bulb to a 25 W bulb? Assume that you have the light on for 6 h per day. (Hint: Refer to the calculations you did in Activity 6-G.)

CHAPTER at a glance

Now that you have completed this chapter, try to do the following. If you cannot, go back to the sections indicated in brackets after each part.

(a) Explain how a hydro-electric dam works to produce electricity. (6.1)

(b) Identify different sources of energy for electricity generation. (6.1)

(c) Describe the advantages and disadvantages of different sources of energy. (6.1 and 6.2)

(d) Explain how to calculate a device's electricity consumption. (6.3)

(e) Describe the safety precautions you should follow when working with electric devices in your home. (6.3)

Prepare Your Own Summary

Summarize this chapter by doing one of the following. Use a graphic organizer (such as a concept map), create a poster, or write a summary to include the key chapter ideas. Here are a few ideas to use as a guide:

- Make a chart listing the different ways that electricity is produced in British Columbia. In columns, identify the advantages and disadvantages of each method.
- Draw a flow chart to describe how electricity gets to your home. Start with the energy source and end with the electric device in your home. Identify as many steps and features along the way as you can.
- Explain the difference between renewable and non-renewable sources of electricity and identify how electricity should be generated in the future.
- Describe how circuit breakers and fuses ensure safety in the home.
- Draw a concept map (see SkillPower 1) to describe ways you can conserve electricity at home and at school.

Figure 6.32 Is this house conserving electricity?

CHAPTER

6 Review

Key Terms

electric generator	solar energy
turbine	wind energy
hydro-electric power	tidal energy
hydro	geothermal energy
hydro-electric dam	biomass energy
fossil fuels	fuses
coal	circuit breakers
oil	electrocution
natural gas	electric meter
nuclear energy	energy consumption
renewable	energy efficiency
non-renewable	energy conservation

Reviewing Key Terms

If you need to review, the section numbers show you where these terms were introduced.

1. Describe the difference between:
   (a) fission and fusion (6.1 and 6.2)
   (b) renewable and non-renewable sources of energy (6.1)
   (c) fuse and circuit breaker (6.3)

2. The following diagram illustrates how flowing water is used to produce electricity. In your notebook, match the function in column A to the number representing the component of the energy station.

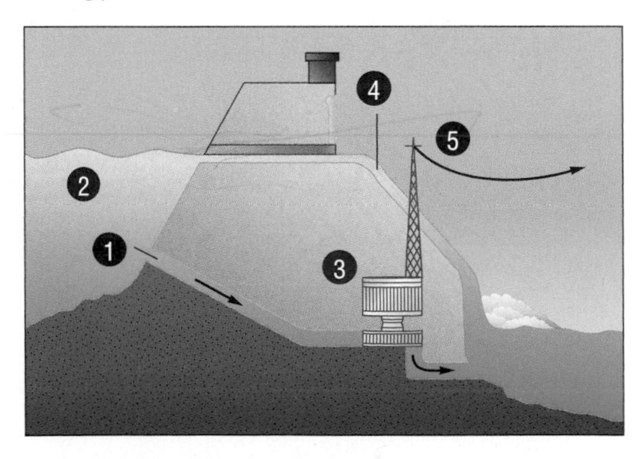

   (a) carries water to the turbine
   (b) transforms the energy of falling water to electrical energy
   (c) stores water
   (d) releases surplus water
   (e) transports electrical energy away from power station

3. In your notebook, match each description in column A with the correct term in column B. Use each description only once.

A	B
(a) can flood land	• hydro-electric energy (6.1)
(b) makes toxic waste	• burning fossil fuels (6.1)
(c) contributes to air pollution	• solar energy (6.2)
(d) can be noisy	• nuclear fission (6.2)
(e) produces almost no pollution	• wind generation (6.2)

Understanding Key Ideas

4. Hydro-electric power stations, coal-fired power stations, and nuclear power stations all contain turbines. In hydro-electric power stations, the movement of water turns the turbines. In coal-fired and nuclear power stations, the movement of high-pressure, flowing steam turns the turbines. What type of energy do all of these turbines harness? (6.1)

Developing Skills

5. Use the data in the table to construct a bar graph of energy use in British Columbia. Include one bar for each source of energy and one bar for the total energy consumed in the province. (A kWh stands for a kilowatt-hour, which is equal to 1000 watt-hours.)

Energy Source	Energy Generated in kWh
Hydro-electric	36 689
Fossil fuels	28 535
Nuclear	0
Renewable (geothermal, tidal, wind, solar)	6 373

Problem Solving

6. Scientists working on solar cells often study green plants. Explain why.

7. You are waxing your skis in the garage on a cold winter day. To melt the wax, you attach an extension cord to a portable heater. After a short period of time, you notice that the extension cord is warm to the touch. Should you be concerned? Explain.

8. Why are the birds in this photo able to perch safely on high-voltage power lines? Can you think of a situation that might make it dangerous for birds to perch there?

Critical Thinking

9. Why is it dangerous to replace a fuse with a copper penny?

10. What do you think might prevent solar-powered cars from becoming common on Canadian roads?

11. Power workers working on high-voltage power lines wear a suit called a Faraday Cage. A Faraday Cage has many conducting fibres woven into the fabric of the suit. Explain how this type of suit would protect the worker from electric shock.

Pause& Reflect

Go back to the beginning of this chapter on page 150, and check your original answers to the Getting Ready questions. How has your thinking changed? How would you answer those questions now that you have investigated the topics in this chapter?

Ask an Elder

Jean Isaac

Williston Lake is the biggest lake in British Columbia. Fifty years ago, it didn't exist. The lake was created in the late 1960s a huge hydro-electric dam was built across the Finlay River. Williston Lake is now the reservoir for British Columbia's largest hydro-electric generating station.

This hydro-electric project did not only create a power supply for British Columbia. It also flooded the traditional lands of the Tsay Keh Dene peoples. Jean Isaac is a Tsay Keh Dene Elder. She grew up at Grassy Bluff beside the Ingenika River, near what is now the north end of Williston Lake. Jean knows first-hand the effect that a big hydro-electric project can have on ecosystems and on people.

Q. What was it like here before the valley was flooded?

A. It was beautiful, beautiful country. This was a deep valley and the Finlay river flowed along the bottom. The water was green—it was probably glacier water—and there were a lot of fish in it.

All along the valley, each family looked after its own territory. My family had a main cabin, and through the year we moved around and set up camps for hunting, trapping and gathering berries. All the families worked together and helped each other. We had a very good life.

Q. Did you know that the flood was going to happen?

A. We were told about the dam, but no one knew that it was going to be so extreme. It was really just a rumour. The water came up fast, about three or four feet the first night. But we still didn't know how high it would go. So we moved our homes to higher ground, and then we moved again and again. At that time the men were away working at the mills, so we women had to do the moving. I had four young children then. It was very scary.

The flooding continued for a year. We moved further and further. When the lake finally reached its level, we were cut off from our own homelands. We had no way of getting back. We had to work hard just to get food and clean water, and our children suffered. They didn't understand how we could let this happen.

Q. How did the lake change your way of life?

A. Before the flood, we knew how to live off the land. We were independent. After the flood we couldn't travel because the rivers were all blocked with debris. We had to find new places to trap and to gather berries. Our livelihood was just devastated and we had to start over.

The flood created a gap between the older people and the younger generation. I can't even show my grandchildren the places where my family lived along the river, because they're all under water. Being able to identify with a place gives you a feeling of belonging. Those places are gone now.

Q. What is it like living beside the lake now?

A. After the first flood, we thought that this would be a lake like any other. We thought it would have a shoreline and stay at the same level. That would be easier to live with. But the water goes up and down about 40 or 50 feet each year. In the winter people use more electricity, so the water level goes down. That leaves big mud flats of sand and silt. In the spring, strong winds create dust storms. Some days we can't even go outside. People have tried to plant grass on the mud flats to stabilize the soil, but then the water comes up again and the waves wash everything away.

We don't even have access to that power. When I turn on the light here, it doesn't come on because of the big hydro-electric development right beside us. Our community gets its electricity from a generator. But it's important not to live with bitterness in our hearts. If things are going to work out for us, that's the part we have to overcome.

Q. What can people learn from what happened here?

A. We all need to learn that nothing comes from nothing. In some way, everything you do has short-term and long-term impacts. It would be a good thing if young people could learn that.

Some things we know we'll have to live with. That lake will never go away. But the worst has passed. Now we're trying to gather up the pieces so we can use them as stepping stones to continue—to balance our own traditional ways and the non-Native ways. And I figure it can be done.

EXPLORING Further

Working in groups, imagine you are part of a community that had to move to make way for a big hydro-electric power plant. Imagine that the people who will use the electricity live far away, and do not know much about your community or your way of life.

Prepare a presentation that you would make to the users of the electricity. What would you want the users of the electricity to know? Are there things you would ask them to do that could reduce the impact on your community? Discuss your presentation with your class.

Fifty years ago this was a forested valley bottom. Williston Lake covers almost 2000 square kilometres of land, and is more than 150 metres deep.

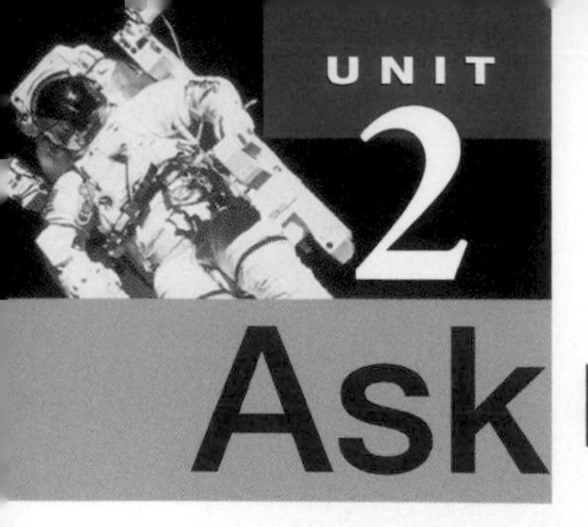

UNIT 2 Ask an Expert

Taras Atleo

Billions of dollars are spent every year in British Columbia on constructing homes, schools, offices, and factories. All those buildings need electric circuits. Without electricity they would be lifeless shells, with no heat, no light, no fresh air, no appliances, and no computers. Taras Atleo knows how to install electric circuits. He is a journeyman electrician, which means he has completed his apprenticeship as an electrician. He knows how to use the power of electricity.

Q. How did you become interested in working with electricity?

A. I think I first got interested as a kid, playing with video games. You know, lots of buttons and little red lights on the screen. But it was my trade school program that turned me on to the wonders of the electric circuit—making the connection between the buttons and the lights.

Q. Tell us about your job.

A. Now that I'm a journeyman (a ticketed electrician) my job requires more problem solving. I have to read and interpret blueprints. I have to be sure the latest legal codes are followed. I work with BC Hydro and the electrical inspector. Sometimes I supervise apprentices who are learning the trade. It's up to me to make sure that the job is done right.

Q. What steps are involved in wiring a house?

A. When the electrician arrives, the house is already framed in. The roof is up and the building is closed against the weather. The first thing we do is look at the blueprint (the house plans) to figure out the electrical needs of that house. How is the house going to be heated? What appliances and lighting are planned? How many electrical outlets are needed in each room? How much electricity do we have to supply—and where?

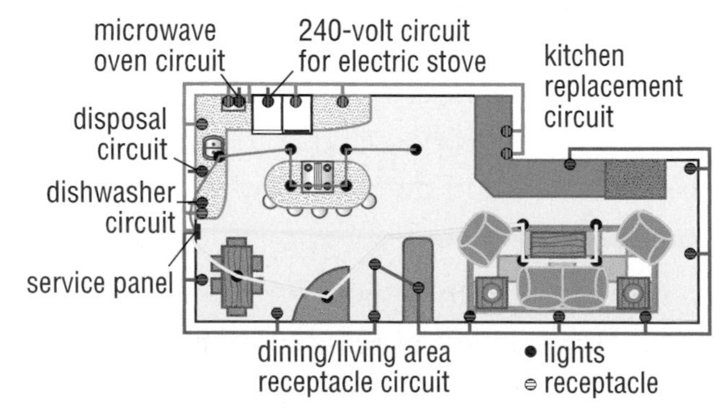

Wiring a house involves many decisions about how to connect the circuits.

Then we map out circuits to meet those demands. It's like a game. We play using the rules laid down by the electrical code. It's important that no circuit carries too much of a load. Major appliances like electric stoves or clothes dryers will need their own circuits. Smaller appliances can share a circuit.

Q. So each circuit must be designed to carry the right amount of electricity for its intended use?

A. That's right. Electrical outlets in the kitchen, for instance, will need to carry more electricity than outlets in the bedroom do. Each part of the circuit needs careful consideration: the heavy wires that carry electricity into the house, the main electricity cut-off switch for the whole house, the service panel, the circuit breakers and wiring, the switches and outlets.

Then we have to install the circuit. The boxes that will carry switches, outlets and lighting fixtures are nailed into place. Wiring for each circuit is run from the service panel to the boxes. Then the electrical inspector is called in to make sure everything has been done according to the code. Electricity won't be sent to the building until the inspector gives permission.

Q. Safety first, right?

A. Always you need to be very careful, very respectful, of the power of electricity. If you make a mistake, electricity can hurt you—or hurt somebody else. You need to think of the consequences. What will happen if I do this? Or that? Such knowledge comes from listening to others. Or from personal experience that you accumulate, day by day, on the job.

Q. What do you like best about your work?

A. I enjoy the problem solving, especially on a difficult job. Sometimes a job seems just about impossible. It's always satisfying to prove that the work can be done, and that it can be done to the proper standard. I'm still thrilled when the switch is pushed and the lights come on.

Q. What's in the future for you?

A. I want to give something back to my First Nations community, the Ahousaht people, at Marktosis on the west coast of Vancouver Island. I can use my skills to help them. I can train apprentices. I'm looking forward to installing the electrical wiring on a house that my grandmother will be moving into. I know that my great grandmother Nan would be proud of my accomplishments as a journeyman electrician.

EXPLORING Further

Often an electrician is asked to solve a problem: something isn't working properly. An electrician is skilled at identifying and eliminating possible causes.

You know that a light bulb needs a complete electric circuit to work properly. Suppose a friend calls you and complains that the light in her bedroom is not working. What are the possible causes? How might you go about testing and eliminating the possibilities? It's a good idea to begin with the causes that are the most likely and that are the easiest to test. (Remember, safety first.)

SKILLCHECK

- Interpreting Data
- Predicting
- Modelling
- Communicating

UNIT 2

Project

Building Communities

When a town is being built, a town council is formed to make decisions. One decision the council must make is how to provide the town with the most effective source of electricity. Experts survey the environment to determine which source of energy would give them the most electricity for the least cost. Models are built before the final decision is made so that the town council can make the best choice for the community.

Challenge

In a group, design a town and decide on a source of energy for this town.

Design Criteria

A. Imagine you are on the council for a new town.

B. Assume that three energy sources are available to your town: solar power, wind power, and hydro-electric power (hydro power).

C. Use information in your textbook as well as other sources (Internet, library books) to produce a list of advantages and disadvantages of each energy source.

D. Design your town with a population between 1500 and 5000 people.

E. Decide how many houses will be built in your town.Assume that four people live in each house.

F. Assume that you cannot store energy (that is, if you have extra energy in February, you cannot use it in May).

G. Assume that the energy usage by buildings (Table 1) does not change each month.

Materials

drawing paper

pens and pencils

Bristol board

Apparatus

1 1.5 V D-cell batteries

3 1.5 V 10 watt light bulbs

10 copper wires with alligator clips

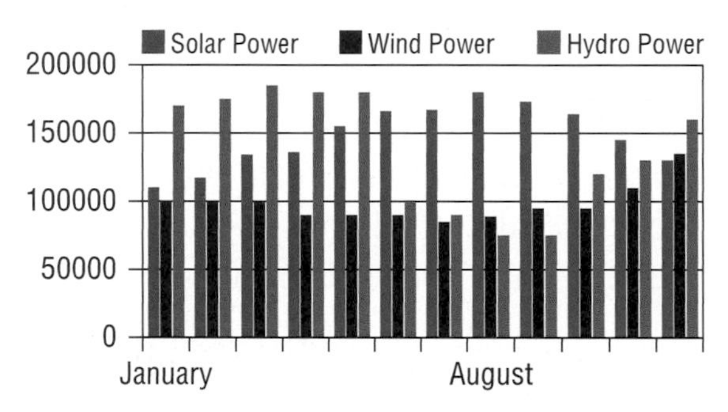

Figure 1 Energy sources available (measured in kilowatt-hours/month).

Table 1 Energy Usage for Buildings

Building	Energy Usage
House	360 kWh/month
School, Store or Office Building	800 kWh/month
Hospital or Airport	850 kWh/month
Recreational Facility	875 kWh/month
Factory or Industry	975 kWh/month

Plan and Construct

1. In your group, create a table to show the advantages and disadvantages of the three energy sources available to your town. Your table should also identify which of the sources is most reliable throughout the year (based on Figure 1).

2. Use the information in step 1 to decide as a group which power source you would prefer to use. Explain why.

3. (a) Decide on the population of your town.
   (b) Estimate the number of buildings (houses, stores, hospitals, facilities, etc.) that will be constructed in your town. Select the buildings you will need from those listed in Table 1.

4. Create a map of your town plan. Use pencil in case you have to make changes to your plan. (Don't try to include every house on your map. You can draw squares to representing groups of houses. Indicate the number of houses each square represents.) Include a power station.

5. Calculate the total amount of electricity required by your town in one month. Do this by adding together the energy usage of all the buildings in your town. Show your work. Table 1 provides the energy usage for different types of buildings in kilowatt-hours/month.

6. Predict whether your chosen energy source will be able to supply electricity for your town throughout the year. Which months, if any, do you think could be a problem?

7. Determine whether your chosen energy source will be able to supply electricity for your town throughout the entire year.
   (a) From Figure 1, use the bars of the graph to estimate how much of your chosen energy source (in kilowatt-hours/month) is available each month of the year.
   (b) Compare these numbers to the total amount of electricity required by your town in one month (step 5).
   (c) Are there any months in which the energy source will not provide enough electricity for your town?

8. If the energy source will not provide enough electricity, you will have to either:
   (a) choose another energy source (repeat step 7 to ensure that enough electricity is available), or
   (b) remove some of the buildings from your town plan until enough electricity is available.

9. Draw a final map of your town on Bristol board. Show each building wired into the power plant. Each building should be labeled with its name and amount of energy usage.

10. Present your model.

Evaluate

1. How would you change the design of your town if you repeated this project?
2. What was the most important information your group used for choosing the source of energy?
3. Were you able to use the source of energy you chose, or did the amount of electricity needed by your town change the decision?
4. Compare your results to those of other groups. How did the size of the towns affect the choices of energy source?

Extend Your Skills

5. Which energy source would you choose if you needed a second one?
6. Imagine you are on the council for a city with 10 000 homes:
   (a) What other ways could you provide electricity to your city?
   (b) Which of these choices is least harmful to the environment?

UNIT 3

Exploring Extreme Environments

High above the coast of British Columbia, a satellite records images of the planet below. The pictures will help researchers predict weather patterns and understand changes in the environment. Meanwhile, far below in the deep ocean, a scientist peers through a tiny window in her underwater vehicle. Nearly 3 km below the ocean surface, she sees a landscape that human eyes have never seen before. By using special tools and vehicles, explorers today are making amazing discoveries about our world.

Humans have always searched for ways to explore new environments. People have crossed the widest oceans and have climbed into the mouths of volcanoes. They have crawled into the deepest caves and have travelled over frozen lands. What problems did they face in each of these environments? How did they overcome these problems? What traditional tools and knowledge did they use?

The knowledge and tools that have helped people explore challenging environments on Earth have also helped us explore the challenging environment of space. In this unit, you will learn how Canadian technologies help people explore the universe beyond our home planet. Why are such explorations important?

Unit Contents

Chapter 7
A Challenge to Explorers **186**

Chapter 8
Technologies for Unknown Worlds **226**

Getting Ready...

- What are the obstacles that challenge explorers in extreme environments?
- How does technology help us learn about unknown environments?
- How have Canadians contributed to exploration technologies?

CHAPTER 7

A Challenge to

Getting Ready...

- What is an extreme environment?
- What would you need to survive in an extreme environment?
- What are some challenges to exploring space?

The first humans stepped onto the Moon in July 1969. What knowledge and tools made their journey possible?

A hundred years ago, most people did not think that humans could leave our planet and travel to the Moon. At that time, nobody had the knowledge or the tools needed to build and fly a spacecraft. People wondered whether space travel was possible. What materials could they use to build a spacecraft? What fuel could a spacecraft use? How could people survive outside Earth's atmosphere? How could space travellers communicate with people on Earth? How could the spacecraft return safely to Earth? One by one, these problems were solved by new scientific discoveries and new tools and procedures.

Travellers in space are just like explorers of long ago. They travel to places no one has been before. In this chapter, you will learn how people survive in space. You will find out how technology is used to solve problems for explorers.

Explorers

What You Will Learn

In this chapter, you will learn

- what kinds of conditions are in extreme environments
- how technology helps people to travel to and explore unknown environments
- what challenges are found in the extreme environment of space
- what knowledge might be gained from distant explorations

Why It Is Important

- Technologies developed for extreme conditions can help people live more comfortably in many different environments.
- Resources and knowledge can be developed from distant explorations.

Skills You Will Use

In this chapter, you will

- construct models of extreme environments and exploration technologies
- research and communicate Aboriginal technologies
- interpret data about planets
- control variables to investigate the force of gravity
- model movement in space

Why might you call spacecraft "the sailing ships of the twenty-first century?"

Starting Point 

Exploring Your Home Province

Exploring the province of British Columbia is not like exploring the Moon. Or is it? You may be surprised by how much technology you use every day.

What to Do (Group Work)

1. In groups, make a list of the items you would need in each of the following situations. Hint: Think of what you need for safety, travel, warmth, and food.
   (a) It is a cold, wet day. You are visiting a friend, but you must return to your home at the other side of town.
   (b) It is a hot summer night. You must get from Vancouver Island to the mainland before morning.
   (c) You are driving in the mountains when a snowslide blocks the road. You must wait 48 h before the snow is cleared.

What Did You Find Out?

1. What were your three most important items for each situation? Explain why you chose each item.
2. Would you be able to survive in each situation without these items? Why or why not?

Extension

3. Imagine you are exploring an unknown environment such as a desert or a large cave. Draw and label what you will take with you on your exploration. How did the environment you chose affect your decisions?

Section 7.1 What Are Extreme Environments?

Key Terms

- environment
- extreme
- technology
- exploration
- astronaut
- life-support
- recycling

Imagine yourself in a place that is as dry as the driest desert. It is bitterly cold and dark. The average temperature is close to −57°C and the night can last for several months. To get to this place, you must travel across an ocean where fierce winds can blow with speeds of 300 km/h.

There is actually such a place on Earth. It is the continent of Antarctica, shown in Figure 7.1. Antarctica is the coldest place on our planet. It is almost entirely covered by ice. The environment there is so extreme that Antarctica is the only continent where humans have never settled to live.

Figure 7.1 People first began to explore Antarctica on foot in the early 1900s. Why do you think people have never settled there?

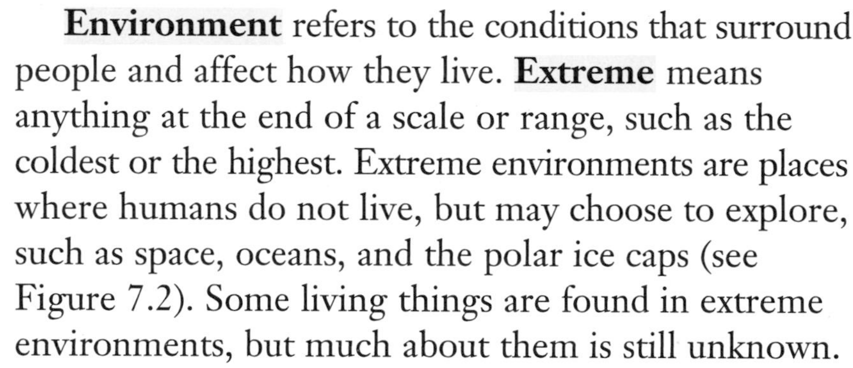

Environment refers to the conditions that surround people and affect how they live. **Extreme** means anything at the end of a scale or range, such as the coldest or the highest. Extreme environments are places where humans do not live, but may choose to explore, such as space, oceans, and the polar ice caps (see Figure 7.2). Some living things are found in extreme environments, but much about them is still unknown.

Special equipment and technology are often needed to explore extreme environments. **Technology** refers to the practical knowledge and tools that people use to make life easier.

READING Check

Why is Antarctica called an extreme environment?

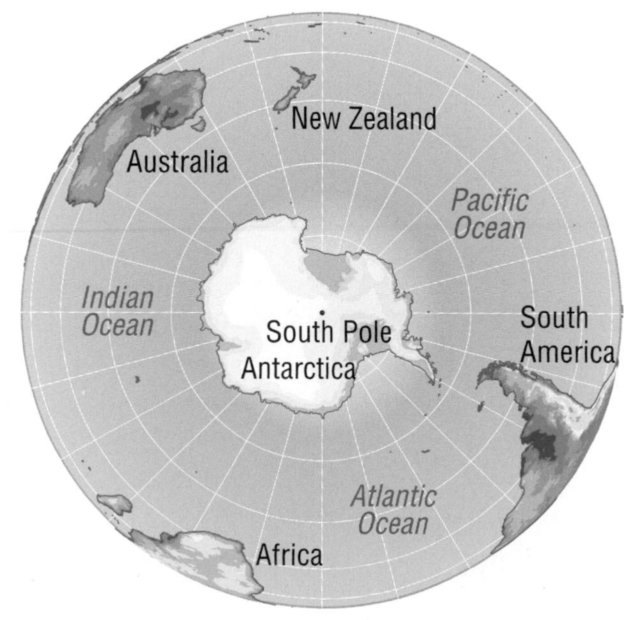

Figure 7.2 Antarctica is located around the South Pole. The Arctic is located around the North Pole.

Find Out **ACTIVITY 7–B**

Model an Extreme Environment

What is an extreme environment? In this activity, you will list as many extreme conditions as you can think of. Where can you find these conditions—either on Earth or in space?

What to Do

Part 1

1. Draw a 2-column chart like the one below. Label the left column "Extreme Conditions." Label the right column "Extreme Environments."

2. In the left column, list at least six extreme conditions. Consider conditions such as rainfall, temperature, and light. The first two are done for you.

3. In the right column, record the environments in which each of these conditions might be found.

Extreme Conditions	Extreme Environments
cold	Arctic, Antarctic, deep ocean, space
no air	deep ocean, space

Part 2

4. Choose one extreme environment that interests you. Make a concept map showing all the details you know about your extreme environment. Include the conditions you identified in Part 1.

5. Add to your map all the questions you have about your extreme environment. For example, how might people travel in a frozen environment? How can people breathe in an environment that has no air?

6. **Research** your extreme environment at the library or on the Internet. Answer the following questions:
   (a) What do people need to survive in this environment?
   (b) How do people explore this environment?

Part 3

8. **Design** and **create** a 3-D model that shows the conditions in your extreme environment. Use a shoebox and art supplies to construct your model. Add labels to identify different parts of your environment. Display your model in the classroom.

What Did You Find Out?

1. Compare the environment you studied with the environment in which you live.
   (a) How are the environments similar?
   (b) How are the environments different?

2. Compare the environment that you studied with the environments studied by other students in your class.
   (a) How are the environments similar?
   (b) How are the environments different?

3. **(a)** Define an extreme environment in your own words.
   (b) Does your definition define all extreme environments? If not, how can you improve your definition?

Skill POWER

For tips on drawing a concept map, refer to SkillPower 1.

Some parts of the Atacama desert in Chile, South America, have had no rainfall for 400 years.

READING Check

What do all extreme environments have in common?

A Guide to Extreme Environments

Extremes on Land

Caves

- little or no sunlight
- some caves are hot, some are cold
- some are flooded, some are dry

Explorers need:

- source of light
- protective clothing

Figure 7.3 The largest cave system in the world is the Mammoth Caves in Kentucky. Before Europeans arrived in North America, Aboriginal peoples had explored parts of these caves. Today, more than 580 km of underground passageways have been explored and mapped in the Mammoth Cave system.

Volcanoes

- some volcanoes produce flows of red-hot melted rock
- some volcanoes send out dust, ash, gases, and rocks

Explorers need:

- protective clothing
- remote sensors to record gases and temperatures

Figure 7.4 A special suit and helmet protect this scientist collecting lava samples.

Deserts

- have less than 25 cm of precipitation per year
- can be very hot in the daytime and very cold at night
- some deserts can be covered in snow and ice year round

Explorers need:

- supply of water
- protection from heat, cold, and drying

Figure 7.5 The only hot desert in Canada is in the southern Okanagan Valley in British Columbia.

Extremes in Ocean

Ocean Trenches

(very deep cracks in sea floor)

- no sunlight
- high water pressure
- cold temperature

Explorers need:

- supply of air
- submarine vehicle
- source of light

Figure 7.6 Very little is known about creatures that live in the deepest parts of the ocean.

Sea Floor Vents

(deep ocean hot springs)

- no sunlight
- high water pressure
- hot temperature

Explorers need:

- supply of air
- submarine vehicle
- source of light
- protection from heat

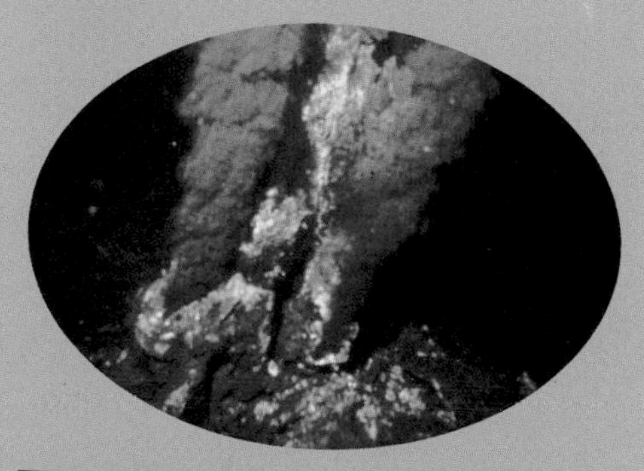

Figure 7.7 Some sea floor vents can be hotter than 300°C. The hot water contains dissolved minerals from rocks below the surface.

INTERNET CONNECT

www.mcgrawhill.ca/links/BCscience6

Would you like to go on an expedition to the sea floor? New species of plants and animals are found on almost every trip. Go to the web site above and click on **Web Links** to find out where to go next.

Extremes in Space

Distance

- distances between objects are so huge they are difficult to imagine

Temperature

- average temperatures on nearby planets can vary between −230°C to 480°C

Explorers need:

- fast vehicle with large supplies of food, water, and oxygen
- protection from hot and cold temperatures

Figure 7.8 Space extends in all directions.

What are three things people need to survive?

What Do Explorers Need to Survive?

Exploration means travelling to discover what a place is like. For example, an **astronaut** is an explorer who travels in space. As well as needing the right knowledge and tools, explorers also need oxygen, water, and food—just as you do.

Oxygen

When you breathe in, you take oxygen and other gases into your body. There is little or no oxygen available to breathe in space or underwater. Explorers in those environments must bring oxygen or air with them (see Figure 7.9).

Figure 7.9B What advantages does this modern diver have compared with the underwater explorer in 7.9A?

Figure 7.9A Before modern diving equipment was developed, underwater explorers were connected by a long tube to air at the surface.

Water

Water is bulky and heavy to carry. However, explorers in deserts can collect water vapour from the air and condense it to make liquid water.

There is plenty of water in frozen regions, but it is in the solid form of snow and ice. Explorers must bring fuel to produce heat to melt snow or ice and make drinking water.

Submarines are surrounded by water, but the salt and other impurities have to be removed before the water can be used for drinking.

Food

In the weightless conditions inside a spacecraft, crumbs of food may float away and damage sensitive instruments. To prevent this from happening, astronauts use “spoon-bowl” packages. The moisture in the food makes it cling to the spoon, so eating in space is more like eating on Earth.

Long ago, the Dunne-za people of northern British Columbia learned to make food for long trips (see Figure 7.10). They cut long, paper-thin strips of moose meat and fish and dried the strips beside a fire. The dried meat was nutritious, lightweight, and not easily spoiled. Why is each of these three factors important for survival?

Figure 7.10 Drying fish

Pause& Reflect

A healthy explorer needs exercise as well as good food. Why might it be hard to exercise in space? Record your ideas in your science notebook. Design exercise equipment that could be used in a spacecraft.

INTERNET CONNECT

www.mcgrawhill.ca/links/BCscience6

Watch an astronaut drink tea with chopsticks. Discover how honey and peanut butter behave in space. Go to the web site above and click on **Web Links** to find out where to go next.

Eating Like an Astronaut

What foods would you pack for an astronaut's lunch? Look at the food you have at home and plan a meal that an astronaut could make in space using only water and heat.

Safety Precautions

- Ask an adult in your home for permission and help with this activity.

What You Need

paper
pencil
packaged food labels
coloured crayons

What to Do

1. Talk about this assignment with an adult in your home.
2. Look at the foods in your home. Which ones could be prepared in space? Read the labels on the packages to help you decide.
3. Create a menu for a meal an astronaut could make and eat in space.
4. Draw a picture of the meal.
5. Present your picture and menu to the class. Explain how the meal could be prepared in space.

What Did You Find Out?

1. What was the easiest part of planning the astronaut's meal?
2. What was the hardest part?
3. What problems do you think astronauts have when they prepare food in space?

Life-Support Systems

Explorers in space must travel over long distances. How could they get enough oxygen, water, and food on missions that last for several years?

To answer this question, space researchers are developing life-support systems. **Life-support** systems are methods of providing air, water, and food using techniques such as recycling. **Recycling** is using the same material over and over again. For example, water is recycled naturally on Earth by the processes shown in Figure 7.11. How could recycling be used on a spacecraft?

Write a sentence that correctly uses the terms life-support and recycling.

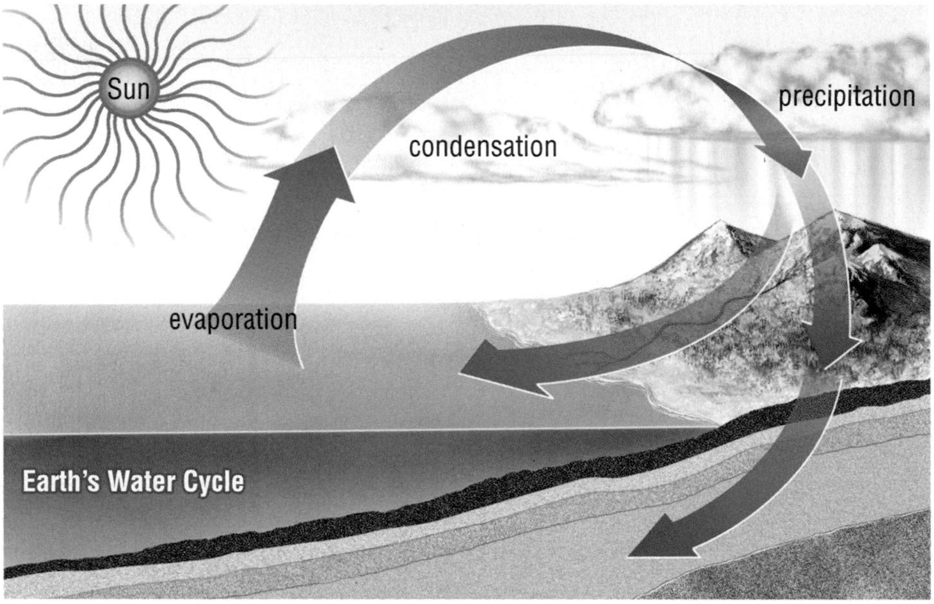

Earth's Water Cycle
Sun
condensation
precipitation
evaporation

Figure 7.11 In the water cycle, the limited amount of water on Earth is used over and over again.

astronaut
food
water
oxygen
grey water
(from washing)
solid and
liquid waste
nutrients + carbon dioxide
waste material
+ oxygen
plants
microorganisms

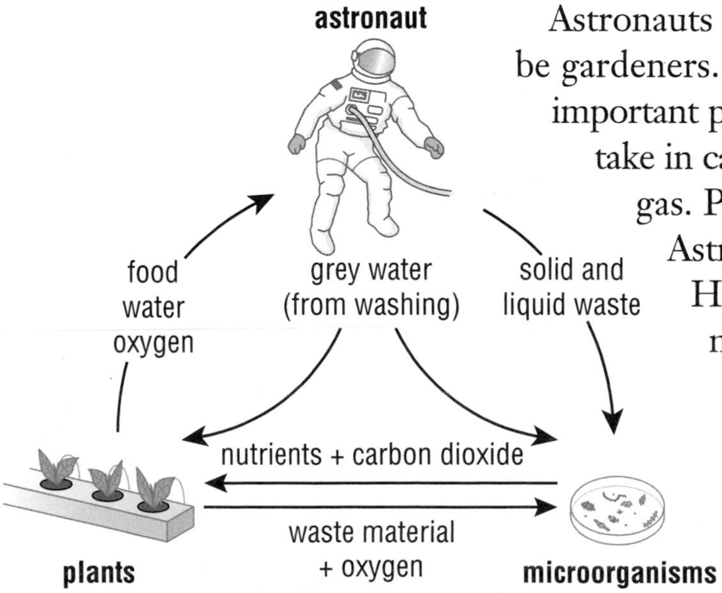

Figure 7.12 What are the three main elements of this life-support system designed for a spacecraft? How is this system like natural systems on Earth?

Astronauts on long trips in space may have to be gardeners. How would growing plants be an important part of life-support technology? Plants take in carbon dioxide gas and produce oxygen gas. Plants can also purify waste water. Astronauts can eat parts of plants for food. Human wastes can be broken down by micro-organisms and used to provide nutrients for more plant growth. Together, astronauts, plants, and micro-organisms make a life-support system that recycles air, water, and food as shown in Figure 7.12.

Transportation

In Starting Point Activity 7-A, you chose transportation for travelling in British Columbia. What transportation would you choose for travelling across the snow? Long ago, the Algonquin and Cree Aboriginal peoples developed snowshoes and toboggans to help them move across deep snow. In the following photographs and activity, you can identify some other examples of technology for travel in extreme environments.

In the past, some Inuit sled runners were made from frozen fish wrapped in caribou hides.

Figure 7.13A What are the advantages and disadvantages of this Inuit sled compared with a snowmobile?

Figure 7.13B A snowmobile

Figure 7.14A What are the advantages and disadvantages of a canoe for ocean exploration compared with a kayak used by the Inuit?

Figure 7.14B A kayak

Figure 7.15A What are the advantages and disadvantages of a helicopter compared with a light aircraft?

Figure 7.15B A light aircraft

At Home **ACTIVITY 7-D**

Technology for Travel

Throughout human history, different cultures have developed vehicles to move through different environments. How does the design of each vehicle help it move through an environment?

What to Do

1. Your teacher will assign you one of the following environments: land, ice and snow, water, air, and space.
2. Work with someone at home to list as many vehicles as you can think of that can travel in your environment.
3. Choose one of the vehicles. Draw and label a picture to show how its design helps it move through the environment.

What Did You Find Out?

1. Prepare an oral presentation using your picture to explain how the vehicle travels in your environment.

Clothing

What would you wear to explore Canada's Arctic region? You would need an insulating jacket, mittens or gloves, footwear, and headwear, so that very little skin is exposed to the freezing air. You would also need sunscreen and sunglasses to protect your skin and eyes (see Figure 7.16).

When you are completely bundled up from head to foot against the cold, you may feel like the astronaut working outside the spacecraft in Figure 7.17. The spacesuit that astronauts use for such tasks is like a mini spacecraft. Not only does it protect the astronaut from the environment of space, it also includes a life-support system. Protective clothing must be flexible and comfortable enough for the explorer to move and work in it.

Figure 7.16A Traditional Inuit snow goggles are made of leather, bone, or ivory.

Figure 7.16B How are modern snow goggles similar to traditional Inuit goggles?

Figure 7.17 An astronaut outside the spacecraft must wear a life-support system.

DidYouKnow?

An astronaut in space has more oxygen to breathe than on Earth. The gas in a spacesuit is 100 percent oxygen. On Earth the air is only 21 percent oxygen.

Shelter

An easier way for an astronaut to work in space is inside the shelter of a space station. The station creates a complete barrier against the environment. Within the station, conditions are like those on Earth, so the astronaut can work in regular clothes.

READING

Why is it important to have the right clothing and shelter in an extreme environment?

Figure 7.18 shows the inside of an igloo, a traditional home of the Inuit in Canada's Arctic region. An igloo can be built in just a few hours from specially packed snow and provides protection from wind chill, blizzards, and drifting ice particles.

What types of clothing and shelter would you need if you were exploring a desert? A cave?

Figure 7.18 When European explorers reached the Arctic, the knowledge and tools of the first inhabitants of the far North helped them to survive. What might have been some of the technology that the Inuit shared with early European explorers?

Find Out **ACTIVITY 7–E**

Aboriginal Technology for Explorers

How does Aboriginal technology help explorers in extreme environments? In this activity, you will research a Canadian Aboriginal technology of your choice and then make a drawing or model of it.

What You Need

library books, magazines, Internet
museum
art supplies to build a model
index cards
pencil
felt pens
paper

What to Do

1. Choose a Canadian Aboriginal technology used for food, clothing, or shelter. Discuss your choice with your teacher.
2. Use a variety of resources to **research** your technology. Answer the following questions.
   (a) How was this technology traditionally used?
   (b) What local materials are used to make it?
   (c) How was this technology developed or improved over time?
   (d) What is the key advantage of this technology? (For example, is it lightweight? Waterproof? Flexible?)
   (e) How might the technology help explorers in an extreme environment?
3. Find or make a picture or model of your technology.
4. On an index card, write about your technology in point form. The information should answer the questions in step 2.
5. Share your model and information with the class.

Section 7.1 Summary

In this section, you learned that:

- extreme environments are places where humans do not naturally live but may choose to explore
- examples of extreme environments are space, polar regions, oceans, deserts, caves, and volcanoes
- life-support systems provide explorers with oxygen, food, and water
- technologies such as vehicles, tools, and clothing help humans to explore extreme environments
- Aboriginal technologies have helped people explore extreme environments

DidYouKnow?

In December 2004, Matty McNair, a resident of Nunavut, became the first female resident of Canada to reach the South Pole on foot without outside help along the way. McNair and her team members dragged sleds with their supplies to the Pole, then kite-skied back across Antarctica.

Check Your Understanding

1. What are six examples of extreme environments?
2. What characteristic is common to all extreme environments?
3. Why is special equipment and technology needed to explore extreme environments?
4. What three things must an effective life support system provide?
5. Draw or describe two examples of Aboriginal technology that could be used for exploring an extreme environment.
6. **Apply** You have found a new life form in an ice cave in Antarctica and you need to take it back to British Columbia to study it further. What conditions do you need to provide to make sure that the life form can survive?
7. **Thinking Critically** Do most extreme environments have more than one extreme condition? Explain your answer.

Pause& Reflect

What resources and knowledge might be obtained from exploring an extreme environment? Choose an environment and record your ideas in your science notebook.

Section 7.2 Observing the Sky

For most of human history, the only way that people could learn about space was by looking at the Moon, the planets, and the stars. Inventions such as the telescope (see Figure 7.19) make it possible to study objects in space in much more detail. A **telescope** is an instrument that uses lenses and mirrors to make distant objects appear larger and closer.

Figure 7.19 How do telescopes help us explore space?

Key Terms
- telescope
- atmosphere
- orbit
- gravity
- planet
- solar system
- Sun

Find Out ACTIVITY 7–F

A Clearer View

How does technology help us learn more about the planets?

What You Need

shoebox
newspaper
clear tape
sheet of thin, transparent blue plastic

What to Do

1. Wrap a sheet of newspaper tightly around the shoebox and tape it in place. The covered shoebox represents a mystery planet.
2. Place the planet at the far end of the classroom.
3. View the planet through the thin sheet of blue plastic. Draw what you see in as much detail as possible.
4. View the planet without the plastic. Draw what you see.
5. Walk halfway to the planet. Draw the planet again in as much detail as possible.

What Did You Find Out?

1. Which of your drawings represents the planet as you look:
   (a) through a telescope from the surface of Earth?
   (b) through a telescope on a spacecraft flying high above Earth?
   (c) from a spacecraft flying close to the planet?
2. Why do you think each of these views of the planet is different?
3. Suggest several ways you could learn more about the planet.
4. Why might it be important to learn about planets?

The Atmosphere

One of the challenges to learning more about space from Earth is the **atmosphere**, the gases that surround our planet. Earth's atmosphere is made up mainly of nitrogen (78 percent), oxygen (21 percent), and small amounts of other gases. The atmosphere also contains water vapour, dust, and pollutants that blur the view of objects in space when seen from Earth, as shown in Figure 7.20.

Earth's atmosphere contains several layers, as shown in Figure 7.21. The lowest layer is where the wind, snow, rain, thunderstorms, and other weather systems occur. This is also the layer where aircraft fly.

In the highest level of the atmosphere, particles of gases are thinly scattered in space. Spacecraft orbit Earth in this layer at altitudes of 250 km or more. An **orbit** is the path of an object in space around a larger object.

Figure 7.20 (A) Saturn, as viewed through a telescope on Earth. (B) Saturn, as viewed from space. Why is image B clearer than image A?

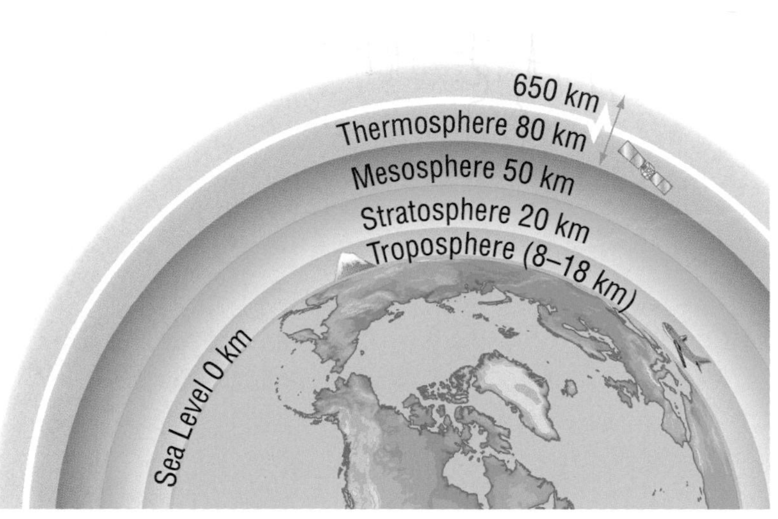

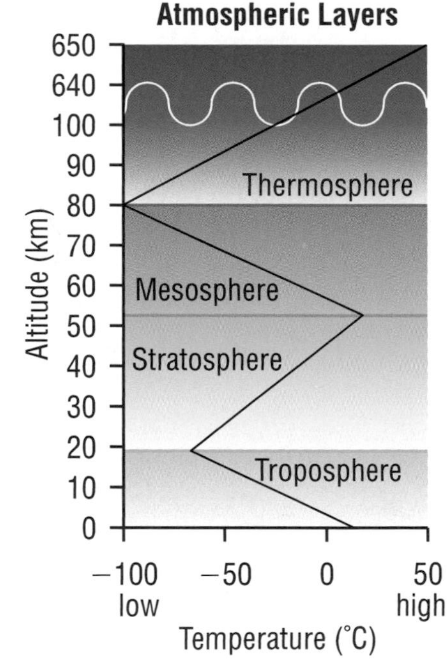

Figure 7.21 How does temperature change as you go higher in the atmosphere?

The Moon has no atmosphere and therefore no weather such as rain and wind. Because of this, footprints left by the first astronauts on the Moon over 35 years ago are still there.

What Is Gravity?

When you throw a ball into the air, why does the ball fall down? The answer is a *force*, which is a push or a pull. The ball is pulled back to Earth by the force of gravity. **Gravity** is a force that pulls together any two objects that have *mass* (the amount of matter in an object). A ball and Earth both have mass, so Earth pulls the ball and the ball pulls Earth, as shown in Figure 7.22.

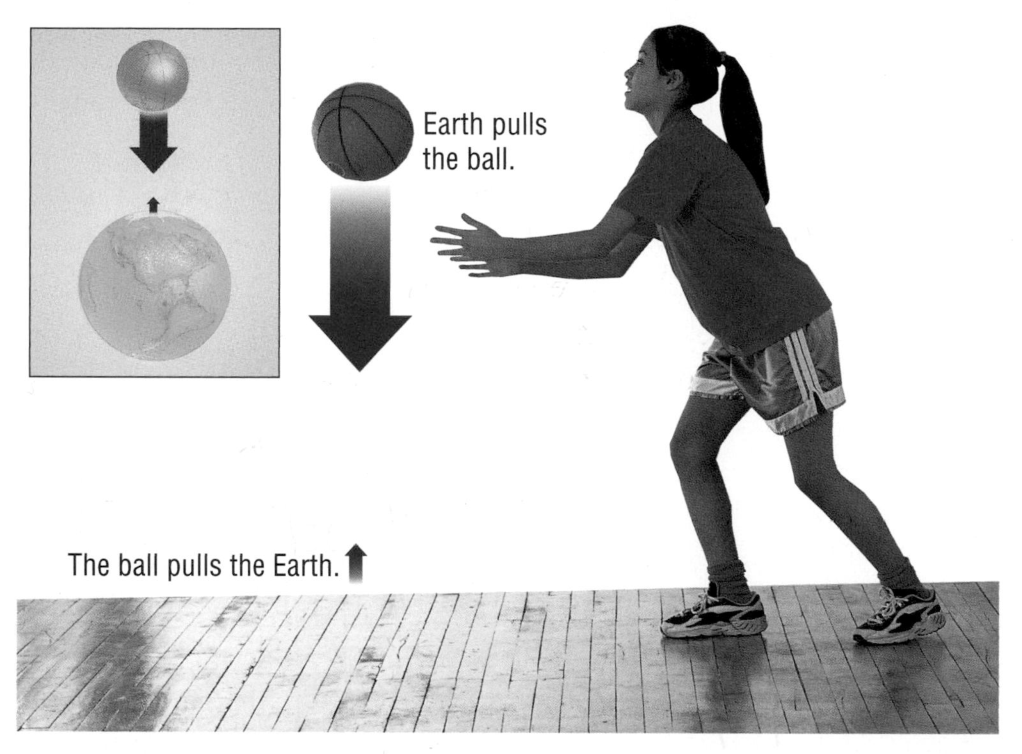

Figure 7.22 Which do you think has the greatest gravitational pull, Earth or the ball?

The strength of gravity depends on the mass of the objects and the distance between them. Because Earth has a huge mass compared with the mass of a ball, we notice Earth's gravity pulling the ball toward it. We do not notice the tiny effect of the ball's gravity pulling on Earth.

How is gravity related to mass and distance?

The strength of gravity also depends on how far apart two objects are. For example, the force of gravity between Earth and a spacecraft in orbit around Earth is greater than the force of gravity between Earth and the same spacecraft when it is much farther away from Earth.

CONDUCT AN

INVESTIGATION 7-G

SKILLCHECK
- Predicting
- Observing
- Communicating
- Inferring

Gravity and Orbits

In this investigation, you will demonstrate how the pull of Earth's gravity keeps a spacecraft in orbit around the planet.

Question

What keeps a spacecraft in orbit?

Safety Precautions

- Make sure the knots are tied firmly. The weights could be dangerous if they fly off.

Apparatus

4 small weights (for example, large washers or metal nuts)

Materials

20 cm length of thick straw or plastic tube

80 cm length of string

Procedure

1. Thread the string though the plastic tube or straw.

2. Tie one washer or nut securely onto each end of the string.

3. Put on your safety goggles. Hold the plastic tube upright in one hand. The upper washer represents a spacecraft. The lower washer represents the force of Earth's gravity.

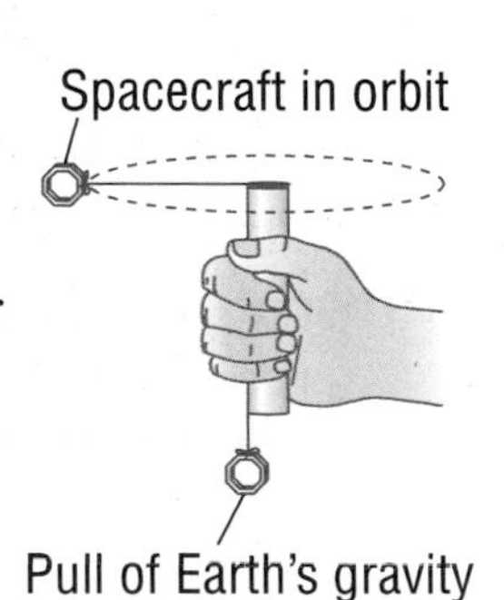

4. Keep the tube upright, and spin the spacecraft as shown in the figure. **Observe** what happens to the lower washer when you change the speed at which you spin it. Can you spin the spacecraft so the lower washer does not move up or down?

5. **Predict** what will happen if you increase the mass of the spacecraft. **Record** your predictions.

6. Attach a second washer or nut to the top string. Spin the spacecraft. **Record** your observations.

Analyze

1. **Summarize** your observations using four or five complete sentences.

2. Was your prediction correct? Why?

Conclude and Apply

3. What must you do to maintain the size of the circle of the spacecraft's orbit?

4. When you spin the spacecraft in a larger circle, must the spacecraft move faster or slower to keep the lower weight at the same level?

5. Does a spacecraft close to Earth have to travel faster or slower than one farther away from Earth to remain in orbit? Why?

Extend Your Skills

6. **Predict** what will happen if you increase the strength of the gravitational pull. Turn the straw around so the greater mass is at the bottom. Spin the spacecraft. **Record** your observations

The Solar System

Long before humans built spacecraft, there was already an object in orbit around Earth. You can easily observe this object moving slowly across the sky almost every night. It is the Moon. The gravitational pull of Earth keeps the Moon in orbit around Earth.

For many centuries people have observed objects other than the Moon moving in the night sky. They noticed that some of these bright, twinkling objects changed position from night to night against the background of stars. These moving objects are the planets. A **planet** is a large body that orbits a star. Most of the planets have several moons in orbit around them. You will learn more about the planets in the next investigation.

The strong gravitational pull of the Sun keeps Earth and the other planets in orbit around the Sun. Together, the planets, their moons, asteroids (chunks of rock), dust, gases, and other objects that orbit the Sun form the **solar system**, shown in Figure 7.23.

Write a sentence explaining the meaning of the word "orbit."

DidYouKnow?

The Sun's diameter is almost 110 times the diameter of Earth. The Sun's mass is about 330 000 times the mass of Earth.

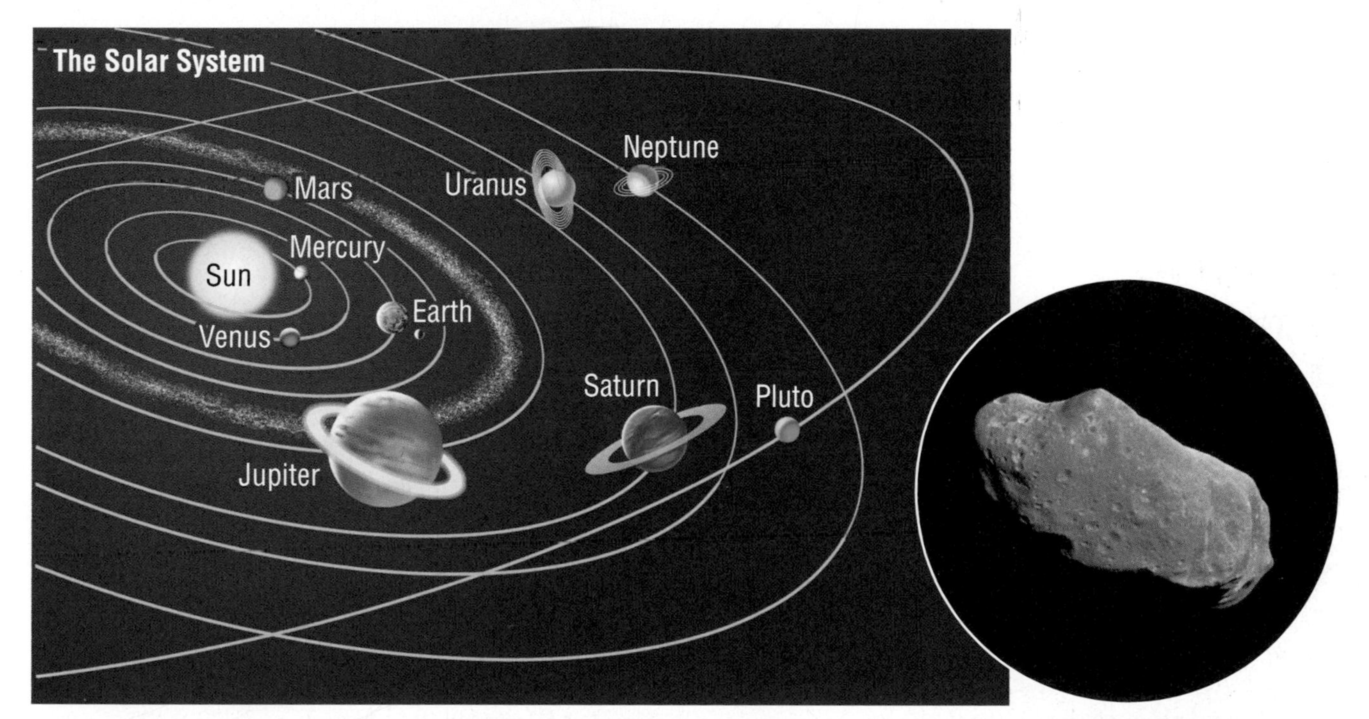

Figure 7.23A How many planets are in the solar system?

Figure 7.23B Asteroids sometimes have moons in orbit around them

CONDUCT AN

INVESTIGATION 7-H

SKILLCHECK
- Classifying
- Inferring
- Interpreting Data
- Modelling

Profile the Planets

We measure time on Earth by the length of Earth's orbit around the Sun. It takes exactly one year for our planet to move around the Sun and come back to the same point to begin the next orbit. How do you think the measurement of time on other planets is different from our own? In this investigation, you will compare the planets in the solar system and evaluate obstacles to exploring them.

Question

Which planets are most similar to Earth?

Apparatus
2 small marbles
1 Ping-Pong™ ball
2 tennis balls
2 baseballs
1 soccer ball
1 basketball
ruler or tape measure

Materials
sheet of paper
masking tape

Procedure

Group Work

1. Examine the sizes (diameters) of the different planets in the table. List the planets in your notebook in order of size from the smallest to the largest.

2. Your teacher will assign your group one planet to study. Compare the sizes of the planets. Is your group's planet the largest, smallest, or in between? Is it similar in size to any other planet?

3. Compare your list of planets with the list of balls above. Select the ball that best represents your planet.

4. **Calculate** the distance between the ball representing your planet and the Sun in your solar system model.

   Use the following ratio or calculate your own ratio based on your study area size: 1000 million km in solar system = 1 m in study area
   (A planet 8575 million km from the Sun would be 8.575 m from the Sun in your model.)

   Measure this distance and mark it with a stick or tape. Move your planet to the mark.

5. Write the name of your planet on a sheet of paper. Place the sheet beside your planet.

Planet Name	Diameter (km)	Maximum Distance from Sun (millions of km)	Average Surface Temperature (°C)	Length of Year (in Earth units)
Mercury	4880	70	–170 to 350	88 days
Venus	12 100	109	480	225 days
Earth	12 756	152	22	365 days
Mars	6787	249	−23	687 days
Jupiter	142 800	816	−150	12 years
Saturn	120 000	1507	−180	30 years
Uranus	51 800	3004	−210	84 years
Neptune	49 500	4537	−220	165 years
Pluto	3000	7375	−230	248 years

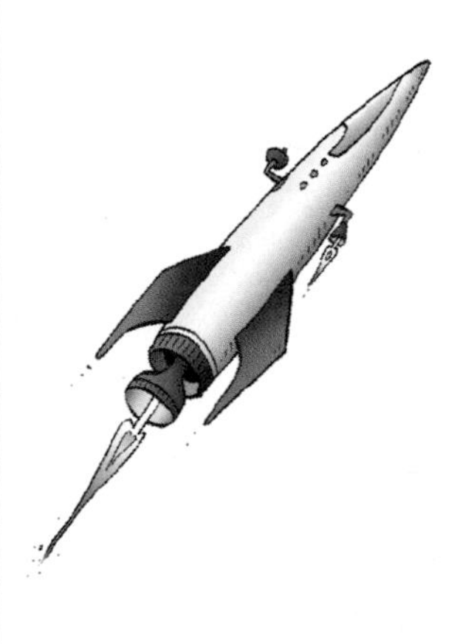

Analyze

1. Which planet is the:
- **(a)** largest?
- **(b)** coldest?
- **(c)** one with the shortest orbit time?
- **(d)** one that is closest to the Sun?
- **(e)** one that is closest to Earth?
- **(f)** one that is most similar to Earth?

Conclude and Apply

2. Why is it important to know the orbits of the planets when planning a space mission in the solar system?

3. (a) Which planet is probably the easiest for humans to visit? Why?

(b) Which planet is probably the most difficult for humans to visit? Why?

4. Write a short story, poem, or song about the planet you would most like to explore.

Extend Your Skills

5. Most planets have several moons orbiting around them. Research to discover which planet has the most moons.

Suppose you were to make a map of space using 1 cm to represent the distance between Earth and the Sun.

- The distance to the next nearest star would be 2.5 km.
- Our Sun is one of over 200 billion stars in a *galaxy* (group of stars) called the Milky Way. To show the size of the Milky Way, your map would have to be about one third the size of Canada.
- There are billions of galaxies in the universe. To show them all, your map would need to be larger than the surface area of our planet!

Pause& Reflect

In Spanish the Sun is called *el sol*. The French term for the Sun is *le soleil*. Notice that both terms use *sol*. The English word "solar" means "connected with the Sun." What terms do you know that have "solar" in them? List the terms in your science notebook.

The Sun

Without one particular object in space, life on Earth would not exist. That object is the Sun. The **Sun** is a huge ball of hot gases, similar to billions of other stars that we see in the night sky. The Sun looks different from other stars only because it is part of our solar system and we are so much closer to it.

Using observations and measurements from different instruments, scientists have learned a great deal about the structure of the Sun. Some of the Sun's features are shown in Figure 7.24.

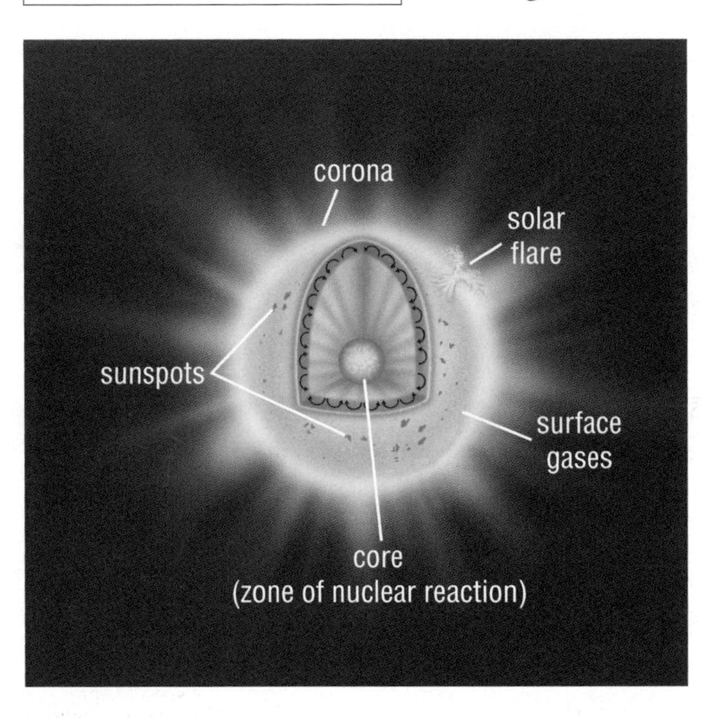

Figure 7.24 The structure of the Sun

- **Corona (outer atmosphere):** This outer layer can be seen as a halo around the Sun during a total eclipse when the Moon passes between Earth and the Sun.
- **Surface gases:** At a temperature of about 6000°C, these gases form the Sun's visible surface and give the Sun its yellowish colour.
- **Core:** Nuclear reactions in the core produce temperatures higher than 15 million °C.
- **Sunspots:** These areas on the surface are cooler and look darker.
- **Solar flare:** When magnetic energy that has built up in the solar atmosphere is suddenly released, there are sudden bright flares. Flares can be as short as a few seconds or as long as 1 h.

READING Check

Why does the Sun look brighter to us than other stars?

The Sun sends out huge amounts of energy into space. For example, the energy in a single solar flare is 10 million times greater than the energy released from a volcanic explosion. On the other hand, a solar flare produces less than one-tenth of the total energy produced by the Sun every second.

Heat from the Sun's core moves outward to the surface, producing temperatures of up to 1 million °C in the corona. The temperature of the corona is so high that the Sun's gravity cannot hold on to it and the topmost layers of the corona flow away from the Sun into space. This produces a solar wind that moves in all directions at speeds of about 400 km/s.

Light from the Sun is so intense that it can damage your eyes. Never look directly at the Sun.

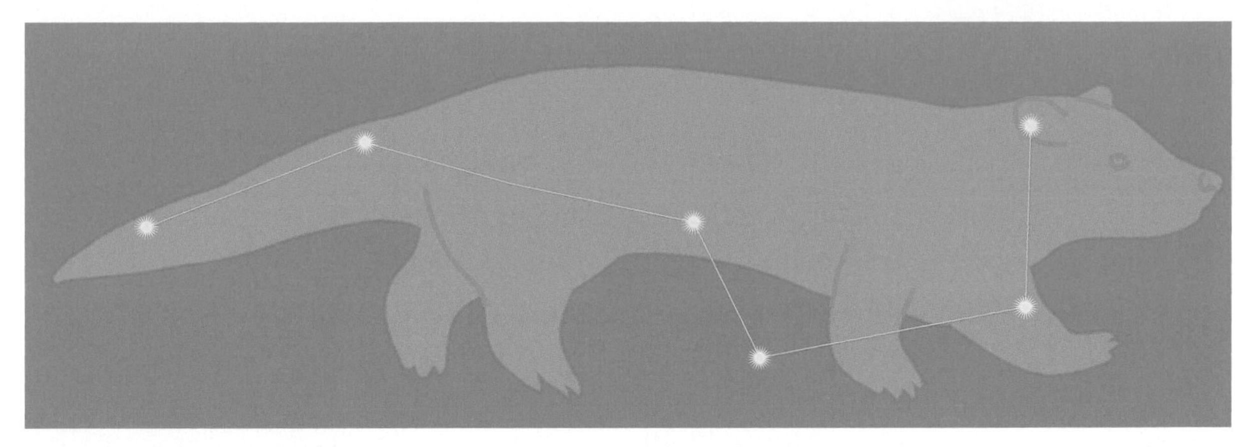

Figure 7.25 The group of stars known as the Big Dipper represents a grizzly bear or a fisher to some Aboriginal peoples.

Section 7.2 Summary

Humans have been observing the sky for centuries to learn more about the Moon, the Sun and other stars, and the planets in our solar system.

The force of gravity is at work throughout the solar system:
- the gravitational pull of Earth keeps the Moon in orbit around Earth
- the gravitational pull of the Sun keeps Earth and the other planets in orbit around the Sun

The solar system includes the Sun, the planets, their moons, asteroids, dust, gases, and other objects that orbit the Sun.

Check Your Understanding

1. In what way is the atmosphere a challenge to viewing objects in space?
2. How does the mass of a spacecraft affect its orbit?
3. What is one way you have experienced gravity in your own life?
4. List or draw the planets in order from the closest to the Sun to the furthest from the Sun.
5. Use a labelled diagram to describe how the Sun produces energy.
6. **Apply** Choose a sport you enjoy. Would the sport be easier or more difficult to play on the Moon? Explain why.
7. **Thinking Critically** What do you think are three challenges for scientists studying the Sun?

INTERNET CONNECT

www.mcgrawhill.ca/links/BCscience6

Aboriginal peoples throughout the world have many legends about the solar system. Find out more about traditional observations and interpretations. Go to the web site above and click on **Web Links** to find out where to go next.

Section 7.3 Leaving Planet Earth

Key Terms
- rocket
- thrust
- payload
- stage
- booster rockets
- vacuum
- airlock
- microgravity
- electromagnetic waves

The rocket shown in Figure 7.26 lifts off from its launch pad and speeds into the sky. Only 10 min after leaving Earth, the rocket is in orbit high in the upper atmosphere. What is a rocket? How can a rocket travel in space?

A **rocket** is a tube with fuel at the lower end. When the fuel burns, the force of the hot gases escaping from the bottom of the tube pushes the rocket upward. You can find out how the movement of escaping gases *propel* (move) a rocket in the next activity.

Figure 7.26 What are some differences between a rocket and an airplane?

Find Out ACTIVITY 7-I

Racing Rockets

You can observe the force that sends rockets into space by making your own rocket with a balloon.

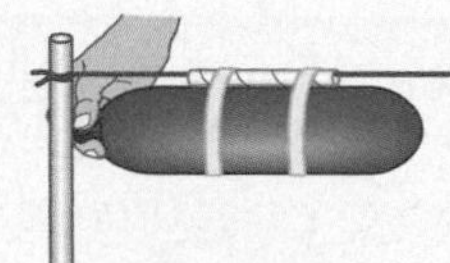

What You Need

pencil balloon
5 cm long drinking straw
masking tape
5 m length of string

What to Do Group Work

1. Thread the string through the straw.
2. Tie each end of the string to a sturdy post about 2 m above the ground. Make sure the string is tight and level.
3. Blow up the balloon. Ask a partner to tape the balloon to the straw while you hold the mouth of the balloon shut.
4. Keep squeezing the balloon end while you pull your balloon rocket to one end of the string.
5. Release the balloon.

What Did You Find Out?

1. Why did the balloon move?
2. What affects the speed of the balloon?

Extensions

3. What happens if you change the volume of air in the balloon?
4. What happens if you change the length of the string?

Action-Reaction

The force that propels a rocket does not push against the ground or against the surrounding air. The rocket's motion is the result of an *action-reaction* pair of forces. You can easily experience these forces. Imagine you are standing on a skateboard holding a heavy ball. What happens if you throw the ball to a friend? The ball goes in one direction and you and the skateboard go in the opposite direction with the same force. You can see this principle shown in Figure 7.27.

Figure 7.27 There is a law of motion that states, "for every action there is an equal and opposite reaction." What is the action shown here? What is the reaction?

Why is the action-reaction principle important? It means that a rocket can move in space where there is no air or anything else to push against. In order to move faster in space, a rocket simply fires its engines. The escaping gases produce a force called **thrust**, which causes the rocket to move. The thrust produced by burning solid fuels, such as gunpowder, is less than the thrust from liquid fuels. In 1926, Dr. Robert Goddard launched the first rocket to use liquid fuel, shown in Figure 7.28.

INTERNET CONNECT

www.mcgrawhill.ca/links/BCscience6

What is the connection between firecrackers and modern spacecraft? You can find out by investigating the history of rocketry. Go to the web site above and click on **Web Links** to find out where to go next.

How is thrust produced in a rocket?

Figure 7.28 Dr. Goddard used a mixture of gasoline and oxygen to fuel this rocket in 1926.

Flying in Stages

You know that if you throw a heavy ball into the air, it will not go as far as if you throw a lighter ball with the same force. How far a rocket flies depends on the weight it is carrying as well as its speed. Much of the weight of a rocket is in the **payload** it carries. Payloads include cargo, measuring instruments, and astronauts. How would you send a heavy payload into orbit? You need a large thrust that lasts long enough for the rocket to reach its orbit. To produce a large thrust you need lots of fuel, which also adds to the weight of the rocket.

To help solve the problem of weight, Robert Goddard designed rocket engines built in several sections. Each rocket engine section, called a **stage**, drops off after all its fuel is burned up. The remaining rocket is lighter and can move faster.

Figure 7.29 shows the space shuttle seconds after liftoff. Notice the hot escaping gases that produce the thrust. You can also see the booster rockets on either side of the large main fuel tank. **Booster rockets** are engines in a space vehicle that provide thrust during a launch. These engines burn up all their fuel within 2 min. Then the booster rockets separate from the shuttle and parachute back to Earth to be used again. The main fuel tank separates from the shuttle about 6 min later. By this time, the shuttle has reached its orbit and is travelling at 8 km/s. The main tank breaks into pieces as it falls back through the atmosphere.

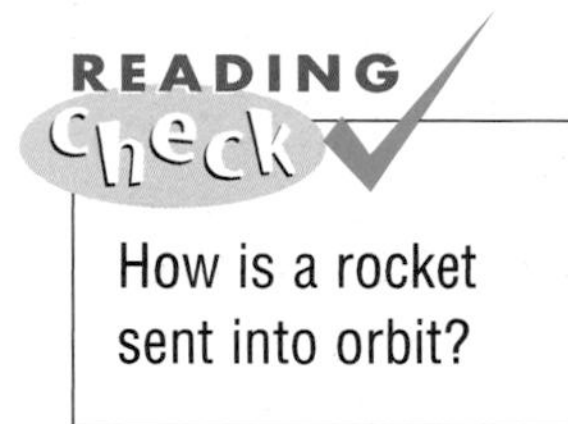

Figure 7.29 Two booster rockets help launch the space shuttle, and then drop off when their fuel is used up.

Working in a Vacuum

Imagine you are an astronaut. As your spacecraft soars upwards, you eventually leave the atmosphere and travel in a **vacuum**, a region where there is very little or no air or other gases. Gravity pulls all the gases and other materials in the universe close to planets and other large objects, leaving areas of empty space in between.

You are protected from the vacuum of space by air pressure inside the spacecraft. If you need to go outside the spacecraft to do work, such as changing the solar panels, then you need to wear a pressurized spacesuit. Once you have your spacesuit and your equipment ready, you enter a part of the spacecraft called an airlock (see Figure 7.30). An **airlock** is an airtight chamber where the air pressure can be reduced until it matches the vacuum of space. From the airlock you can go into space.

You could not survive outside your spacecraft without your spacesuit. In the vacuum of space, there is no pressure outside to balance the pressure of gases inside your body. Your blood and tissues would swell like balloons and burst out of your skin.

Figure 7.30A The astronaut puts on a spacesuit in the airlock, then connects the suit to the airlock using a safety strap.

DidYouKnow?

Although there are radio waves in space, there is no sound. Sound waves produce sound by vibrating the substances they pass through, such as air, water, or your eardrums. Because deep space is an empty vacuum with very little material, it does not transmit sound.

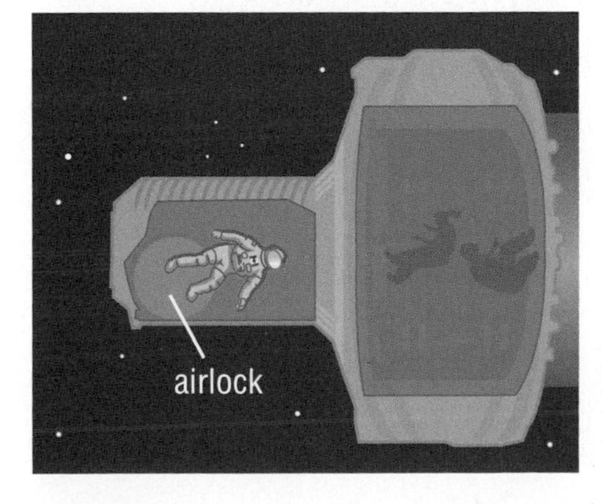

Figure 7.30B The airlock is located between the inside and the outside of a spacecraft.

CONDUCT AN

INVESTIGATION 7-J

SKILLCHECK
- Observing
- Measuring
- Interpreting Data
- Modelling

Working in a Spacesuit

Take the challenge! Make yourself a spacesuit, then see how well you can work without the suit compared with wearing the suit.

Question

How well can you carry out different tasks while wearing a spacesuit?

Safety Precautions

- Take care using scissors.

Apparatus

clock or stopwatch
scissors
pair of hockey gloves or oven mitts
school bag or backpack with zipper
CD or DVD in hard plastic case
jar with screw-top lid filled with dry macaroni
pen or pencil
woollen socks
pull-on boots or shoes

Materials

6 2 L empty milk or juice cartons
duct tape
3 sheets of paper
Ziploc™ bag
brown paper grocery bag

Procedure

Making a Space Helmet

1. Cut an 18 cm × 10 cm rectangle out of the middle of one side of a brown paper bag.

2. Cut a semicircular piece from opposite sides of the top of the bag so that the bag will fit around the shoulders when placed over the head.

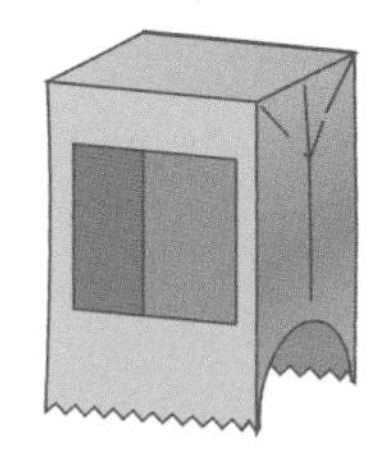

paper bag helmet

Making Spacesuit Arms

3. Cut off both ends of the milk cartons.

4. Join three cartons together with duct tape for each arm. Put your arms and helmet in a safe place.

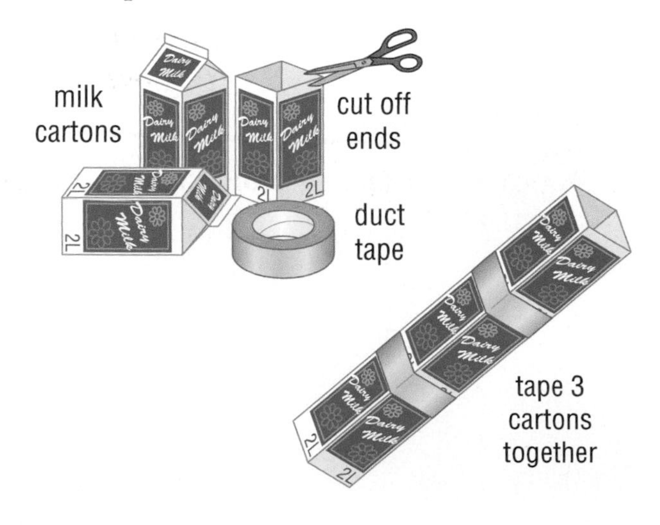

Preparing the Challenge

5. Place the jar of macaroni, Ziploc™ bag, sheets of paper, pen or pencil, socks, boots, and CD or DVD in the school bag and close the zipper.

6. Prepare a data table to record the results of the five tasks described below.

Skill POWER

For tips on making a data table, turn to SkillPower 5.

Taking the Challenge

7. Working with a partner, carry out all five tasks (see below). **Record** the time it takes to complete all the activities.

Task 1	Remove all the contents of the school bag.
Task 2	Remove the CD or DVD from its case and exchange it with your partner's. Place your partner's CD or DVD in your case.
Task 3	Pour the macaroni from the jar into the Ziploc™ bag. Seal the bag. If any macaroni spills, you must pick it up and place it in the ZiplocTM bag.
Task 4	Write your name on one sheet of paper. Fold the paper in half.
Task 5	Put on the socks and boots.

8. Repeat step 5.

9. Put on the space helmet, spacesuit arms, and space gloves (the hockey gloves or oven mitts).

10. Repeat all five tasks. **Record** the time it takes.

Analyze

1. How long did it take you to complete the challenge during each trial?
2. Which tasks were most difficult to carry out when you were wearing a spacesuit?
3. How did having a partner help you or make tasks more difficult?

Conclude and Apply

4. What recommendations would you make to help an astronaut carry out tasks more efficiently?
5. How do you think astronauts hold and grip tools?

Extend Your Skills

6. Draw a design of a spacesuit including features that would make it easier for an astronaut to collect rock samples. Label your design.

Did You Know?

Astronauts train for microgravity conditions in a jet that flies at high speed in sharply curving arcs, producing weightless conditions for up to 30 s at a time. The jet has also been used in filming weightless scenes for movies, and in designing space tools and experiments. How do you think the jet got its nickname, the "Vomit Comet?"

What Is Microgravity?

Many people think there is no gravity aboard a spacecraft. However, if you are in orbit 400 km above Earth, the gravitational pull is still nearly 90 percent of its strength at Earth's surface. In fact, Earth's gravity is what keeps the spacecraft in orbit. Figure 7.31 explains this relationship.

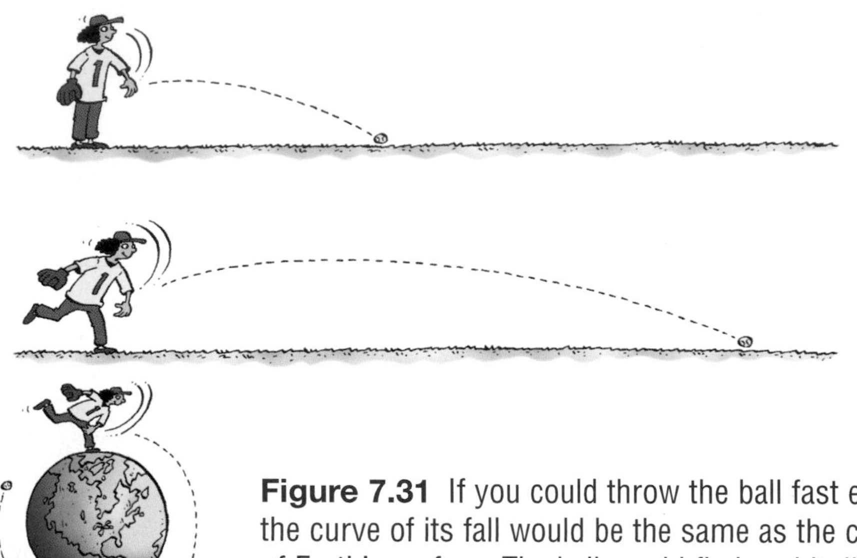

Figure 7.31 If you could throw the ball fast enough, the curve of its fall would be the same as the curve of Earth's surface. The ball would fly in orbit, like a spacecraft.

Why do astronauts in space seem weightless?

If gravity is pulling an orbiting spacecraft, why do objects inside the spacecraft seem weightless, like the astronaut in Figure 7.32? Imagine an astronaut on a spacecraft dropping an apple. The apple falls under the pull of gravity. However, the astronaut and the spacecraft are falling together with the apple at exactly the same speed. Therefore, the apple does not appear to be falling. It seems to float in space.

Objects falling together at the same rate seem to float in what is sometimes called zero gravity, or **microgravity**. A common term for objects floating in orbit is "free fall."

You may have experienced the feeling of free fall yourself without realizing it. Roller coasters produce brief periods of free fall when the cars finish climbing then suddenly speed down steeply rolling rails. Jumping on a trampoline also creates brief experiences of weightlessness while you are in the air.

Figure 7.32 Canadian astronaut Chris Hadfield in microgravity. The prefix "micro" means very small.

Find Out ACTIVITY 7–K

Floating in Space

These simple tests give you an idea of the problems of working in a microgravity environment in a spacecraft.

What You Need

an office chair that spins around
a partner
an empty jar with a screw-top lid

What to Do

Part 1

1. Sit on the chair. Make sure your feet do not touch the floor.
2. Imagine you are on the crew of a space shuttle in flight. You cannot touch the ceiling, floor, walls, or any other fixed object. Try to turn yourself around to reach something behind you.
3. Try to turn all the way around in a complete circle.

Part 2

4. Close your eyes and put your head down on your chest as if asleep.
5. Have your partner spin you on the chair at a constant, safe speed as demonstrated by your teacher for four or five complete rotations.
6. Have your partner stop the chair. Predict which wall you are facing. After a few seconds, open your eyes.

Part 3

8. Have your partner hold the jar about 30 cm above your head with the lid downwards.
9. Try to unscrew the lid.

What Did You Find Out?

1. Which tests showed action-reaction forces?
2. Was your prediction in Part 2 correct? How did you feel when you opened your eyes?
3. How would you make it easier for an astronaut to open a jar in a microgravity environment?

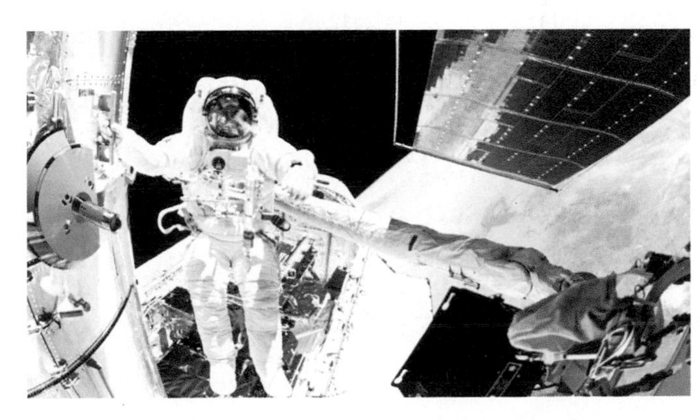

Figure 7.33 What are some of the challenges to working in space?

DidYouKnow?

Radio waves also carry signals from television stations to your television set. Your television converts the signals to pictures as well as sounds. Other radio wavelengths carry signals for cellular phones and for communications between Earth and spacecraft or satellites.

Communicating with Earth

"Space to Earth. Can you hear me?" Messages travel back and forth through space as radio waves, like those that carry music from your favourite radio station to your radio at home.

Radio waves, light waves, X-rays, and microwaves are all types of waves of energy called **electromagnetic waves**. Like the waves on an ocean, electromagnetic waves differ in size, or wavelength, as shown in Figure 7.34. Radio waves have the longest wavelengths.

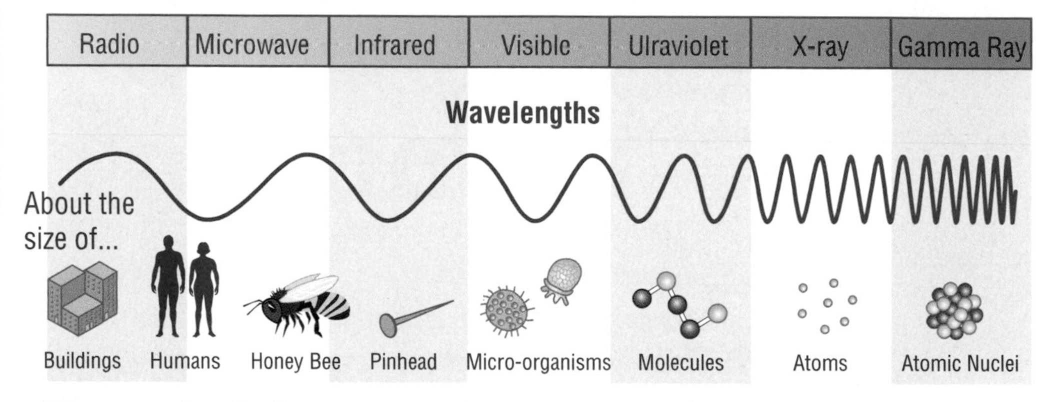

Figure 7.34 Radio waves may be as long as a building or as short as a football. Other electromagnetic waves are much shorter.

How are messages transmitted between Earth and a spacecraft?

Each radio station *transmits* or sends signals using radio waves with a particular range of wavelengths. The waves travel from an antenna at the radio station until they meet an antenna in your radio *receiver*. Your radio changes the signals into sounds. To tune into a different radio station, you adjust your receiver to pick up a different range of wavelengths.

Radio waves sent from spacecraft must travel much longer distances than those used by radio stations on Earth. Over distances, radio waves spread out and become weaker. To overcome this, a radio transmitter in space focusses its signals in a narrow beam directed by its antenna to a receiver on Earth (see Figure 7.35).

Figure 7.35 These radio telescopes near Penticton are used to observe gases between stars, the remains of exploded stars, and the Sun.

How can scientists on Earth communicate with a spacecraft in orbit above the opposite side of the planet? Figure 7.36 shows how this is done. By having three large receiving stations spaced apart at 120° around the globe, stations and spacecraft can be in touch with each other at all times.

Figure 7.36 Why is it necessary to have three receiving stations spaced apart at 120°?

"Seeing" Radio Waves

Human broadcasters are not the only source of radio signals from space. Just as stars send out energy in the form of light waves, they also send out radio waves. These waves carry information about processes that produce energy in stars and other bodies in space. To help analyze these radio waves, scientists use computers to convert different radio wavelengths into different colours. Figure 7.37 shows an example of a radio image of a galaxy made in this way.

Figure 7.37A A photograph of the galaxy Centaurus A, taken using visible light.

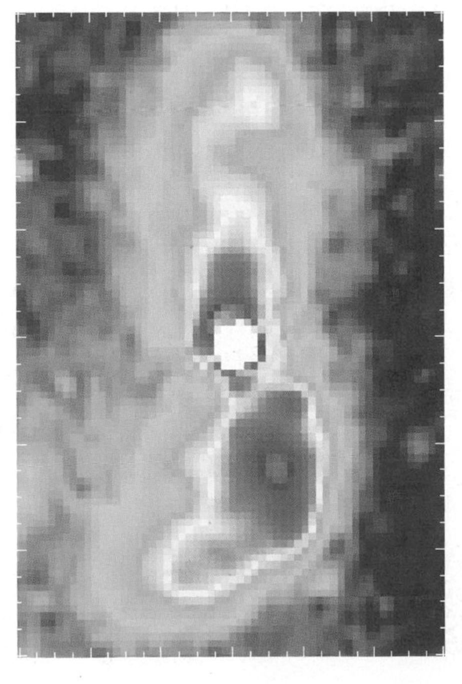

Figure 7.37B A radio image of Centaurus A. The white section in the centre covers the same area seen in 7.37A. Notice how radio energy from the galaxy extends into space much farther than visible light energy.

INVESTIGATION 7-L

SKILLCHECK
- Communicating
- Inferring
- Interpreting Data
- Predicting

Space Report

Think About It

What do you know about the extreme environment of space? For example, how far would you have to travel to explore the nearest planet? How would you get oxygen, food, and water? What hazards would you face during your space journey?

What other questions about space would you like to find answers to?

What to Do

1. As a class, brainstorm all the information you have learned about space so far. **Record** your ideas as single words or phrases on a large piece of paper.

2. In pairs or small groups, choose one topic that interests you. Your group will become class experts on that topic.

3. When you have made your choice, think about three or four questions related to the topic that you would like to answer. **Record** your questions in your notebook.

4. Use your library and the Internet to **research** the answers.

5. As you carry out your research, you may think of more questions or ideas. You may find out unexpected and interesting facts about your topic, including legends or stories about space from different cultures. **Record** any additional information that will fit with your final report.

6. With your group, **analyze** all the data you have researched. Organize the data into different sections according to the question that it answers about your topic.

7. Prepare a report of your findings for the class. Your report may be a model, display, poster, radio show, or other means approved by your teacher.

Analyze

1. After all the students have presented their reports, discuss what factors limit the exploration of space today. How might humans overcome these limitations?

2. What is the value to people on Earth of exploring space?

Voyage to the Edge

On September 5, 1977, a spacecraft named *Voyager 1* was launched from Earth. The spacecraft, shown in Figure 7.38, was beginning its spectacular voyage to the edge of the solar system and beyond. *Voyager 1* is now in a violent zone of high-energy particles unlike any it has met before. In this region, the solar winds slow down suddenly as they meet the area of thin gas between the stars.

INTERNET CONNECT

www.mcgrawhill.ca/links/BCscience6

What would you tell another world about life on Earth? Both *Voyager 1* and *Voyager 2* carry messages in case they someday reach a distant civilization. Each spacecraft has a copper disc with scenes, greetings, music, and sounds from Earth. Find out what is included on the disc. Go to the web site above and click on **Web Links** to find out where to go next.

Figure 7.38A *Voyager 1* continues its journey today as the most distant human-built object in the solar system.

Figure 7.38B A photograph of Jupiter taken by *Voyager 1*

Voyager 1 and its twin, *Voyager 2*, have flown near four planets (Jupiter, Saturn, Uranus, and Neptune) and 48 of their moons (see Figure 7.39). The two spacecraft have enough power to continue sending data until about 2020.

Where is *Voyager 1* travelling?

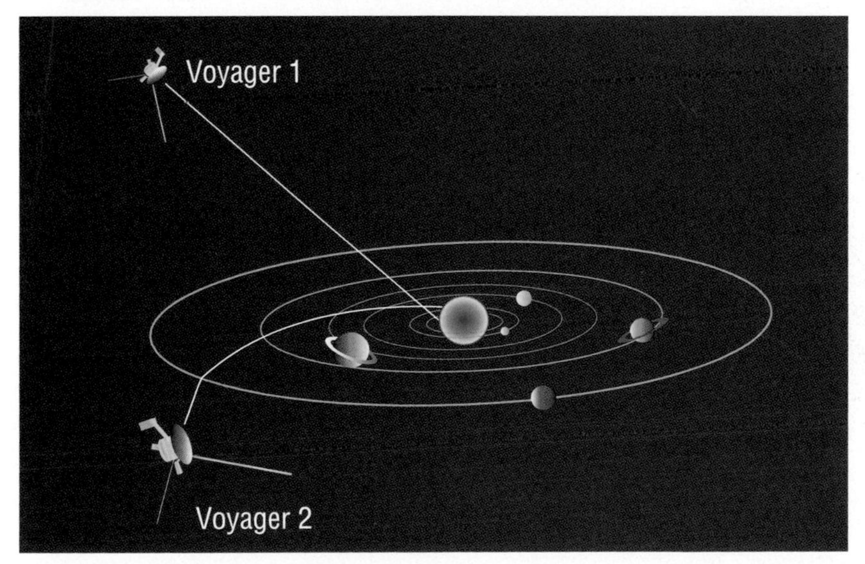

Figure 7.39 After leaving our solar system, both *Voyager 1* and *Voyager 2* may continue to wander forever in space.

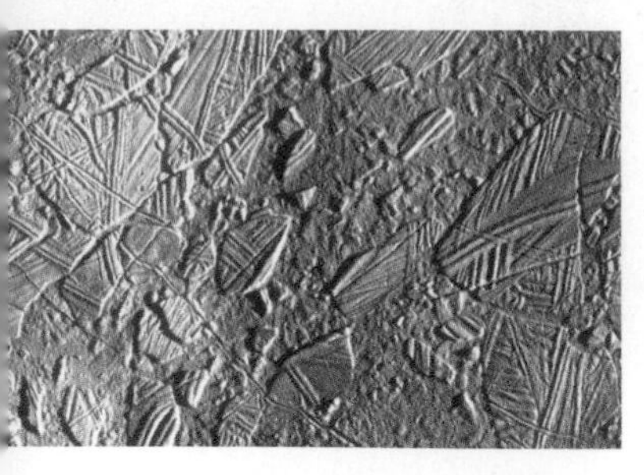

Figure 7.40 Maybe there is a liquid ocean beneath the icy crust of Europa.

Why is a lake in Antarctica of interest to space scientists?

Life in Space

Where would you look for evidence of life in space? What would you look for?

Strange as it seems, some good places to start searching are in the most extreme environments on Earth. Exploration in extreme environments helps us define the limits of life. For example, Antarctica is sometimes thought to be like a laboratory for the extreme conditions on Mars and in other parts of the solar system.

In 1996, scientists made an amazing discovery in Antarctica. Using radar and sound waves, they detected a large warm-water lake under nearly 4 km of solid ice. This lake has been sealed off from the rest of the world for at least half a million years. Why should this lake interest space scientists? On a mission to Jupiter, the spacecraft *Galileo* sent back images of Europa, one of Jupiter's moons (see Figure 7.40). The images suggest that Europa may have a liquid ocean beneath an icy crust. Space scientists are now developing a small robot to drill through 4 km of ice in Antarctica and sink into the water below with instruments to detect life. This venture is like a rehearsal for a mission to Europa without leaving the planet.

In the next chapter you will learn more about robots and other technologies for exploring extreme environments. In the next investigation you will design some technology of your own.

Pause& Reflect

Some micro-organisms thrive in extreme environments. There are micro-organisms that live in Antarctica (as shown in the photograph) and in the very salty water of the Dead Sea. There are even micro-organisms that live under great water pressure and high temperatures by sea floor vents. How might the study of these extreme organisms help scientists understand where life could be found in other parts of our solar system?

PROBLEM-SOLVING

INVESTIGATION 7-M

SKILLCHECK
- Identify the Problem
- Decide on Design Criteria
- Plan and Construct
- Evaluate and Communicate

A Visit to Venus

Challenge

Design a model that meets the design specifications below to transport astronauts on the surface of Venus.

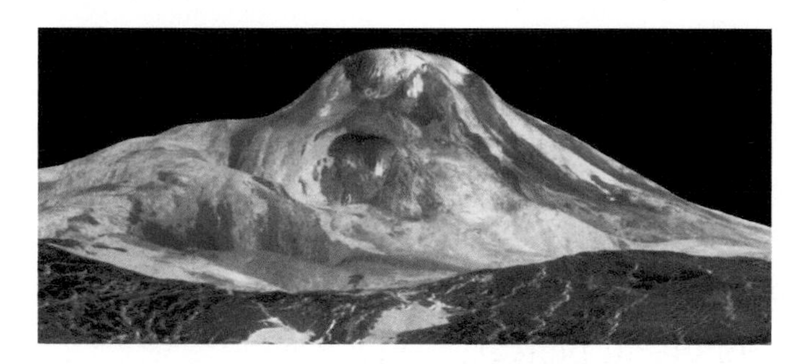

Apparatus
library books
Internet access
scissors

Materials
paper for designing and planning
art supplies
assorted recycled materials for building your model

Design Specifications

A. Your model must show how two astronauts can be carried for 1 km on the surface of Venus. Be sure to indicate what fuel and what source of fuel your design would use.

B. Your model must consider the following conditions on the surface of Venus:
- very high temperatures (close to 450°C day and night, hot enough to melt lead) and extremely high pressure
- the atmosphere is mostly carbon dioxide; thick, white clouds made from poisonous acid cover the planet
- gravity is slightly less than on Earth (about 91 percent of Earth's gravity)

C. Your model must not be larger than a shoebox.

Plan and Construct 

A. With your group, discuss the different types of vehicles that could be used. Consider the extreme conditions on Venus. What human needs will you have to meet? How are these needs already being met in space travel? What factors make it difficult to travel on and explore Venus? Plan several possible solutions.

B. Make a list of the materials you could use for your model. Consider the limitations of materials now used in space travel—could you invent something new? Draw some possible designs.

C. As a group, select the best design. Draw a labelled sketch of the model showing what materials you will use.

D. Obtain your teacher's approval. Then construct the model.

E. Present your model to the class, explaining the features that will help astronauts explore Venus.

Evaluate

1. What improvements did you make to your original design?
2. How would you change your design if you were to build the actual vehicle?
3. Consider your model and your classmates' models. What ideas did other groups use that you would like to use? Explain why.

Earth's Past and Future

Scientists think that conditions on Venus today may be similar to conditions on Earth in the very distant past. By studying Venus, scientists may better understand how certain materials formed on Earth and better predict where precious minerals may be found. Venus may also hold clues about Earth's future. Scientists study the atmosphere on Venus to help them predict changes that could occur on Earth if the amount of carbon dioxide in our atmosphere increases.

Section 7.3 Summary

In this section you learned:

- A rocket can reach its orbit because the force of the hot gases escaping from its base pushes it upward.
- Challenges to astronauts in space include working in a vacuum and in microgravity conditions.
- Radio waves are used to send data between Earth and spacecraft.
- Exploring extreme environments on Earth can help our understanding of space. Exploring space can help our understanding of environments on Earth.

Check Your Understanding

1. Why do rockets have stages?
2. Laboratories in space can be used to carry out experiments that could not be done on Earth. What condition(s) make it possible to carry out such experiments?
3. How do scientists on Earth communicate with spacecraft in space?
4. What is happening as the exhaust leaves a rocket?
5. How can studying other planets help us understand Earth's environment?
6. **Apply** How could you increase the speed of a rocket in space?
7. **Thinking Critically** You are on Earth communicating with one astronaut on Mars and another on the Moon. What differences in communication might you notice? Explain your answer.

CHAPTER at a glance

Now that you have completed this chapter, try to do the following. If you cannot, go back to the sections indicated.

(a) Define an extreme environment. (7.1)

(b) Explain what explorers need to survive in an extreme environment. (7.1)

(c) Draw an example of a life-support system. (7.1)

(d) Draw a variety of Aboriginal technologies that have been used for exploration. (7.1)

(e) Explain how the atmosphere affects our view of objects in space. (7.2)

(f) Use the example of a ball to help you explain gravity. (7.2)

(g) Draw a labelled diagram of the solar system. (7.2)

(h) Describe the different parts of the Sun. (7.2)

(i) Describe the different parts of a rocket and explain the role each plays. (7.3)

(j) Use the example of a ball and a skateboard to help you explain the action-reaction principle. (7.3)

(k) Use the example of a ball to help you explain how Earth's gravity keeps a spacecraft in orbit. (7.3)

(l) Describe some of the extreme conditions that astronauts experience while in space. (7.3)

(m) Explain why exploring our solar system can help us understand more about Earth. (7.3)

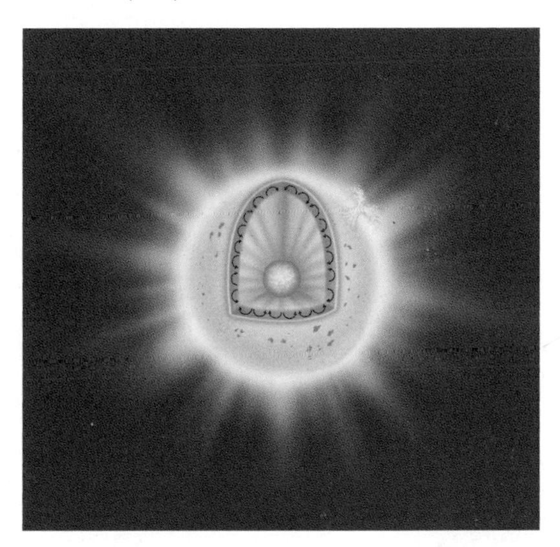

Prepare Your Own Summary

Summarize this chapter by doing one of the following. Use a graphic organizer (such as a concept map), produce a poster, or write a summary to include the key chapter ideas. Here are a few ideas to use as a guide:

- Write a plan for exploring an extreme environment. Include the technology you would need, and the conditions you would face.
- Create a skit to show how astronauts work in space. Include a description of their life-support systems.
- Draw a poster to show the different parts of a rocket. Show how it reaches its orbit and how it can travel in a vacuum.

CHAPTER

7 Review

Key Terms

environment
extreme
technology
exploration
astronaut
life-support
recycling
telescope
atmosphere
orbit
gravity
Sun
solar system
rocket
thrust
payload
stage
booster rockets
vacuum
airlock
microgravity
electromagnetic waves

Reviewing Key Terms

If you need to review, the section numbers show you where these terms were introduced.

1. Describe the difference between:
   (a) life-support and recycling (7.1)
   (b) atmosphere and planet (7.2)
   (c) thrust and payload (7.3)

Understanding Key Ideas

2. Why do you think space is called an extreme environment? (7.1)

3. **(a)** How is gravity related to mass?
   (b) How is gravity related to distance? (7.2)

4. Explain why astronauts experience the sensation of weightlessness in space. (7.2)

5. Describe how astronauts in a spacecraft communicate with scientists on Earth. (7.3)

6. Why would you need to enter an airlock before going from a spacecraft into space? (7.3)

Developing Skills

7. Describe two difficulties that a student could face creating a scale model of the solar system.

8. Draw a circle to represent the Sun. Draw and label the following:
   (a) a solar flare
   (b) corona
   (c) core
   (d) a sunspot

9. **(a)** Draw an example of an Aboriginal technology used for exploring an extreme environment.
   (b) Explain the key advantages of this technology.

Problem Solving

10. Choose an extreme environment other than space. Design a life-support system for an explorer working in the environment. Label your design.

11. Astronauts use spacesuits to allow them to work in space.
    (a) In what other extreme environments could spacesuits be used to help people explore?
    (b) Explain the advantages and disadvantages of using a spacesuit in each of the extreme environments you identify.

Critical Thinking

12. To live on Mars, we would have to deal with the thin atmosphere, lack of water, and cold temperatures. Suggest or draw a solution for each problem.

13. **(a)** What would happen to your mass if you went to Mars?
(b) Why?

14. The picture below shows the inside of a spacecraft orbiting Earth. Could the dog do this? Explain.

15. Why do objects float around in a spacecraft orbiting Earth even though they are not weightless?

16. If a malfunction occurs in *Voyager 1*, why would it not be known until after almost one full day?

17. **(a)** What are some advantages of space travel over using telescopes to explore the solar system?
(b) What are some disadvantages?

Pause& Reflect

Go back to the beginning of this chapter on page 186 and check your answers to the Getting Ready questions. How has your thinking changed? How would you answer these questions now that you have investigated the topics in this chapter?

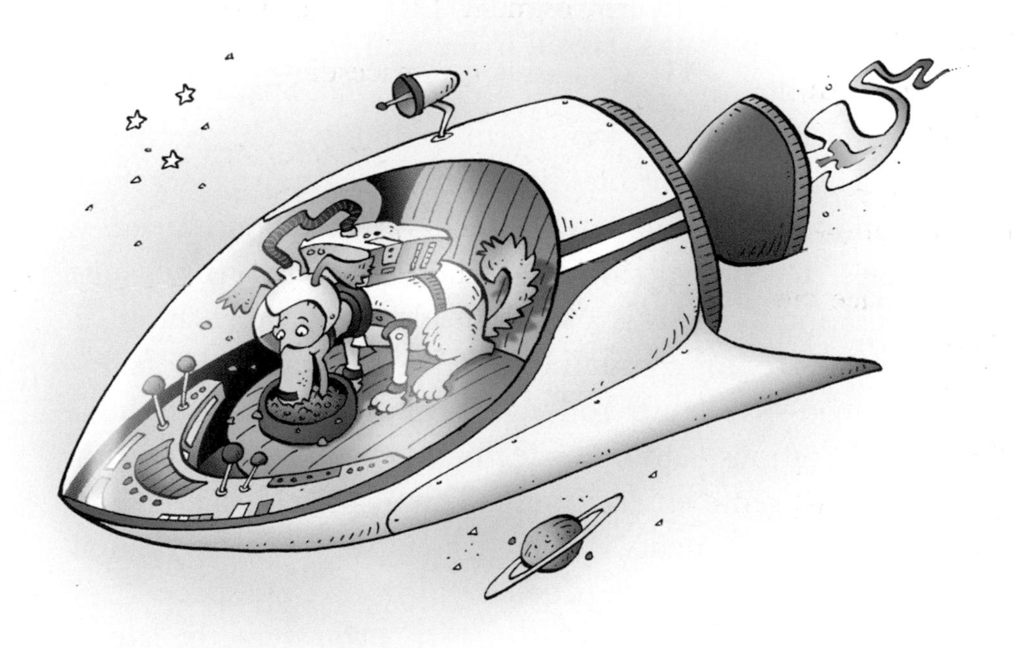

CHAPTER

8 Technologies for

Getting Ready...

- How do robots help us explore unknown worlds?
- What Canadian space technologies do you use?
- What are the advantages and disadvantages of new exploration technologies?

The International Space Station orbits 360 km above Earth.

The year 2000 was the start of a new era in space exploration. For the first time, humans started living and working in a permanent base built in space. The International Space Station is an orbiting laboratory for long-term space research, and may in future become a site for building and launching spacecraft.

Explorations of extreme environments such as space need more advanced technologies. Every year, new technologies change space exploration. Canada is a world leader in space research, developing satellites for communication and designing robots to make repairs on spacecraft. You will learn more about satellites, robots, space stations, and other technologies in this chapter.

Exploring extreme environments brings problems and challenges as well as benefits. Are the benefits of space exploration worth the huge cost of these projects? How do new space technologies change your life and society? As you study the latest developments in space research, you will learn about some of the benefits and problems of technology.

Unknown Worlds

What You Will Learn

In this chapter, you will learn

- what technology is used to explore extreme environments
- how new tools and technologies are developed
- how Canada contributes to space technology
- why technology has both benefits and drawbacks

Why It Is Important

- The study of space technology can help us understand technological design and problem solving. We can use this knowledge in our everyday life.
- Understanding how technology is developed can help us understand new technologies as well as familiar ones.

Skills You Will Use

In this chapter, you will

- investigate a mystery planet
- communicate with a robot on "Mars"
- design and construct models of space technology
- interpret satellite photographs
- test solutions to a problem

Starting Point ACTIVITY 8–A

Extremely Useful Technology

How many different types of technology for exploration do you know?

What to Do Group Work

1. Your group will be assigned one of these extreme environments: space, polar regions, oceans, deserts, caves, or volcanoes.
2. Brainstorm a list of technologies used by explorers in your extreme environment.
3. Choose three items. Sketch a picture of each item, or find pictures in magazines or on the Internet. Add the pictures to your notebook.
4. How can each of the three items be adapted for use in another extreme environment? Using the pictures, label the changes you would need to make to the items so they can be used in another extreme environment.
5. Share your designs with the class.

What Did You Find Out?

1. What was the most difficult part about adapting a technology for an extreme environment?
2. How could you use ideas from other groups to improve your technologies?

Section 8.1 Rovers and Robots

Key Terms
- rovers
- lander
- robot
- cryobot
- frogbot
- snakebot
- spiderbot

Have you ever seen or used a remote-controlled toy, such as a racing car or airplane? The movements of the toy are directed by radio signals sent from a small hand-held control. Scientists use a similar technology to explore the surface of the Moon and the planets. Scientists send signals from Earth to operate exploration vehicles called **rovers**.

Mission to Mars

In January 2004, two rovers like the one shown in Figure 8.1 began exploring Mars. How did the rovers get to Mars? Each rover arrived on the planet inside a special spacecraft called a **lander**, which was launched from Earth many months earlier. The lander was designed to land safely on the planet's surface without damaging the rover or its equipment (see Figures 8.2 and 8.3).

DidYouKnow?

The rover and lander together weigh about 530 kg on Earth, but only 200 kg on Mars because of the difference in gravity.

Figure 8.1 What are the advantages of a rover over a human explorer? What are the disadvantages?

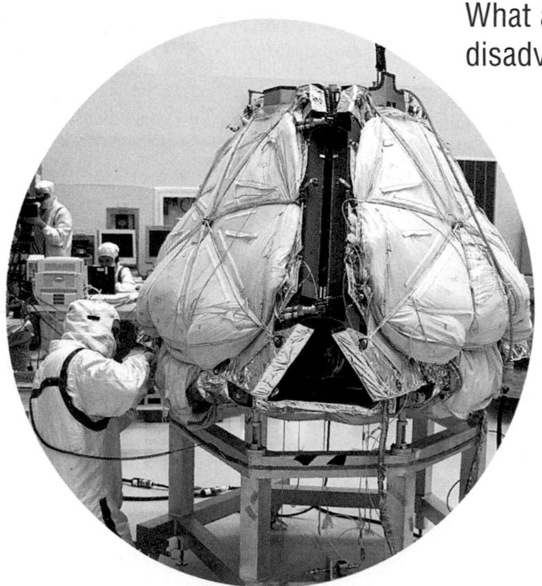

Figure 8.2A Why do you think the lander has a pyramid shape?

Figure 8.2B The pyramid shape allows the lander to open no matter which way up it lands. The side of the lander touching the planet surface becomes the base. The other three sides, or "petals," fold back on hinges.

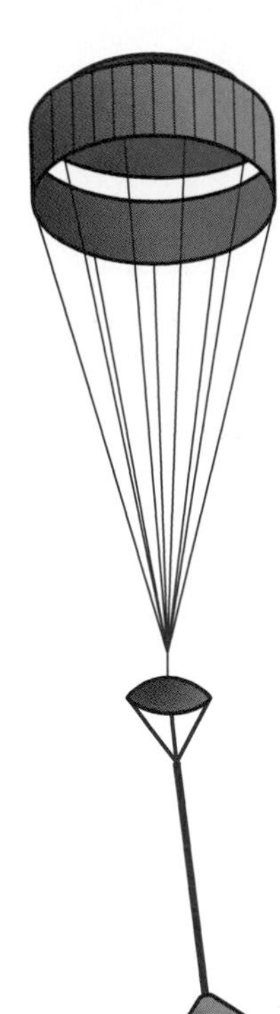

Figure 8.3A As the lander fell through the Martian atmosphere, a parachute opened to slow its speed.

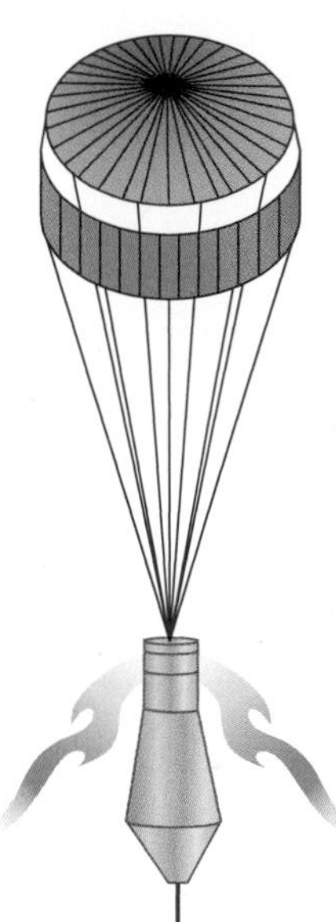

Figure 8.3B Small rocket engines, called *retro-rockets*, produced thrust in the opposite direction, slowing the lander even more.

Figure 8.3C Seconds before the lander touched the planet, large airbags inflated. The lander bounced over the Martian surface like a big beach ball.

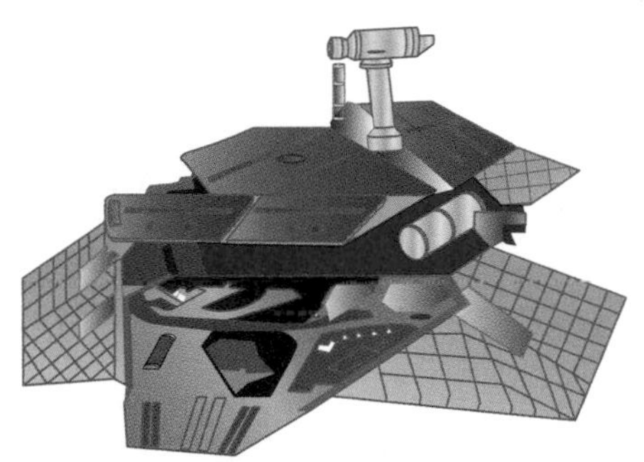

Figure 8.3D After the lander rolled to a stop, the airbags deflated and were drawn in. The petals folded back and the rover rolled out across one of the open petals.

Rovers at Work

A rover is equipped with some of the same tools a scientist would carry, such as a scraper to remove samples from the surface of rocks, and a camera to take and send pictures and other data to a team of scientists and engineers on Earth. After studying the data, the scientists decide what they want the rover to do next.

Scientists compare data sent by rovers with data sent from spacecraft in orbit above the planet. Because the rover data are more detailed, scientists can interpret the spacecraft data with greater accuracy. You can compare data from "spacecraft" and "rovers" in the next investigation.

INTERNET CONNECT

www.mcgrawhill.ca/links/BCscience6

To learn more about the Mars Exploration project, go to the web site above. Click on **Web Links** to find out where to go next.

READING Check

Why were rovers sent to Mars?

SKILLCHECK
- Observing
- Communicating
- Interpreting Observations
- Modelling

Mission to An Unkown Planet

You have learned that new technologies such as the lander and rover have been developed during the history of space exploration. Each new technology leads to new understandings, adding details to our knowledge of space and leading to new inventions. In this investigation you will model different stages of a space program over time. Use the information you gather at each stage to create a model of an unknown planet.

Question

How does new technology lead to new understandings?

Apparatus

ruler

Optional
binoculars
telescope or spotting scope
digital camera

Materials

various materials to make a model, such as paper, cardboard, felt pens

Mission Stage	Observations
10 m away	
10 m away with binoculars	

Procedure

1. Prepare a two-column chart with the headings "Mission Stage" and "Observations."
2. Your teacher has suspended a mystery Planet X from the far end of a sports field or school hallway. **Observe** Planet X from at least 10 m away. **Record** your observations. Include details such as the shape, size, and colour of the planet.
3. (Optional) Use binoculars to **observe** the planet from the same location as in step 2. **Record** any new details you observe. **Infer** whether the planet has an atmosphere, water, or life.
4. (Optional) Use a telescope to **observe** the planet. **Record** your observations. Discuss any changes in your ideas or inferences about the planet.

5. Have one student walk halfway to Planet X and **make observations**. The student returns and shares observations. **Record** any new information.

6. Have several students walk quickly to 1 m from the planet and walk back without stopping. Share and **record** observations.

7. Have one student take notes while walking around the planet five times. Share and **record** observations.

8. (Optional) Have one student walk by the planet taking several photographs with the digital camera. The class views the pictures on a computer and makes observations, notes, and diagrams based on the pictures. As a class, discuss the new information. How did the photographs change your ideas about Planet X?

9. Have two students visit the front side of the planet. Share and **record** observations.

10. Have two students visit the back of the planet. Share and **record** observations.

11. Use the data you have collected to design and construct your own small model of Planet X. Compare your model with the original planet.

Analyze

1. **(a)** What step of the investigation was similar to viewing a planet from Earth?
   (b) What step was similar to flying past the planet?
   (c) What step was similar to putting a spacecraft in orbit around the planet?
   (d) What step was similar to a first landing on the planet?

2. **(a)** How did your observations, notes, diagrams, and inferences change as the tools and technology changed?
   (b) What caused these changes?

3. **(a)** How is your model similar to the original object?
   (b) How is it different?

Conclude and Apply

4. How are the mission stages you used similar to exploring Mars or other planets and moons?

5. **(a)** What are three challenges of exploring other planets?
   (b) What technology would help to meet these challenges?

Robots

The Mars rover is a type of robot. A **robot** is a machine or device that works automatically or by remote control (see Figure 8.4). Simple robots such as a microwave or a washing machine can work by being programmed in advance. Other robots such as an airplane autopilot and a computer can recognize and respond to problems.

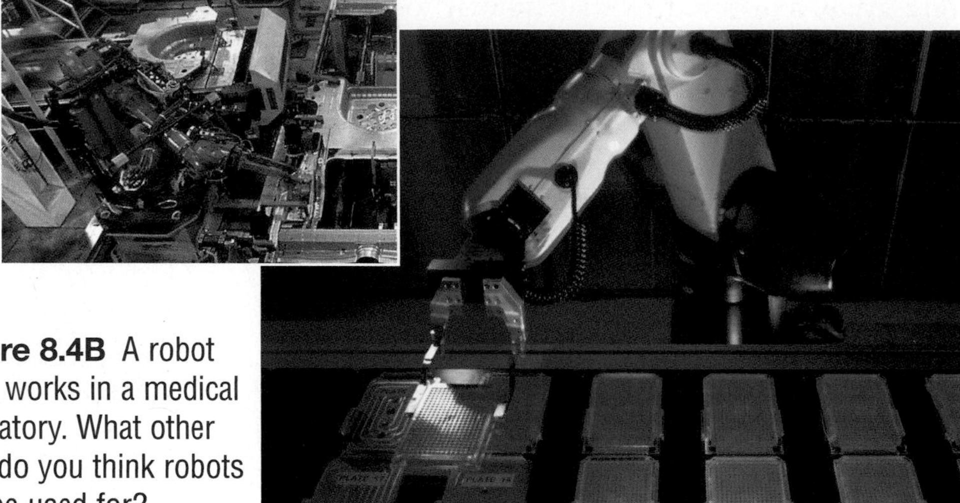

Figure 8.4A Robots are well suited for repetitive jobs. Unlike people, robots do not get tired or bored.

Figure 8.4B A robot hand works in a medical laboratory. What other jobs do you think robots can be used for?

At Home **ACTIVITY 8–C**

Robots in Your Life

Every time you turn on a dishwasher, press an elevator button, or set an alarm clock, you use a robot. How many robots do you use every day?

What You Need

notebook
pencil

What to Do

1. Make a three-column chart with the headings "Robot," "What It Does," and "How to Tell It What to Do."
2. During one week, record all the robots you use.

What Did You Find Out?

1. Analyze your list. What surprised you the most?
2. Which robots perform a job that you could not do yourself?
3. Which of the robots do you think your parents used when they were your age?
4. Which of the robots do you think your grandparents used when they were your age?

Robots can explore extreme environments in conditions that would be dangerous to people. On Earth, robots travel into the poisonous gases of volcanoes, under huge water pressure on the ocean floor, and through the extreme temperatures of deserts and polar ice caps.

How do robots help us explore extreme environments?

Parts of a Rover

To carry out the tasks of an explorer, a rover needs many of the same parts and functions that a living organism has. For example, the Mars rover has the parts and functions shown in Figure 8.5.

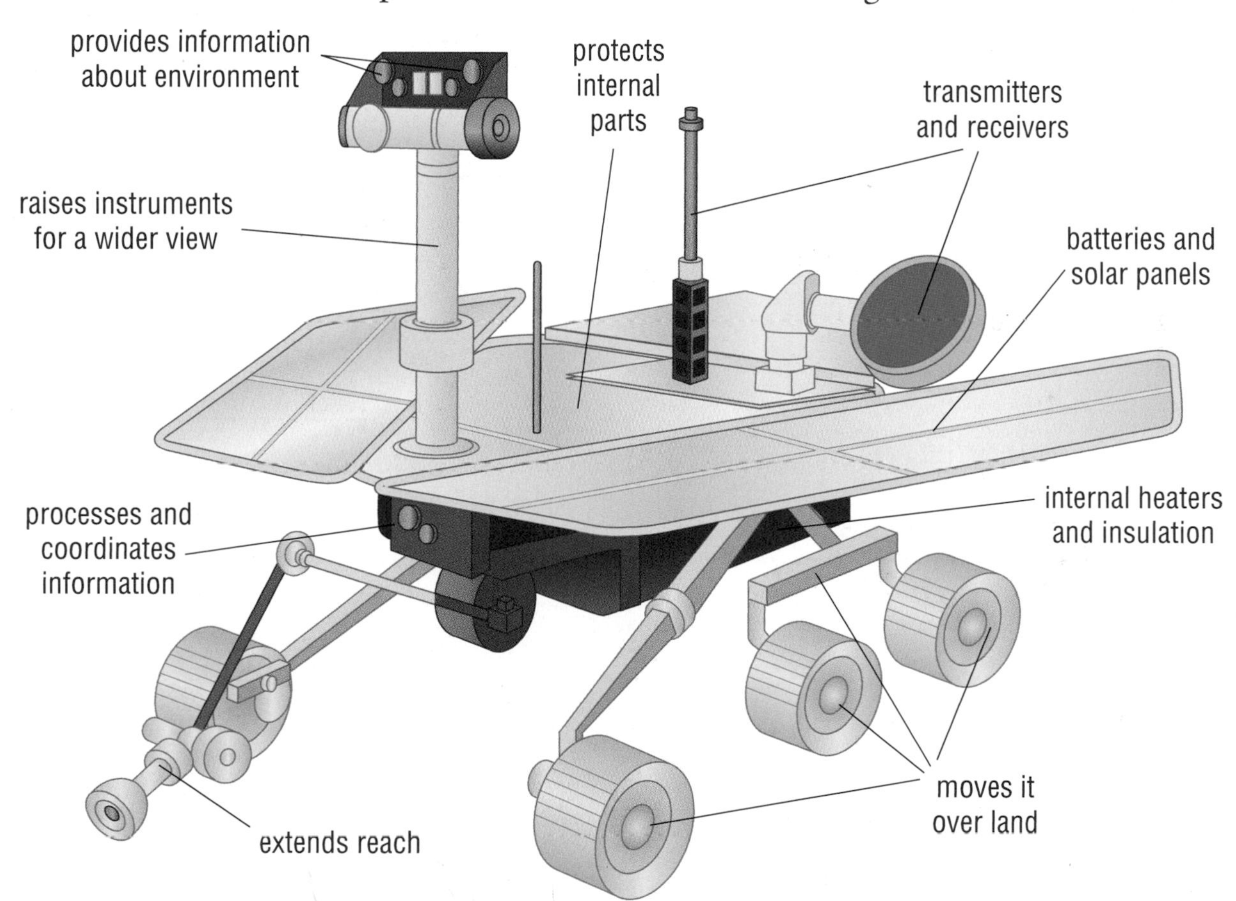

Figure 8.5 Which part of the rover do you think is like a body? Which part is like eyes? Which part is like a brain?

Before sending a rover into space, scientists and engineers test their design. They use locations on Earth that resemble where the rover will be sent. Why is this a necessary stage of developing a rover? Where do you think the rovers are tested on Earth?

CONDUCT AN

INVESTIGATION 8-D

SKILLCHECK
- Communicating
- Inferring
- Interpreting Data
- Modelling

Communicating with a Robot on Mars

How would you communicate with a robot on Mars?

Question

What are some of the difficulties in communicating over long distances?

Safety Precautions

- Blindfolded students ("robots") should move slowly. Partners ("sensors") should watch out for robots' safety at all times.

Apparatus

obstacles in a room (such as three chairs and a table)
several objects placed in room (such as pencil sharpener, eraser)
ruler
compass
blindfold

Materials

index cards
pencils
felt pens (dark colours)
a small notepad with tear-off pages

Procedure Group Work

1. Organize in groups of four. With your group, create a map of the room with a route for the robot to follow. The purpose is to move the robot from the starting point to the finishing point while performing various tasks on the way.
2. As a group, design and create command and response cards for the robot and sensor. Number the command cards in the order in which they will be given. Study your map to decide which commands to write. Each card should have just one command or response.

Sample Commands and Responses

Commands from Robot Operator to Robot
• Move forward _____ metres • Turn right • Move backward _____ metres • Turn left

Commands from Robot Operator to Sensor
• Take a photograph • Beep three times when approaching a collision • Beep once when object cleared • Retrieve sample from surface (pick up small object on table)

Responses from Robot to Robot Operator (to be completed on mission by Communications Link)
• Collision detected (show location on map) • Command completed for command # _____ (number on card) • Command failed for command # _____ (number on card)

Responses from Sensor to Robot Operator
• Photograph taken • Sample retrieved

3. Each student in the group has a different role:

   Robot operator: Sits in a corner of the classroom and facing away from the robot

   Communication link: Takes communication cards to and from the robot operator and the robot and sensor

   Robot: A blindfolded student; only able to move slowly while following directions

   Sensor: Accompanies the robot, but cannot communicate directly with the robot; is able to beep when approaching obstacles, draw pictures, and retrieve samples when directed by the robot operator

4. Place the robot operator in the mission control room where he or she is not able to see the robot.

5. Place the blindfolded robot and the sensor at the starting point.

6. The robot operator gives command cards to communication link one at a time.

7. The mission is complete when the robot has collected samples and pictures and reached the finish point.

Analyze

1. List some of the difficulties your robot had. What caused each difficulty?
2. How could you redesign the robot to avoid some of these difficulties?
3. Why do you think the robot was blindfolded in this activity?

Conclude and Apply

4. **(a)** How does this activity model the use of a real robot on the surface of a planet or moon?
   (b) How is this activity different?

Figure 8.6 This cryobot is a 1 m long torpedo-shaped robot.

Types of Robots

Cryobot

The **cryobot** in Figure 8.6 is a robot that can penetrate thick ice. The prefix *cryo-* means freezing. A cryobot uses heat to melt the ice, and gravity to sink downward. Instruments in the robot send data back to the surface using radio waves. A cryobot uses less power than a drill and does not pollute the environment. Scientists are planning to send cryobots into space in the next few years to explore icy regions on Mars and Europa.

Frogbot

The **frogbot** is a robot with a single leg that has a spring at its jointed "knee" (see Figure 8.7). When the spring releases, the bent leg straightens and the frogbot hops. A frogbot that takes a 1.8 m hop on Earth could jump 6 m in low-gravity conditions. The frogbot is powered by solar panels and has onboard cameras, sensors, and computers. It can operate independently, and is able to right itself if it falls over. Scientists hope to use frogbots to explore comets and asteriods.

What are some differences between a cryobot and a frogbot?

Figure 8.7 Frogbots can be used in places where wheels might not work.

Find Out ACTIVITY 8–E

Robot in a Rabbit Hole

There are three holes in a field. Each hole is about 25 cm in diameter and leads into a tunnel. Design a robot to explore the tunnels.

What to Do

1. With a partner, list several questions you want to answer. (Example: Are the tunnels connected?)
2. Design your robot. Consider these points:
   - How will the robot move?
   - How will it measure the length and direction of each tunnel?
   - How will you communicate with the robot?
   - How will you retrieve it if it gets stuck?
3. Draw and label a sketch of your robot. Explain how each part of your robot works. Give your robot a suitable name.

What Did You Find Out?

1. What was the most difficult part of the robot to design? Why?
2. Can your robot measure changes in the depth of a tunnel? If not, how could it do this?

Extension

3. Suppose your robot comes across an obstacle such as a rock or a living creature that blocks a tunnel. How can you find out what the obstacle is?

4. Computers are used to design and test robots. How could you use a computer to help you make a robot?

Skill POWER

For tips on solving technological problems such as this one, turn to SkillPower 6.

Pause& Reflect

The design of some robots is copied from nature. What other technologies are copied from nature? Record your ideas in your notebook.

Snakebot

What shape of robot did you design in Find Out Activity 8-E? Engineers have developed a 2 m-long flexible **snakebot** to investigate pipes beneath city streets (see Figure 8.8). Shaped like a string of sausages, the snakebot has cameras at both ends to collect data about cracks in the pipes. Operators direct its movements using remote-control radio signals. One day snakebots might be used to explore loose piles of rock or underground spaces on other planets.

READING Check

What can a snakebot do that a spiderbot cannot? What can a spiderbot do that a snakebot cannot?

Spiderbot

The **spiderbot** is a multi-legged robot that can climb over a variety of difficult surfaces (See Figures 8.9A and B). Like other robots described here, the spiderbot's design is copied from nature. Spiderbots carry sensors that they can put in different places to create an interconnected web. Information webs allow scientists to set up a large network to gather data at a low cost.

Figure 8.8 A snakebot moves on small wheels.

Figure 8.9A A spiderbot is shown with one of its inventors, Robert Hogg.

Figure 8.9B In what kind of environment do you think a spiderbot could be used?

Section 8.1 Summary

- The stages of exploring space depend on tools and technologies.
- A robot on a planet's surface provides more detailed information than does a spacecraft orbiting the planet.
- Robots help explore extreme environments that would be dangerous for humans.
- Some examples of robots are rovers, cryobots, frogbots, snakebots, and spiderbots.

Check Your Understanding

1. (a) What are the advantages of building a lander that is shaped like a pyramid?
   (b) What are the disadvantages?
2. What are the main parts of a rover?
3. (a) Identify three different robots you have in your home.
   (b) How does each of these robots help you?
4. **Apply** If you were designing a program for a Mars rover to search for signs of life, what would you have it look for?
5. **Thinking Critically** What qualities do humans have that could never be replaced by robots?

The Personal Satellite Assistant (PSA) is a robot camera the size of a softball that floats inside a spacecraft. The PSA can check air quality and onboard systems, and provide access to computer data. Motion sensors keep it from bumping into things. Its design was inspired by a similar sphere in the movie *Star Wars*.

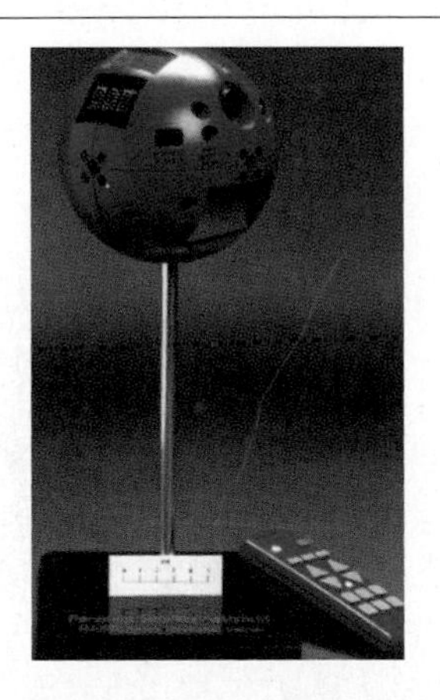

Section 8.2 Satellites and Space Stations

Key Terms
- satellite
- geosynchronous

Have you watched a live hockey game on television while it is happening in another part of Canada? If so, you have used a space satellite. A **satellite** is any object in orbit around another object in space, such as Earth or another planet. Earth has one natural satellite, which is the Moon.

There are hundreds of human-made satellites in orbit around Earth. People all over the world depend on satellites for communications, research, observation (see Figure 8.10), monitoring, scientific mapping, and other purposes. Like robots and rovers, satellites help us explore environments from the ocean floor to space.

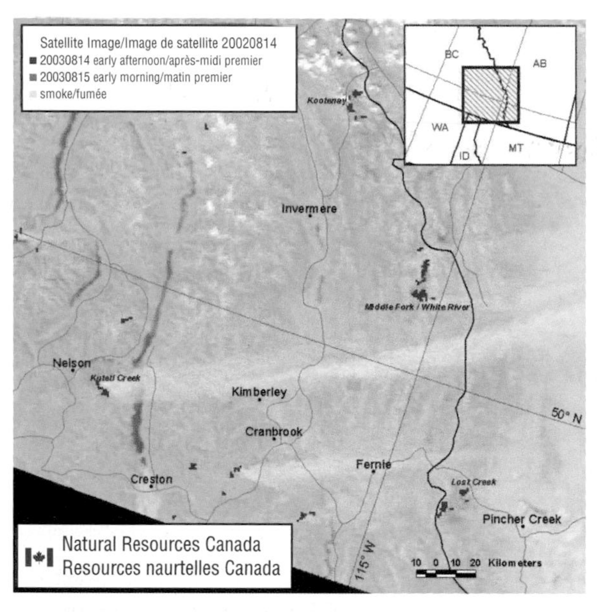

Figure 8.10 Forest fires burned through many parts of British Columbia in 2003. How does a satellite image such as this help firefighters? In what other ways might satellites help firefighters?

Communications Satellites

Canada was the first country in the world to put a satellite into orbit to help people communicate across long distances. The *Anik 1* communications satellite was launched in 1972. Many more *Anik* satellites, like the one shown in Figure 8.11, have been launched since then. The newest versions relay almost all of Canada's television broadcasts, as well as high-speed Internet services.

Figure 8.11 The *Anik E2* satellite. "Anik" means "brother" in the Inuit language Inuktitut. Why might this be a good name for a communications satellite?

Some communications satellites orbit Earth at low altitudes of 200–800 km above the planet. These low-flying satellites take about 90 min to circle the planet. At night, you may see one of these satellites as a bright light moving quickly across the sky.

Other satellites are in orbit at a much higher altitude, about 36 000 km above the equator. These higher satellites are in a **geosynchronous** orbit, which means they move at the same speed as Earth rotates, making one complete orbit in 24 h (see Figure 8.12A). As a result, each satellite appears to be motionless above a point on Earth's surface. Satellite receivers, like the one in Figure 8.12B, are fixed in a position that points directly toward a geosynchronous satellite.

Geo- means Earth, and *synchronous* means at exactly the same time. What is a geosynchronous orbit?

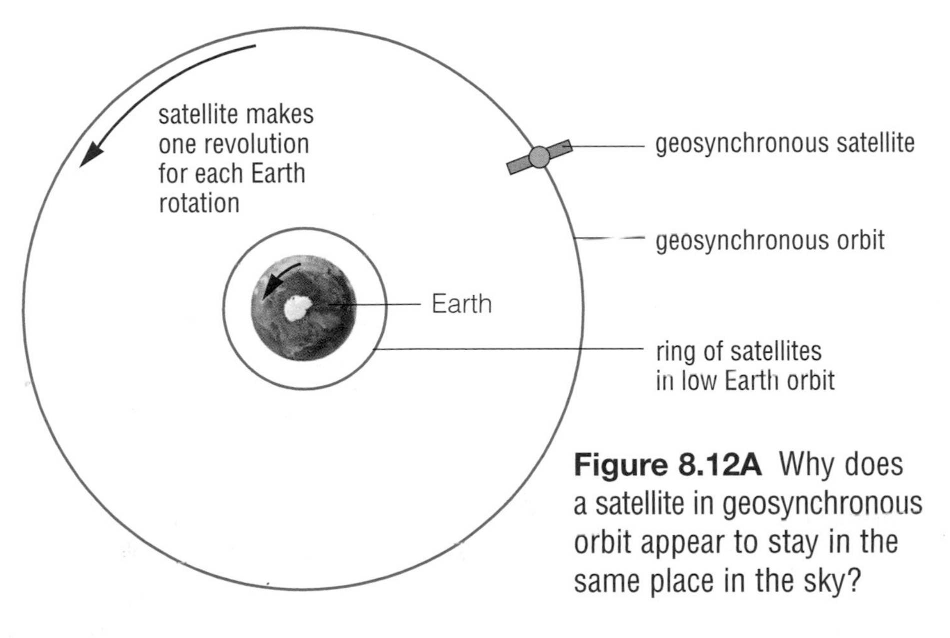

Figure 8.12A Why does a satellite in geosynchronous orbit appear to stay in the same place in the sky?

Figure 8.12B In addition to TV broadcasts, satellite communications can be used in remote areas for learning and health programs and Internet services.

Global-Positioning Satellites

Explorers can travel anywhere on the planet and know exactly where they are using a global-positioning system (GPS) unit (see Figure 8.13). This unit receives radio signals from GPS satellites and compares the signals to calculate an explorer's position on the planet. They are accurate to within a few metres.

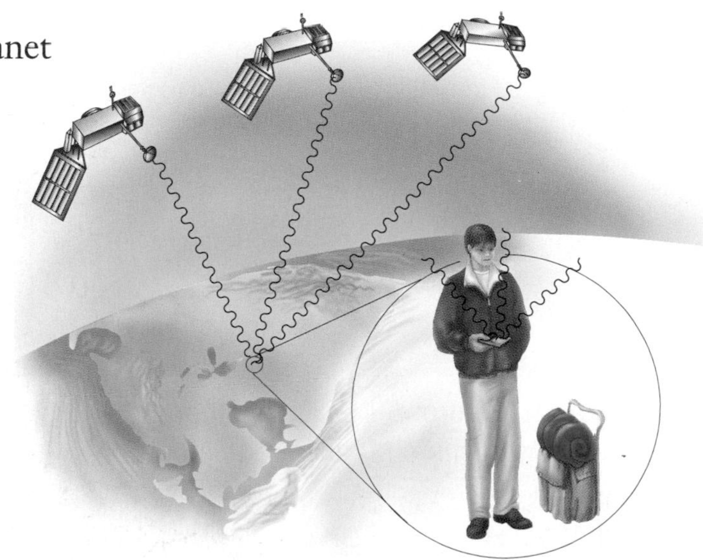

Figure 8.13 More than 24 GPS satellites circle the planet. A GPS unit needs at least three signals to calculate your position.

DidYouKnow?

Space-exploration satellites (also called space probes) leave Earth's orbit and travel into space to observe other planets and moons. These satellites take measurements of the surface and atmosphere, and send data and pictures back to Earth.

Observation Satellites

We use a huge variety of satellites to observe changes in the world around us. One that you may know is the weather satellite. Looking down from high above Earth, weather satellites track the movements of storms and other weather systems, such as hurricanes. Because clouds may block views of Earth's surface, many observation satellites record information using various forms of energy such as radar, infrared, ultraviolet, and X-rays (recall Section 7.4).

We use observations of Earth from space in many ways. For example, farmers can use satellite data to monitor the amount of moisture in soils or to detect crop infections at an early stage. Satellites can track pollution in the ocean and follow the movement of icebergs. Satellites can measure the effects of volcanic eruptions, forest fires, or tsunamis, as shown in Figure 8.14. Prospectors can use satellite images to explore for oil, gas, minerals, and water deposits.

What do observation satellites observe?

Governments use reconnaissance satellites, also called spy satellites, to gather information on military activities such as missile launches, troop movements, or weapon construction. Spy satellites can also pick up communications signals from any country on Earth.

Figure 8.14A Before: A satellite photograph of the Banda Aceh shore of Indonesia before the tsunami of 2005.

Figure 8.14B After: What details does this satellite photograph show of the destruction caused by the tsunami?

Find Out ACTIVITY 8–F

Weather Forecasts from Space

How do satellite images help people forecast the weather? Different satellites produce different images. In this activity, you can interpret two different satellite photographs.

What to Do

1. Study the two photographs. Photograph A shows a chain of three storms over the Atlantic Ocean. Photograph B shows the middle storm one week later.

What Did You Find Out?

1. **(a)** Which photograph was taken from a satellite in geosynchronous orbit?
   (b) How do you know?
2. What information does photograph B have that A does not?
3. What information does photograph A have that B does not?
4. How can the information in these photographs be used?

INTERNET CONNECT

www.mcgrawhill.ca/links/BCscience6

Design a satellite to study hurricanes, or to help lost hikers, or to observe the surface of a planet. Or perhaps you'd like to design a satellite to help Antarctic explorers or Canadian peacekeepers communicate with friends back home. Go to the web site above and click on **Web Links** to find out where to go next.

Pause & Reflect

Most Canadians in the space industry today work on projects involving satellites. What space jobs can you name? Which jobs are of interest to you? Record your ideas in your notebook.

DidYouKnow?

Hubble's future will be decided sometime in 2005. There may be a completely robotic servicing mission in 2007 or 2008. However, the telescope might not receive any further servicing and instead someday be replaced with another telescope.

Hubble Space Telescope

The Hubble space telescope, shown in Figure 8.15, orbits Earth, but looks out into space. Launched in 1990, it provides stunning views of objects in space that cannot be seen using ground-based telescopes. The telescope was designed to allow astronauts to replace its worn-out parts from time to time and to upgrade instruments. This ensures that the telescope has the latest technology.

Every day, the Hubble space telescope sends huge amounts of data to astronomers all over the world. The telescope has given scientists much better views of comets, planets, and other objects, and observed the birth and death of distant stars.

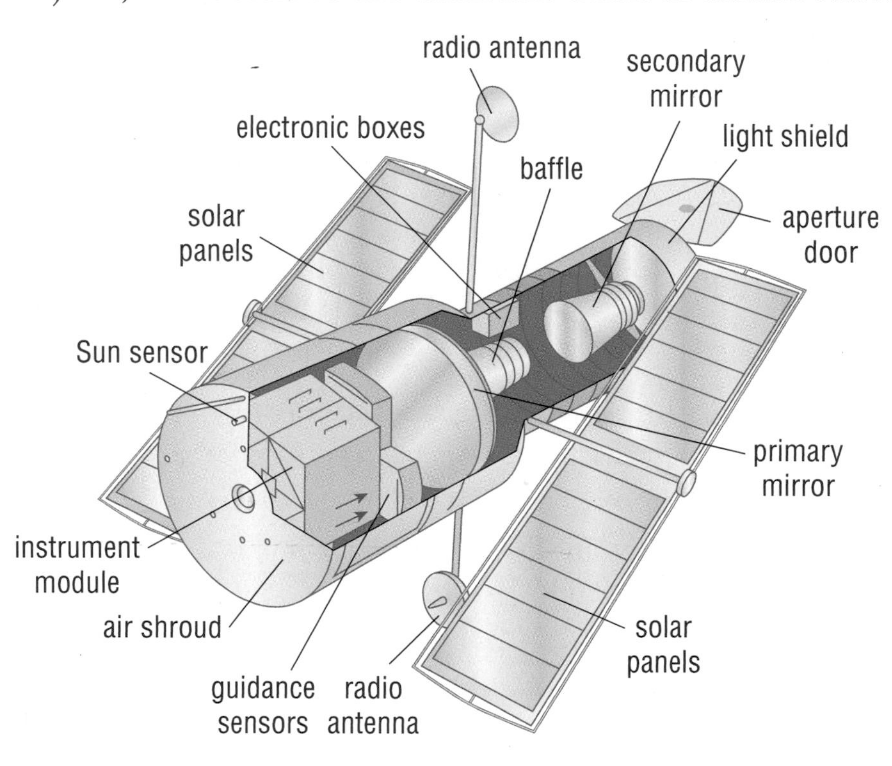

Figure 8.15 The Hubble telescope data have provided a much clearer understanding of black holes—objects with such strong gravity that not even light can escape them.

Figure 8.16 Four pairs of solar panels power the ISS. About 40 percent of the energy generated is used for conducting experiments.

International Space Station

The International Space Station (ISS), shown in Figure 8.16, was built from modules (self-contained units) launched into Earth orbit and connected together in space by astronauts. The first module, *Zarya*, was launched in 1998. When it is fully completed, the station will measure 108 m long and have six laboratories for scientific research. Rotating crews of at least two astronauts have occupied the ISS since November 2, 2000. Over 130 astronauts, including a few tourists, have visited the ISS.

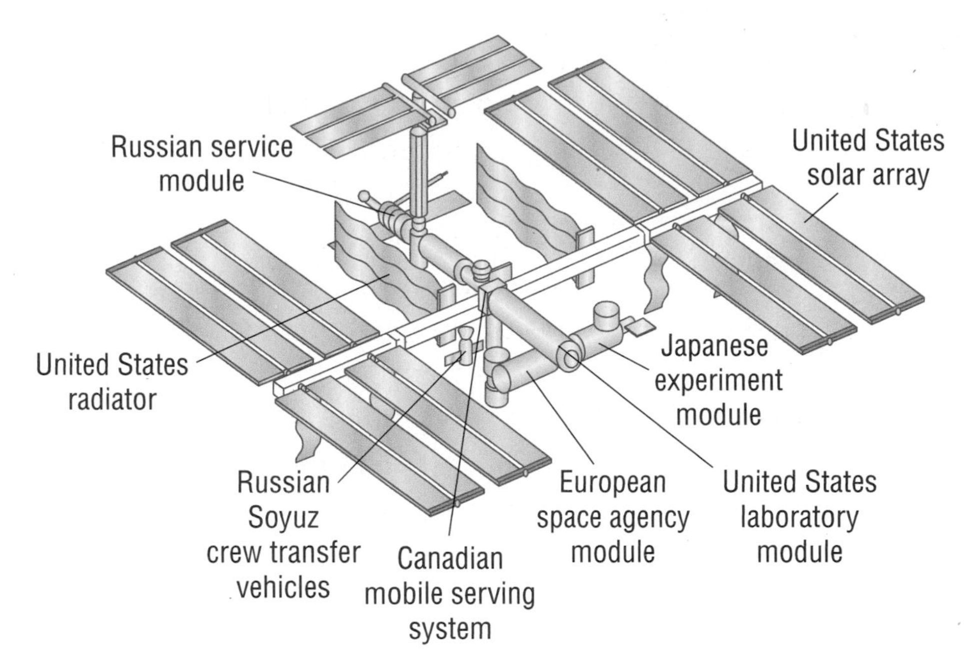

Figure 8.17 The International Space Station is the most complex international scientific project ever undertaken. The 16 nations participating include Canada, the United States, Russia, Japan, Brazil, and 11 European nations.

An important step in designing and building a space station began in the early 1970s, when the United States developed plans for a reusable spacecraft—the space shuttle. Until then, costly spacecraft were used for only a single space journey each. The space shuttle can fly out to the space station with supplies and new crews, then return and be reused for other flights.

The United States launched a smaller space station called *Skylab* in 1973. *Skylab* eventually fell from orbit in 1979 and burned up in the atmosphere. The Soviet Union developed a very successful space station named *Mir*, which remained in orbit from 1986 until 2001.

A Laboratory in Space

The International Space Station allows scientists to carry out experiments that they cannot do on Earth. For example, how do long periods in space affect the growth and development of different kinds of living things? Scientists can grow pure crystals in the microgravity of space and make new types of materials that may one day be used on Earth. A station in space also allows researchers to study gravity, which can lead to a better understanding of how the universe developed.

The ISS circles Earth at a speed of almost 30 000 km/h. It sees 16 sunrises and sunsets within a day.

INTERNET CONNECT

www.mcgrawhill.ca/links/BCscience6

What do tiny flame balls, new scents for flowers, sandcastles, and thousands of tomato seeds have in common? They are all part of experiments in space. Go to the web site above and click on **Web Links** to find out where to go next.

READING Check

What is the purpose of the ISS?

Find Out ACTIVITY 8–G

Technology Timeline

Exploration technologies go through different stages of development as new materials are created and new discoveries are made. Find out the history of a technology.

What You Need

research materials
display materials such as poster paper and markers

What to Do

1. In pairs or in small groups, choose a technology used to explore an extreme environment. For example, you might select a robot, submarine, or GPS.
2. Use a library, the Internet, or other sources to research the developments and major events that have taken place in your technology during the last 100 years. Find pictures as well as facts.
3. With your group, discuss how you will design a timeline to display the information and pictures you have found. Be creative. Decide what time intervals you will use to show dates and decades. For example, the timeline on the right shows the highlights of Canadian satellite launches.
4. Prepare your timeline and present it to the class.

What Did You Find Out?

1. What were you surprised to learn about your technology?
2. What do you think was the most important stage of its development? Explain.
3. How could you improve your timeline?

Extension

4. What developments do you predict for the future of your technology?

Key Canadian Satellite Launches

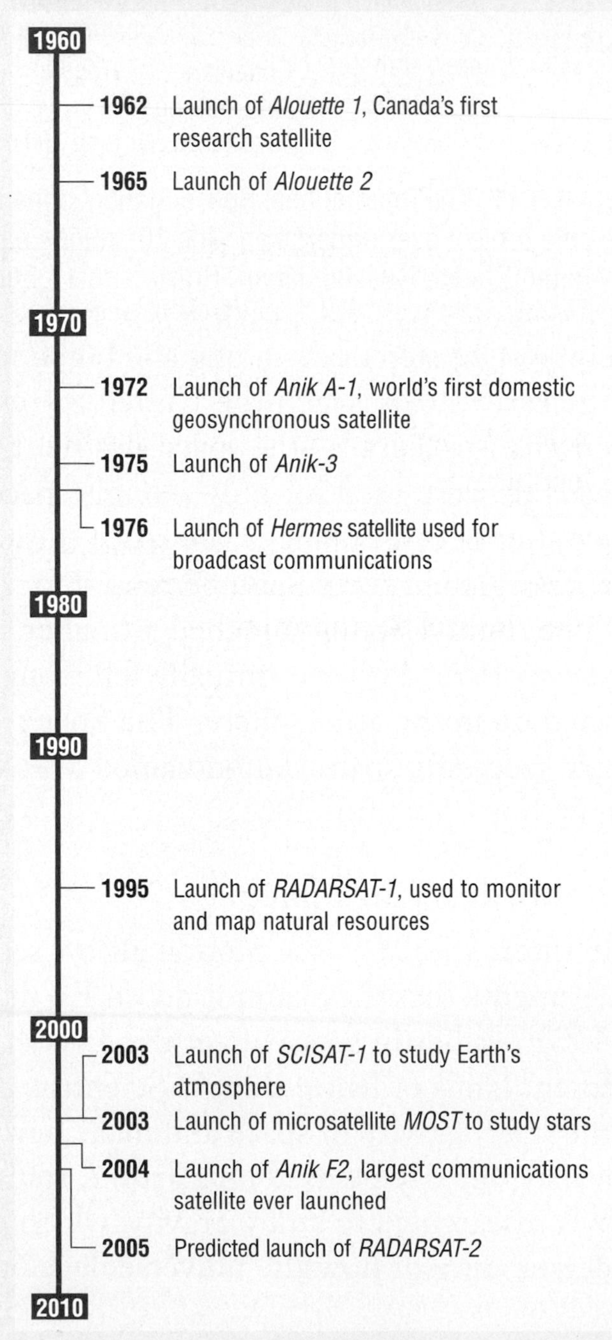

Section 8.2 Summary

There are hundreds of human-made satellites in orbit around Earth, including the following:

- satellites such as those used for global-positioning, communications, and observation
- the Hubble space telescope
- the International Space Station

The development of new technologies happens in different stages.

Check Your Understanding

1. Satellites are used to observe weather. What are two other ways that satellites are used to help collect information?
2. How is a satellite in geosynchronous orbit different from other satellites?
3. How does a GPS unit work?
4. What does the Hubble space telescope observe?
5. How was the International Space Station constructed?
6. **Apply** What are two ways that satellites could be used to find lost hikers?
7. **Thinking Critically** Scientists conduct microgravity experiments in the International Space Station. Why might microgravity experiments be important?

DidYouKnow?

Some cars are equipped with GPS units that can report accidents and provide location information. Almost all ocean-going ships use GPS for navigating the oceans of the world.

More than 130 giant planets (the size of Jupiter and Saturn) have been discovered in orbit around nearby stars by using telescopes. A new space mission to search for planets in other solar systems will begin in 2007.

Section 8.3 Canadian Technologies

Key Terms
- Dextre
- Canadarm
- Exosuit
- Deepworker
- NEPTUNE
- fibre optics

You have learned that Canada is a world leader in satellite technology. Canadian scientists and engineers have also developed robot arms for working in space, a diving suit for undersea exploration, and a network of sensors on the sea floor that will revolutionize studies of the ocean. You will discover these and other Canadian technologies next.

Robot technology that was originally developed for space is now at work in movie studios. Dinosaurs and monsters move as if they are alive, thanks to robot technology.

Dextre

The robot in Figure 8.18 weighs nearly 1000 kg and has arms 3 m long. Despite its huge size, the robot can carry out tasks with a precise and gentle touch. This robot, named **Dextre**, is part of Canada's contribution to the International Space Station (ISS). Dextre is designed to carry out jobs normally done by astronauts during space walks, such as replacing small batteries, assembling parts, and manipulating scientific instruments. Having the robot do these jobs will allow astronauts more time to conduct scientific experiments.

Figure 8.18A An artist's model of Dextre in action on the International Space Station.

Figure 8.18B Dextre has arms that can turn, reach, and grip in seven different ways.

Canada to the Rescue

Figure 8.19 An astronaut is attached to Canadarm as he is moved towards the Hubble space telescope.

Dextre is the latest version of an earlier Canadian remote manipulator named **Canadarm**. The Canadarm was first used on board the space shuttle *Columbia* on November 13, 1981. Since then it has been used on over 40 missions and has never failed.

One of Canadarm's tasks was to help capture and hold a five t docking module, then turn the module 90° and lock it onto the space station. The Canadarm has also been used on missions to repair the Hubble space telescope and several communication satellites. Without the Canadarm, these satellites would have become unusable. There are plans to use Dextre on the ISS to help engineers develop and test new technology for future space exploration.

Operating Canadarm

Canadarm is operated by an astronaut using joysticks, as shown in Figure 8.20. All Canadian astronauts are trained to operate Canadarm as well as to take *space walks* (work outside of the spacecraft or space station) and conduct scientific experiments. In 1999 astronaut Julie Payette operated the Canadarm in orbit and became the first Canadian to board the International Space Station (see Figure 8.21).

What are three tasks that Dextre can do?

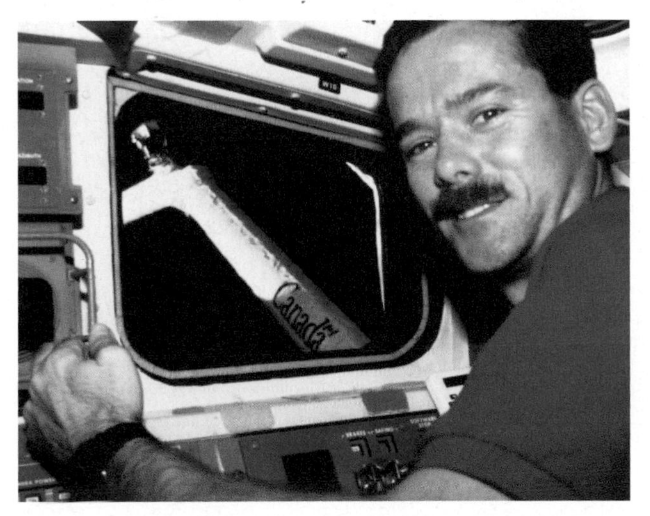

Figure 8.20 Canadian astronaut Chris Hadfield grips two joysticks that control the moves of the Canadarm, much like the controls you use to play video games.

Figure 8.21 Captain Julie Payette was Canada's second woman in space.

PROBLEM-SOLVING

INVESTIGATION 8-H

SKILLCHECK

- Identify the Problem
- Decide on Design Criteria
- Plan and Construct
- Evaluate and Communicate

Build a Space Arm

To build Canadarm, engineers had to develop a tool with the skills of the human arm and the strength of a machine. In this investigation, you will design and create a mechanical device that can be used to collect samples in space.

Challenge

Design and build a device to grasp and hold objects in space.

Safety Precautions

- Wear safety googles to build and test your device.

Materials

paper
pencils
ruler
books and magazines
found materials for model construction

Design Specifications

A. Your drawing must show how an operator would control the device. Label each part.

B. Your device must be able to pick up and hold an object.

Skill POWER

Turn to SkillPower 6 for tips on solving technological problems like this one.

Plan and Construct

1. With your group, discuss different technologies that already exist to extend the reach of people and to collect objects. How can these technologies be adapted for use in space? What extreme conditions of space do you need to consider?
2. Design and draw a device. Include different views of the device to show how the working parts interact.
3. With your group, construct a model of the device using materials such as wood, elastic bands, and string. Give your device a name.
4. Test your device. Make further adjustments as needed.
5. Present your drawing and model to the class. Explain how your device works.

Evaluate

1. Could your model work in the extreme environment of space? Suggest why or why not.
2. What special considerations would be needed to transport your device into space?
3. What ideas did other groups have that you could use for your model?
4. How would you improve your model if you could use other materials to build it?

Extension

5. What challenges do you think the designers of Canadarm faced?

Other Canadian Technologies in Space

Canadian astronaut Bjarni Tryggvason invented the microgravity vibration isolation mount or "MIM," (see Figure 8.22). MIM uses magnets to make a vibration-free floating platform for conducting experiments. Canadian researchers hope to use this platform on the ISS to melt and mix various materials to make new metals, computer parts, and other products for Canadian industry.

DidYouKnow?

Bjarni Tryggvason completed high school in Richmond, British Columbia.

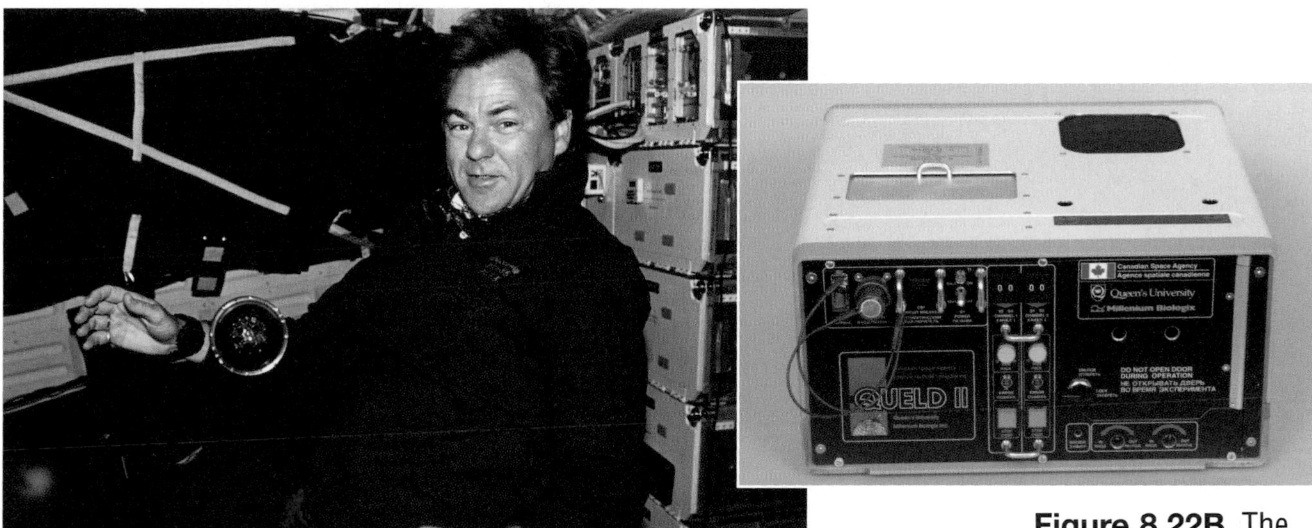

Figure 8.22A Astronaut Bjarni Tryggvason has MIM fluid on his wrist, about to perform an experiment in space.

Figure 8.22B The microgravity vibration isolation mount is about the size of a small suitcase.

Another Canadian technology is the thermal plasma analyzer (see Figure 8.23A) This instrument was designed to gather samples of the Martian atmosphere in order to determine how the atmosphere was formed and what it is made of. Its first mission was to be on board the Japanese spacecraft *Nozomi* (see Figure 8.23B). Unfortunately, one of the spacecraft's thruster valves was damaged and *Nozomi* was forced to fly past Mars in 2004 instead of orbiting the planet as planned.

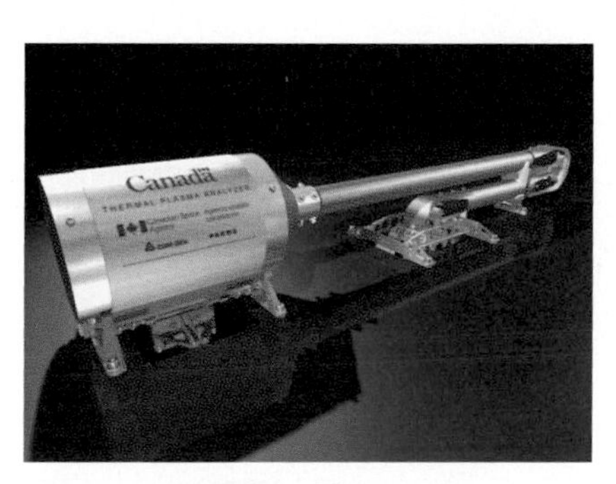

Figure 8.23A Thermal plasma analyzer

Figure 8.23B Nozomi

Canadian Ocean Technology

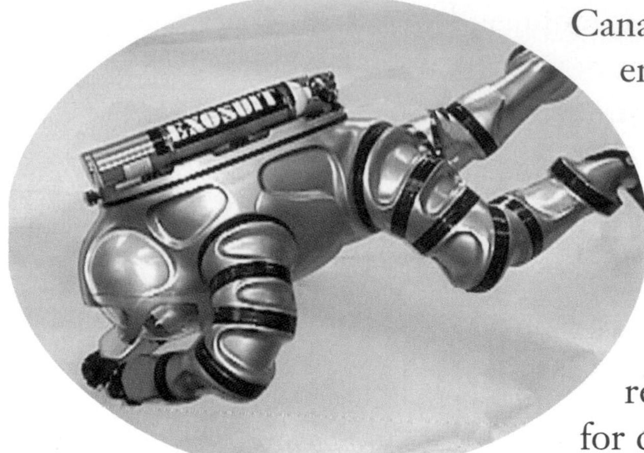

Figure 8.24 How are the Exosuit and a spacesuit similar?

Canadian technology has applications in extreme environments on Earth as well as in space. Two-thirds of planet Earth is covered by water, but scientists know more about deep space than they do about the deep ocean. Below a depth of about 40 m, divers need the protection of a vehicle or a special diving suit. Figure 8.24 shows a recent version of a revolutionary diving suit for deep-ocean exploration designed by Vancouver engineer Phil Nuytten. The **Exosuit** weighs only 54 kg and can be used at ocean depths up to 200 m. Pincers in the hands allow the diver to work underwater.

Another of Nuytten's inventions is a single-seat submersible called **Deepworker**, shown in Figure 8.25. A Deepworker can explore to a depth of 600 m and can be fitted with a pair of manipulator arms. Deepworkers can be used to recover valuable booster rockets that have fallen from their spacecraft into the ocean.

The original "Newtsuit," developed by Nuytten in the 1970s, is a hard diving suit that can be worn to a depth of 300 m. It is sometimes called "the submarine you can wear."

Figure 8.25 A micro-submersible called Deepworker is shown on its way into Vancouver harbour.

How is a Deepworker different from an Exosuit?

The NEPTUNE and VENUS Projects

Even with the best diving apparatus, divers can only explore small areas of the ocean and remain underwater for only a few hours or days at a time. A different technology will soon allow researchers to study large areas of the ocean continuously. The **NEPTUNE** project, illustrated in Figure 8.26A, will lay a 3000 km network of powered fibre optic cable on the seabed off the coast of British Columbia, Washington, and Oregon. **Fibre optics** is a technology of using very long thin tubes of glass or plastic to transmit data (see Figure 8.26B).

This cable network will include 30 or more seafloor "laboratories," spaced about 100 km apart. Land-based scientists will monitor and control sampling instruments, video cameras, and remotely operated vehicles based at the laboratories. Because the instruments in the network are interactive, scientists will be able to carry out experiments. Scientists can also observe events as they happen, such as fish migrations and underwater volcanic eruptions. This information will be available to scientists around the world in real time through the Internet.

A smaller test network called VENUS has begun observations off the coast of southern British Columbia. The project is led by the University of Victoria. This new Canadian technology allows explorations of the ocean environment that are not possible using ships, submersibles, or even satellite imaging.

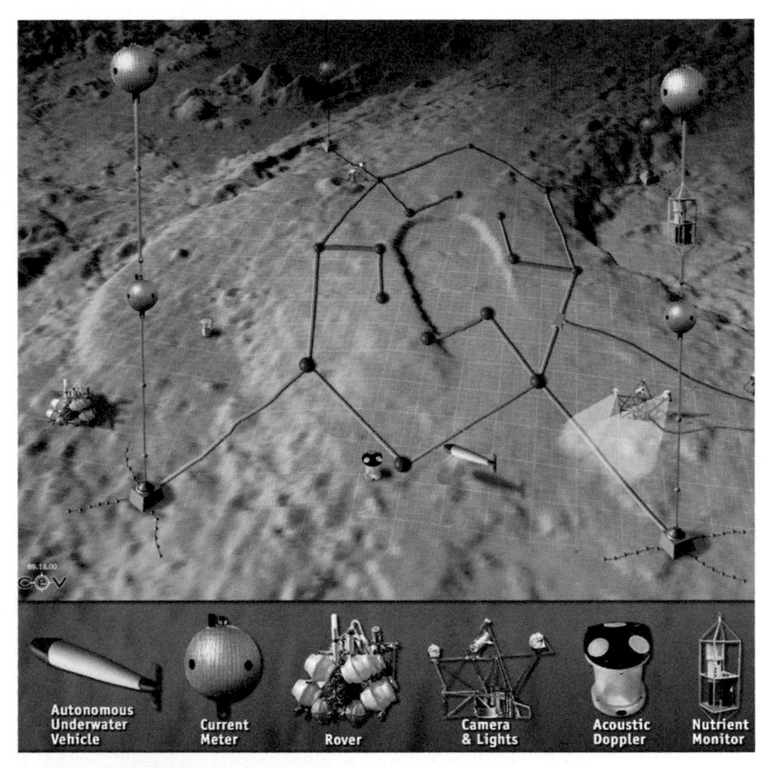

Figure 8.26A NEPTUNE is expected to begin operation in 2007.

Figure 8.26B There are many fibre optic tubes in a fibre optic cable.

DidYouKnow?

In the NEPTUNE project, NEPTUNE stands for the **N**orth-**E**ast **P**acific **T**ime-series **U**ndersea **N**etworked **E**xperiments. The name VENUS in the VENUS project stands for **V**ictoria **E**xperimental **N**etwork **U**nder the **S**ea. A third project off the coast of California is called MARS (Monterey Accelerated Research System).

Find Out **ACTIVITY 8–I**

Canadian Technologies

From satellites to diving suits, Canadians have pioneered technologies for exploring extreme environments. Research some of these inventions and the people involved.

You can use this information in the following Find Out Activity 8-J, Extreme Board Game.

What You Need

research materials

What to Do (Group Work)

1. Your teacher will assign your group one of these extreme environments:
   - polar regions
   - volcanoes
   - caves
   - oceans
   - deserts
2. Prepare a class chart with these four columns:

Name and Description of Technology	Date of First Use	People/Groups Involved	Extreme Environment Where It Is Used

3. Your group will use the library or Internet to research Canadian technologies used in your environment. Include traditional Aboriginal technologies (see section 7.1) as well as recently-developed technologies.
4. Add the information you find to the class chart.

What Did You Find Out?

1. **(a)** What information did you learn that surprised you?
   (b) Why did it surprise you?
2. **(a)** What technologies would you like to learn more about?
   (b) How could you find out more about them?

DidYouKnow?

Canadian Joe MacInnis designed and built a "subigloo"—an underwater resting place for scuba divers working in polar waters.

INTERNET CONNECT

www.mcgrawhill.ca/links/BCscience6

What other technologies have Canadians invented? Go to the web site above and click on **Web Links** to find out where to go next.

Find Out ACTIVITY 8–J

Extreme Board Game

Design and create a "Canadian Extreme Exploration Board Game."

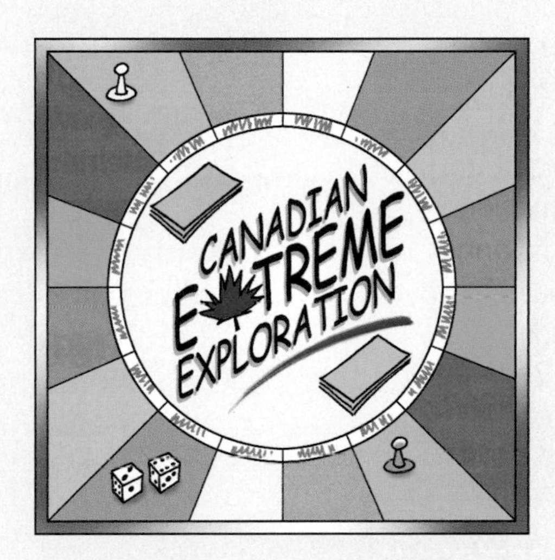

Safety Precautions

- Be careful when using sharp objects.

What You Need

construction materials such as paper, cardboard, scissors, and markers

Optional
dice

What to Do Group Work

1. Work with your group to design and create a board game. This board game will teach the development history of Canadian technologies for exploring extreme environments. Use information you have learned in this unit, including your research from Find Out Activity 8-G and the class chart for Find Out Activity 8-I. Conduct more research if necessary. Include these key features:
   - the conditions found in extreme environments
   - different stages in the development of the technologies
   - the pros and cons of each technology
   - important inventors
   - events that influenced the technology
   - resources needed to build the technologies
2. Create a list of rules for playing the game. Test your design to ensure it works as you planned. If not, change it as necessary.
3. Trade games with another group and play.

What Did You Find Out?

1. What new information did you learn about Canadian technologies by making and playing the game?
2. What evidence did you find in your research that shows new technology is the result of many previous developments?

DidYouKnow?

When *Apollo 11* landed on the Moon in 1969, it landed on Canadian-built landing gears.

Section 8.3 Summary

In Section 8.3 you learned:

- Canadian space technology is being used on the space shuttle, the International Space Station, and other spacecraft.
- Dextre is the latest version of the Canadarm.
- Canadian ocean technology includes diving suits, underwater exploration vehicles, and underwater laboratories.

Pause& Reflect

How do you think Dextre could be adapted to work under water? How could it be adapted to work in polar regions? Draw your ideas in your notebook. Show what Dextre could be used for in each environment.

Check Your Understanding

1. What tasks is Dextre designed to do?
2. What is the microgravity vibration isolation mount used for?
3. What are some differences between the Exosuit and a spacesuit?
4. What is the purpose of the NEPTUNE project?
5. **Apply** Why would it sometimes be necessary to use the Exosuit rather than a robot, even though it is safer to use a robot? (Recall your experiences communicating with your robot in Conduct an Investigation 8-D).
6. **Thinking Critically** Canadarm is designed to work in microgravity. How do you think it would work on Earth?

Section 8.4 Putting Technology to the Test

Before using technology in space, scientists and engineers test their design in locations on Earth that resemble the planet where the technology will be sent (see Figure 8.27). Technologies are not always successful the first time they are tested. Sometimes a new design, or a new material, or a new procedure (way of doing something) may fail. As a result, missions to extreme environments such as space can result in injury, loss of life, or the loss of very expensive equipment.

Key Terms

space junk

Figure 8.27A Dr. Pascal Lee crosses the Mars-like landscape of Nunavut to test the new design of a planetary suit.

Figure 8.27B Dextre is tested many times before travelling to the International Space Station.

Twenty-four astronauts have died in space accidents between 1961, when the first human flew in space, and 2004. Because of the risk to human lives, some people argue that explorations in space should be carried out using robots rather than people. Robot missions not only save lives, they are usually less costly than missions using human explorers. On the other hand, robots can only collect data they are programmed to collect. Humans are more adaptable and can collect a wide range of information.

READING

Why does technology need to be tested many times before use?

INTERNET CONNECT

www.mcgrawhill.ca/links/BCscience6

Find out more about space "spinoffs"—inventions that have changed human life thanks to space technology. Go to the web site above and click on **Web Links** to find out where to go next.

Benefits of Space Technology

Why do humans explore space if there are so many dangers? There are many benefits to humans that have resulted from the development of space technology (see Figure 8.28). The Canadarm technology has been used for robots that handle explosive and hazardous materials on Earth. Firefighters' air tanks, voice-controlled wheelchairs, improved car brakes, and better school bus design have all resulted from space technology.

Figure 8.28 Bicycle helmets, ski boots, golf balls, flat-screen TVs, and the bar codes used in grocery stores have been improved because of space technology.

Hazards of Space Travel

One of the biggest hazards to travellers in space is the risk of collision with tiny particles that break off from meteors and from orbiting space vehicles. **Space junk** refers to the old satellites, parts of spacecraft, tools, and other garbage travelling in orbit around Earth at very high speeds. Even a chip of paint moving at thousands of kilometres per hour can damage sensitive equipment or puncture a spacesuit. Astronauts in space also need to be protected from harmful radiation such as ultraviolet rays and X-rays.

SKILLCHECK
- Identify the Problem
- Decide on Design Criteria
- Plan and Construct
- Evaluate and Communicate

Suited for Space

When astronauts work outside their spacecraft, they must wear spacesuits to protect them. Engineers must design spacesuits to protect astronauts from space junk.

Challenge

Design and make a jacket to protect a potato from being punctured.

Safety Precautions

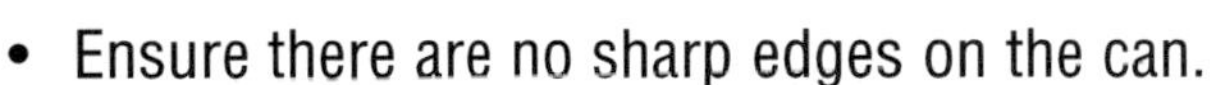

- Ensure there are no sharp edges on the can.

Materials

plastic milkshake straws
potatoes (red skinned are best)
empty, clean tuna or similar sized can to hold the potato upright
materials for construction of suit

Design Specifications

A. Your spacesuit must protect the potato from being punctured by the straw.

B. Your spacesuit must be made of flexible material.

C. Your spacesuit cannot be more than five layers thick of wrapping material.

Plan and Construct

1. Demonstrate the risk of a potato being punctured by a straw in the following way.
   - Place the potato firmly in the empty can.
   - Hold the straw in one hand and stab the potato with a *slow* motion. Does the straw penetrate the potato?
   - Repeat the experiment, but this time stab the potato with a fast motion.
   - Compare the two results. As a class, discuss how this demonstration illustrates the effects of high-speed impacts of tiny particles in space.
2. In small groups, discuss how you could make a "spacesuit" to protect the potato from being punctured by a straw. With your group, discuss the different types of materials you could use. Which materials give most protection against the straw? How will you make your spacesuit? How will you attach it to the potato?
3. Design and construct your spacesuit.
4. Test the spacesuit with the straw. Adapt or change your design if necessary.
5. Present your spacesuit to the class, explaining the features that help protect the potato from being punctured.

Evaluate

1. Consider all the designs of spacesuits presented.
   (a) What are the advantages of the different designs?
   (b) What are the disadvantages of the different designs?

2. How would you improve your spacesuit if you were to repeat the investigation?

Extension

3. What other technologies have scientists designed to protect people from impacts? Make a collage of different technologies that protect people.

Making Choices

In the last activity, you discovered that there are different ways of solving the same problem. In the same way, explorers can often achieve their goals using different kinds of technology.

For example, what source of energy would you use to power a submarine for underwater exploration? Some submarines, like the one in Figure 8.29, burn diesel fuel, which needs oxygen to burn. A diesel submarine is an example of a *hybrid vehicle*, which uses two types of fuel. While on the surface of the ocean, a diesel submarine uses power from its engine to charge large batteries. Once the batteries are fully charged, the submarine can head underwater and use electrical power obtained from its batteries. The amount of charge in the batteries determines how long the submarine can stay underwater.

Figure 8.29 A diesel submarine

An alternative source of power used in some submarines is nuclear power. Nuclear submarines, like the one shown in Figure 8.30, use small nuclear reactors to produce heat. The heat converts water to steam, and the steam drives a mechanism that turns the submarine's propeller.

Nuclear reactions do not need oxygen, so a nuclear submarine can stay underwater for long periods of time. Nuclear fuel lasts much longer than diesel fuel. While a diesel submarine must return to land to replace its fuel, a nuclear submarine can remain at sea without refuelling for years if necessary.

Figure 8.30 What risks does this nuclear submarine have compared with a diesel submarine?

INVESTIGATION 8-L

SKILLCHECK
- Communicating
- Classifying
- Inferring
- Questioning

Using Technology Responsibly

Think About It

Almost every new technology has both benefits and risks. Sometimes there may be benefits for one group of people and risks for another group. How would you make decisions about using such technology?

Procedure

1. As a class, brainstorm different types of technology that you have learned about in this unit and others that you know. **Record** your examples in a three-column chart like the one below.

Name of Technology	Benefits/ Advantages	Problems/ Disadvantages

2. Choose one technology and create a large T-chart on the board to record the pros and cons of this technology. Discuss how each item on your chart could affect decisions about using the technology.

Pros	Cons

3. Working in pairs or small groups, choose one technology from the list. Make a list of the key features of your technology. Some questions to get you started are:
   - What does it do?
   - What extreme environment is it used in?
   - What energy source does it use?
   - How does it affect the environment?
   - What happens to any waste material it creates?
   - How long will it last?

4. Create a Venn diagram to illustrate the key features of the technology. What are the benefits? What are the costs and risks? Which key features are benefits to both science and humans? Suggestions for the labels on your diagram are: Costs for Science, Risks for Humans, Benefits to both Science and Humans.

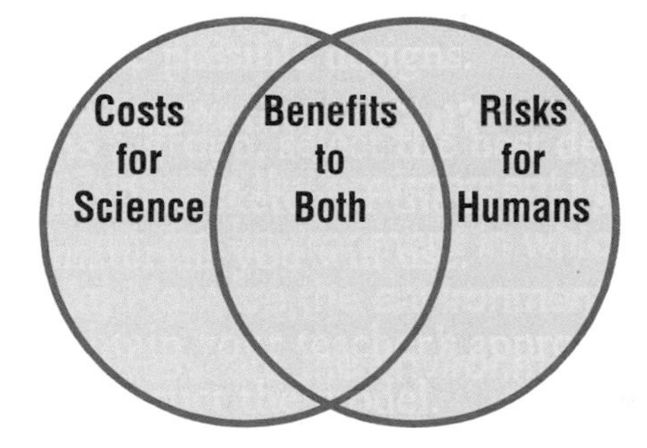

5. As a group, present your Venn diagram to your class.

Analyze

1. What difficulties did you have completing this investigation?

2. What are the most important considerations when deciding whether to use a new technology in an extreme environment?

3. Who do you think should make the decisions about the use of technologies to explore or conduct research in extreme environments?

Section 8.4 Summary

In Section 8.4 you learned:
- Technology has to be tested many times before being used.
- There are both benefits and hazards that come from using technology.
- There are many different points to consider when choosing a technology.

Check Your Understanding

1. (a) What are the advantages of sending robots rather than people to space?
   (b) What are the advantages of sending people rather than robots to space?
2. List three benefits of space technology.
3. List three hazards of space travel.
4. What is meant by responsible use of technology?
5. **Apply** How would you redesign an exploration technology from this chapter to be useful for your environment?
6. **Thinking Critically** What are the advantages and disadvantages of international co-operation for producing exploration technologies?

Pause& Reflect

Would you like to ride a space elevator to the Moon? Engineers are designing a paper-thin, super-strong ribbon of carbon between Earth and outer space. An elevator attached to the 100 000 km-long ribbon could haul equipment—and even people—up into space.

Draw a picture of what this elevator might look like. Record your ideas about the view you would have from the elevator.

CHAPTER at a glance

Now that you have completed this chapter, try to do the following. If you cannot, go back to the sections indicated in brackets after each part.

(a) How are a rover and a lander related? (8.1)

(b) Describe the main parts of a rover. (8.1)

(c) What is a geosynchronous satellite? (8.2)

(d) How do global-positioning satellites help people determine their location? (8.2)

(e) Describe three ways in which people benefit from satellite technology. (8.2)

(f) What is the Hubble space telescope used for? (8.2)

(g) What is the purpose of the International Space Station? (8.2)

(h) What are three Canadian contributions to space research and technology? (8.3)

(i) Why is it important to test technology? (8.4)

(j) What are some benefits of space technology on Earth? (8.4)

(k) What are some hazards of space travel? (8.4)

(l) What does it mean to use technology responsibly? (8.4)

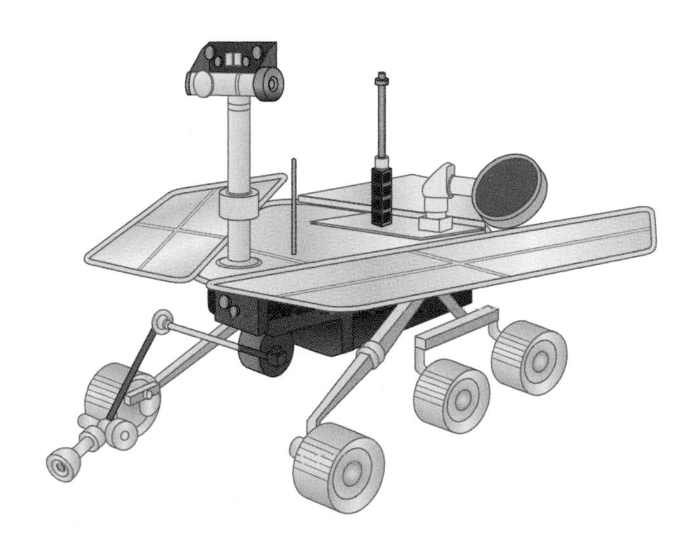

Prepare Your Own Summary

Summarize this chapter by doing one of the following. Use a graphic organizer (such as a concept map), produce a poster, or write a summary to include the key chapter ideas. Here are a few ideas to use as a guide:

- Make a model of a rover.
- Make a poster illustrating the different types of satellites and their uses.
- Design a robot to perform a space job.
- Select a Canadian technology. Write a brief speech explaining its benefits, costs, and risks.

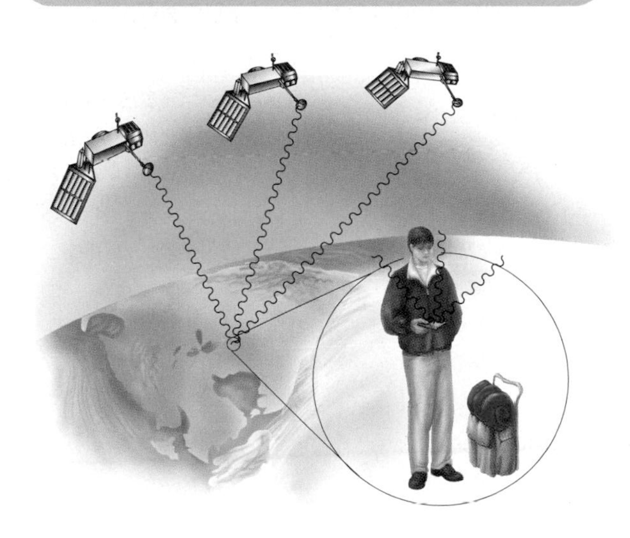

CHAPTER 8 Review

Key Terms

rovers
lander
robot
cryobot
frogbot
snakebot
spiderbot
satellite
geosynchronous
Dextre
Canadarm
Exosuit
Deepworker
NEPTUNE
fibre optics
space junk

Reviewing Key Terms

If you need to review, the section numbers show you where these terms were introduced.

1. Describe the difference between:
   (a) rover and lander (8.1)
   (b) robot and satellite (8.1, 8.2)
   (c) Exosuit and Deepworker (8.3)
   (d) Canadarm and Dextre (8.3)

2. In your notebook, match each description in column A with the correct term in column B. Use each description only once.

A	B
(a) a receiver that picks up radio signals from GPS satellites in orbit above Earth	• *Anik*
(b) robot designed to penetrate through thick ice	• GPS
(c) telescope that orbits Earth, but looks out into space	• Hubble space telescope
(d) a mechanical device that does some of the work of humans	• Dextre
(e) communications satellite	• cryobot
(f) Canadian-made robot used on the International Space Station (ISS)	• Deepworker
(g) single-seater submersible	• robot

Understanding Key Ideas

Section numbers are provided if you need to review.

3. The global positioning system is used around the world. Describe three different occupations or activities where using a GPS unit would be helpful. (8.2)

4. What is the advantage of the Hubble space telescope over other telescopes? (8.2)

5. What are at least four considerations that should be discussed before technology is used? (8.4)

Developing Skills

6. Study the diagram of the International Space Station, then answer the following questions:
   (a) What section is probably solar collectors?
   (b) Where are laboratory tests done?
   (c) Where does the crew sleep?
   (d) Where should technicians make any repairs?
   (e) Name two countries involved.

1 Research Modules
2 Service Module (behind)
3 Power Modules
4 Lab Module
5 Logistics Module
6 Experiment Module
7 Accomodation Module
8 Solar Arrays

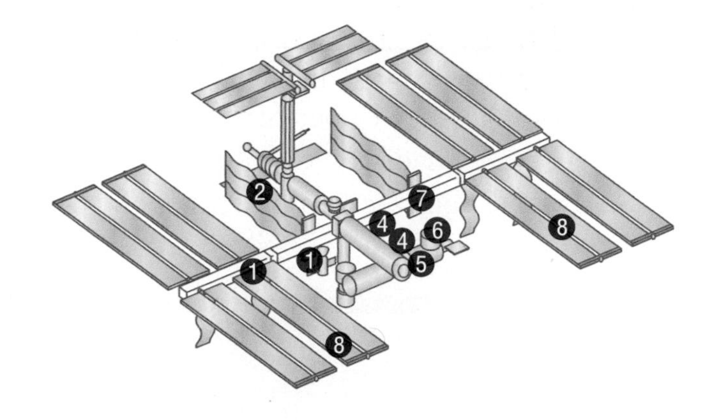

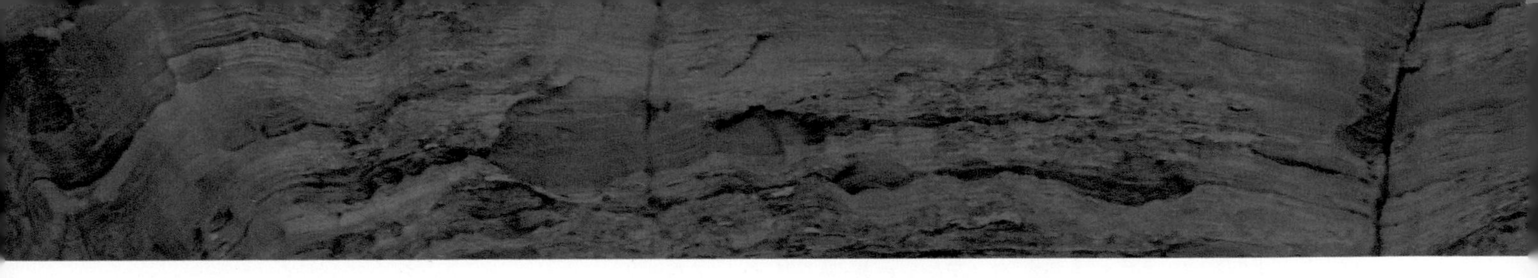

Problem Solving

7. A geosynchronous satellite is about 35 600 km above Earth's surface. How much time does it take a radio signal, travelling at 300 000 km/s, to reach an overhead geosynchronous satellite and go back down to the receiving station? Use the formula $t = \frac{d}{s}$, where t = time, d = distance, and s = speed.

Critical Thinking

8. There have been many movies in which a navy ship uses sonar to hunt a submarine. Sonar is a type of sound wave that is emitted by the ship. It rebounds off the submarine, producing an echo that the sonar operator can hear. What are some limitations to this technology?
9. Why is satellite communication technology especially important to Canada?
10. The photo at right shows a new type of space robot, called a Tumbleweed rover, being tested in Greenland. What features of this rover make it suitable for travelling great distances on other planets and their moons?
11. Why do humans need complicated technology to survive in and explore extreme environments?

Pause & Reflect

Go back to the beginning of this chapter on page 226, and check your original answers to the Getting Ready questions. How has your thinking changed? How would you answer those questions now that you have investigated the topics in this chapter?

UNIT 3 Ask an Elder

Wilfred Jacobs

Years ago, the Kootenay region of British Columbia was made up of a series of lakes, rivers and marshlands. The first European explorers to arrive in this area noticed the unique canoes used by the Aboriginal people who lived there.

Wilfred Jacobs is an Elder from the Ktunaxa (too-NA-ha) Nation. He is an expert at building the traditional sturgeon-nosed canoe. Wilfred's Ktunaxa name is *Ni‡sik Ak‡am* (nuk-sik ACT-sam), which means Buffalo Bull Head. Wilfred has travelled around the world to share his knowledge of Aboriginal cultures. He has spoken to many audiences, including the Queen of England. He lives in Yaqan Nukiy (Creston, B.C.) with his wife and grandchildren.

Q. How did you come to be involved in making this canoe?

A. I was fascinated when I first saw this structure and how it was put together. What really intrigued me was the scientific method that was applied to the construction. My ancestors who originally built this canoe didn't use the term 'geometry,' but if you apply the formulas you can see that it's all based on geometry.

I'd like to make a tribute to all the people who taught me to build the canoe. My wife Agatha and her mother and father more or less kept this handicraft alive. They brought it back. There are other people who are very knowledgeable too. Isabel Louie gave me a lot of information. Some of these people, like Thomas White and Isaac Basil, are gone now. These are the people who taught me how to build this canoe.

Q. How important was the canoe for your people?

A. The canoes were basically used for travelling and gathering. Horses were here then, but it was much easier to jump in a canoe to do your hunting or fishing or gathering. We gathered materials for building, for clothes, and for food.

Only 40 years ago the bottomlands here were all under water. There were dozens of canoes on the shores down here. Everyone owned a canoe, even children. The canoe was a thing you had to have, just like today we have to have our vehicles. The canoe was adapted to the reeds and the high grass in the water. It was very functional and of course it has another, spiritual significance. Its shape shows respect to the sturgeon fish, which is important to this tribe.

Q. Would you have seen this environment as being extreme at the time, before the settlers came?

A. Yes, in the sense that if it wasn't for the canoe people would not have been able to travel around. They would not be able to get through the marshland or to gather their food.

Q. What are the steps involved in making the canoe?

A. The construction of the canoe involves planning, gathering, and knowledge. You have to know how to harvest the different parts of the canoe, and how to assemble them. The canoe is a mix of materials: white pine bark for the cover, cedar planks, cedar roots for sewing, and black cherry bark for binding. These materials only come up at certain times of the year.

In the fall you'd build the frame. You would use long cedar planks, split into about 3 cm widths. The next step is to get cherry bark for the binding. You have to harvest that in May or June. By July the bark starts to harden and you can't work with it any more. Maple is used for the rings and for the ribs. You have to split and shape the maple in the spring too. The wood is easier to work with then. Then there's the pine—you would know where the pine tree was that you were going to use. You would just leave it until you needed it. So you would have to have it all planned.

Q. Do people still use the canoes?

A. No, I'm about the only person who still uses one. I have students asking me how to build them, and I share information with them. Some students have built canoes from books and videotapes, and they are pretty good.

I've been asked many times by the younger generation, why should we learn how to make the canoes? There are a lot of reasons why I learned to make them. I wanted to reaffirm the ingenuity of my people. This is the way they adapted to their environment. They survived for thousands of years using this technology. We pass that knowledge on to our young people saying, "This is the same knowledge that your ancestors had."

Our tribal council used to have a poster that said, "Go into the future with a computer in one hand and a drum in the other." I say this to our young people. Keep your tradition, but yes, learn modern technology too. If you combine those two things, there will be nothing stronger than that.

In what way is the sturgeon-nosed canoe adapted to the wetland environment?

EXPLORING Further

Imagine that you are building a sturgeon-nosed canoe. Write a story or series of journal entries, or create a poster to describe how you build it. Use Wilfred's directions, and research using the library or the Internet if you need more information. Include how the design of this technology is particularly suited to its environment and purpose.

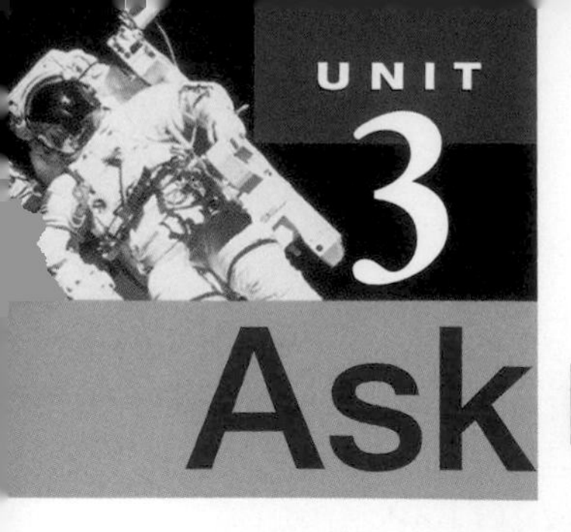

Ask a Robotics Engineer

Mark Fricker

Conditions in space are extreme. There is almost no air pressure or oxygen, and the extreme temperatures and radiation levels make exploration difficult and dangerous. Robots are used to minimize the hazards for astronauts and cosmonauts (Russian astronauts). Marc Fricker is a computer and robotics engineer. He works as a Robotics Instructor with the Canadian Space Agency. His work is part of Canada's contribution to the construction of the International Space Station.

Q. What is your job with the Canadian Space Agency?

A. In lectures and simulator sessions, I teach astronauts, cosmonauts, and ground support personnel how to operate the Canadarm2. There are eight instructors on the Canadian Space Agency astronaut training team. I'm the robotics, electronics, and computer expert. Other instructors are expert with mechanical systems or optics or physics. We combine our knowledge to teach the astronauts what they need to know. One person could not do the job. It takes the full team.

Q. How did you become interested in space robotics?

A. I've been thinking about space robots most of my life. The elementary school I attended in Salmon Arm recently opened a 25-year time capsule. When I was a student, I contributed a letter to that capsule. I wrote of my plans for building robots that would build me a house on Mars. I suppose it was my father who got me interested. He likes working with electronics and computers.

Q. Why do we need robots like Canadarm2?

A. A robot can withstand much harsher environments than humans. Astronauts and cosmonauts make it look easy, but working in a spacesuit outside the space station is cumbersome and very dangerous. Robots can be controlled from the safety of the space station or from the ground. Robots are also much stronger than humans. The maximum payload that the Canadarm2 can move in space is approximately 116 000 kg—about the weight of the space shuttle. No human could hope to move that mass, unaided.

Q. What's in the future for space robots?

A. We'll probably see more space robots being operated from the ground. Eventually we'll need to develop robots with artificial intelligence. Robots with enough intelligence to make their own decisions will be essential for exploration on faraway planets. Because of the vast distance, Earth control will be impossible. Even now, sending a radio signal to Mars can take up to 20 minutes. The Mars rovers—Spirit and Opportunity—under the control of personnel on Earth take up to 3 days just to turn around. If they were smart enough to make their own decisions, that operation would take just a few minutes.

Q. What advice would you give to students who might be interested in space robotics as a career?

A. You're fortunate that space robotics is one of Canada's specialties. Robots are the future of space exploration. Even in life on Earth, robots are important. For example, using robots being developed by Canadian engineers, a surgeon in one hospital will be able to operate on patients in other hospitals, all over the world. The sky is no longer the limit for robot technology!

On February 24, 2005, Canadarm2 on the International Space Station was operated for the first time in a new way. The ground crew on Earth controlled the arm motion and camera controls.

EXPLORING Further

Canadarm2 is just one element of the Canadian-designed Mobile Servicing System (MSS). The MSS is a group of robots that will assist astronauts and cosmonauts in assembling, maintaining, and servicing the International Space Station. Canadarm2 differs from the original Canadarm in several important ways. It's bigger, more flexible, and more mobile. The original Canadarm was permanently attached to the Space Shuttle. Canadarm2 can move around the outside of the International Space Station like a mechanical acrobat, walking on its hands.

- The original Canadarm performed its space shuttle missions very well. What extra demands in the International Space Station's environment required the changes in design?

When either end of the Canadarm2 is firmly fixed onto a control point, the other end is free to do the necessary work.

UNIT 3

Project

SKILLCHECK
- Communicating
- Inferring
- Solving Problems
- Modelling

Home Away from Home

Use your knowledge of the extreme environment of space and what living things need to survive to create a model of a habitat in space. Your model should focus on the following:
- How the basic needs of astronauts in space can be met while in the habitat
- What are the extreme conditions in space that you need to consider?
- Will the habitat be supplied from Earth or will it be self sufficient?
- Have display cards explaining the function and purpose of different parts of your model, for example—how the air supply is produced or how waste is disposed of.

Challenge

Work in a group to follow the design criteria below to design and build a model of a habitat in space that can sustain four astronaut researchers for six months.

Materials

Art supplies; cardboard boxes, plastic bottle and cups, paper cups, straws, aluminum foil, plastic wrap, paints, glue, scissors, felt pens, paint, modelling clay, index cards

Design Criteria

A. The size, shape and style of your model are for you to decide. It must designed as either a cut away model or be able to open up to show the interior of the habitat.

B. Your model must show how the astronauts obtain food and prepare it.

C. Your model must show how the astronauts obtain and recycle water.

D. Your model must show how astronauts deal with waste products and garbage.

E. Your model must show how the astronauts obtain and recycle air.

F. Your model must show how the habitat is designed to protect the astronauts in space from the extreme environmental conditions such as extreme heat or cold, radiation, no pressure or air, and microgravity.

G. You must include display cards to explain the features you have included in your model that show how the basic needs are met.

Plan and Construct

1. With your group, discuss what your habitat might look like and how the needs of the astronauts could be met.
2. Draw a labelled sketch of your habitat.
3. Obtain your teacher's approval and construct the model.
4. Improve your model until you are satisfied with the way it looks and that you have considered all the necessary things to keep the four astronauts alive for 6 months.
5. Write brief descriptions on the index cards that explain the different features of your habitat.
6. Demonstrate your model to your class.

Evaluate

1. What difficulties did you encounter when designing and building your model?
2. (a) What features did you have to change as you built the model?
   (b) Why did you have to change the features?
3. What knowledge of technology did you use when designing the habitat?
4. After viewing the other groups' model presentations, what ideas could you use to improve your model?

Tips and Hints

✓ Consider how plants could be used for food production, air recycling, and waste recycling.

✓ Consider how the astronauts will enter and leave the habitat.

✓ Consider other features you can add to your habitat to make life enjoyable for the astronauts during their six-month stay.

USING GRAPHIC ORGANIZERS

Drawing diagrams and charts helps you make mental pictures of the concepts you are learning. When you look at a diagram or a chart, you can see how different ideas are connected. Diagrams and charts are examples of graphic organizers. In this SkillPower, you will find out how graphic organizers can help you learn and plan in science.

Venn Diagram

A Venn diagram helps you compare objects or ideas. Venn diagrams have two or more overlapping circles. Each circle stands for an object or main idea.

- Where the circles overlap, you write the *similarities* between the objects or ideas.
- You write the *differences* where the circles do not overlap.

Examine the Venn diagram in Figure S1.1. The diagram compares the Arctic fox and the Arctic hare.

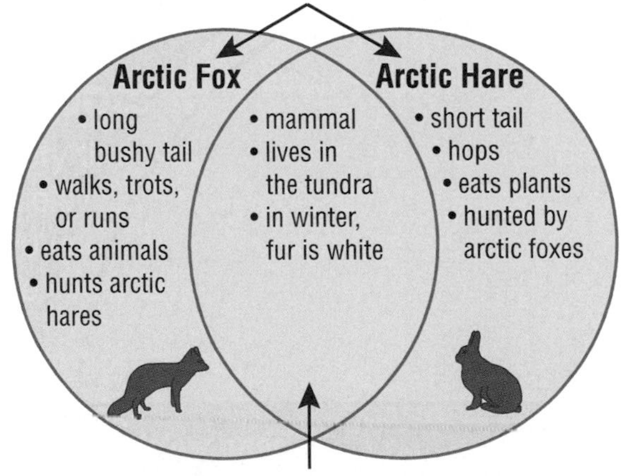

Figure S1.1 Arctic foxes dig burrows for shelter but Arctic hares do not. Where would you write "lives in burrows" on this Venn diagram?

Flow Chart

A graphic organizer called a **flow chart** shows the order of events in a process. For example, the process might be directions for carrying out an experiment. In another flow chart, the process might be changes that happen to an organism over time.

For example, the flow chart in Figure S1.2 shows how a monarch butterfly grows from an egg to an adult butterfly. The word "metamorphosis" means change.

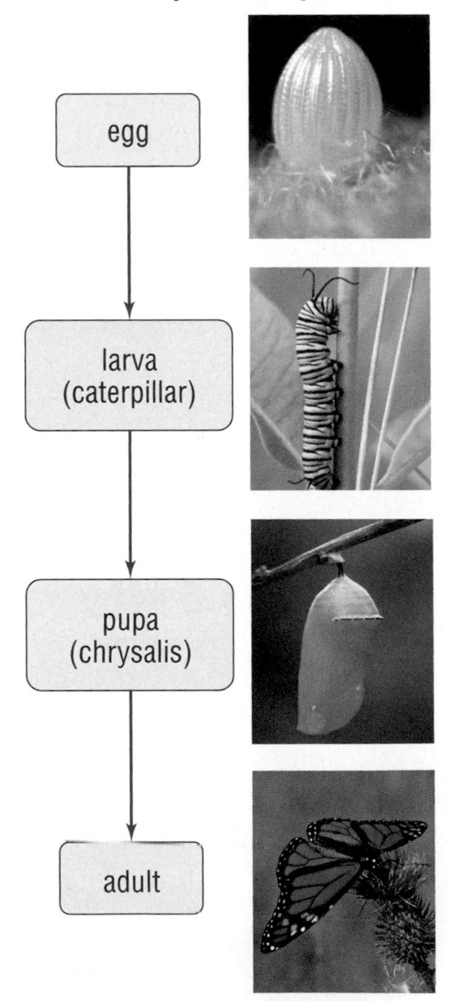

Figure S1.2 Most insects go through four stages of development: the egg, the larva, the pupa, and the adult stages. In butterflies, the larva is called the caterpillar and the pupa is called the chrysalis.

Network Tree

A **network tree** helps you see how one main idea or object breaks down into smaller parts. The network tree starts with the main idea or object. The branches of the tree show how the main idea or object is made up of smaller parts. Network trees are usually drawn "upside-down." The main idea is at the top and the branches go down. The network tree in Figure S1.3 shows you the different groups of protists. (Protists are organisms made of only one cell).

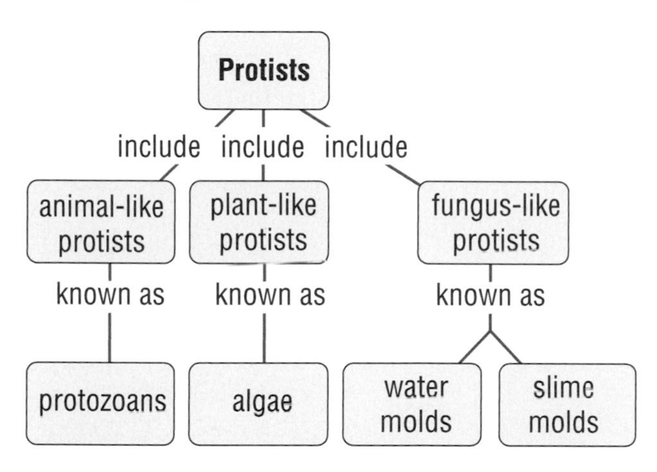

Figure S1.3 This network tree shows you how protists are divided into three main groups.

Concept Map

A **concept map** consists of boxes or shapes that are connected by lines. You write key terms or ideas in the boxes. The lines show how the terms or ideas are connected. You can print words or short phrases on the lines to explain how the ideas are connected. Flow charts and network trees, described above, are both examples of concept maps.

Time Line

A **time line** shows you the order in which events occurred. It also shows you how much time passed between the events. To make a time line, draw a vertical line near the left side of a piece of paper. Make a scale of time on the line. Write the event that occurred beside the year that it occurred. For example, the time line in Figure S1.4 shows events in the development of transportation technologies.

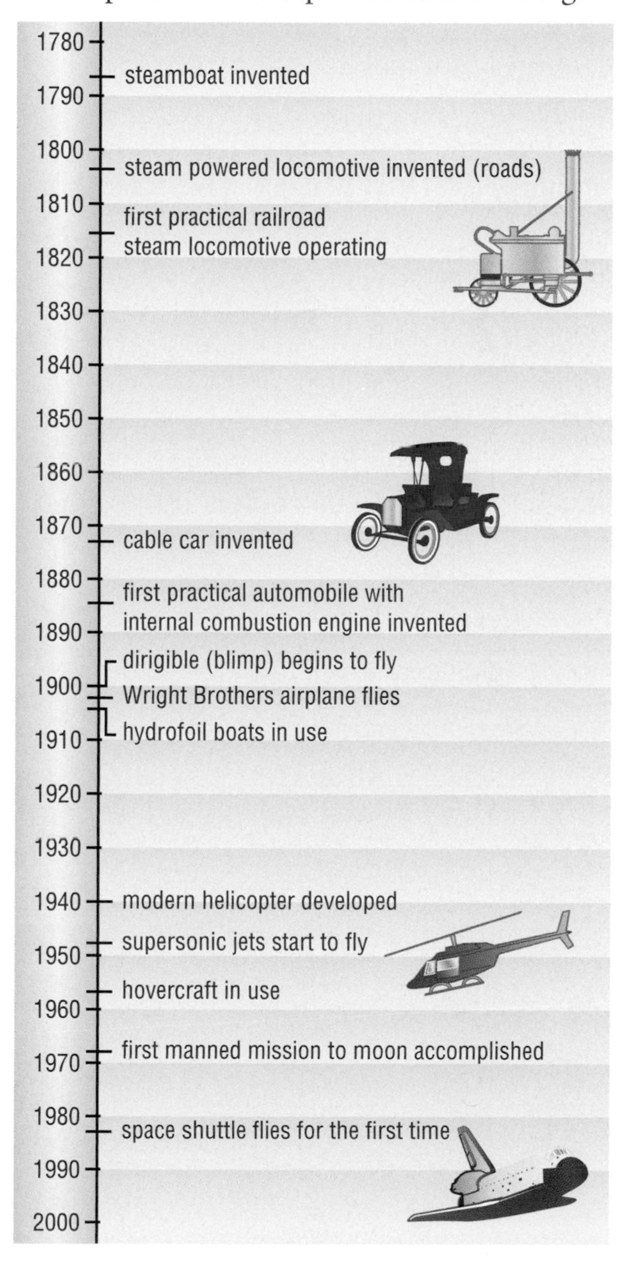

Figure S1.4 You may wish to add illustrations to your time line, as shown here.

Know-Wonder-Learn Chart

A **Know-Wonder-Learn chart** helps you plan and record your study of different science topics. Know-Wonder-Learn charts have three columns.

- In the first column, labelled "What I Know," you write things that you already know about the subject.
- In the second column, labelled "What I Wonder," you write what you would like to know about the topic. While you are learning, you look for answers to the "Wonder" questions.
- As you find answers, you fill in the last column, labelled "What I Have Learned."

Table S1.1 shows a Know-Wonder-Learn chart for hot springs.

Table S1.1 Know-Wonder-Learn Chart for Hot Springs

What I Know	What I Wonder	What I Have Learned
Hot water comes from deep underground.	What causes the water to become hot under the ground?	Cracks in Earth let water seep far below the surface. The water runs over very hot rocks. The hot water expands and pushes up through other cracks.
Boiling hot water kills organisms such as disease-causing bacteria.	Can anything live in the hot water?	Scientists have recently discovered bacteria that can live in boiling water.
Many hot pools have bright colours.	What causes the colours in hot pools?	The bacteria cause the yellow and orange colours.

Mind Map

A **mind map** is a tool that you can use for note-taking, brainstorming, planning a presentation, and many other purposes. In the centre of a large piece of paper, you write a key word that represents the main theme. As you think of ideas or concepts related to the main theme, you write them down or draw a picture on the paper. Then you draw lines to show how all of the ideas relate to each other and to the main theme. It is often helpful to make a mind map when you are brainstorming as a group.

Suppose your group is going to make a presentation about polar bears. You might use a mind map similar to the one in Figure S1.6 on the next page to record a brainstorming session.

Figure S1.5 Hot springs are an extreme environment. In the past, scientists believed that no life could exist in hot springs.

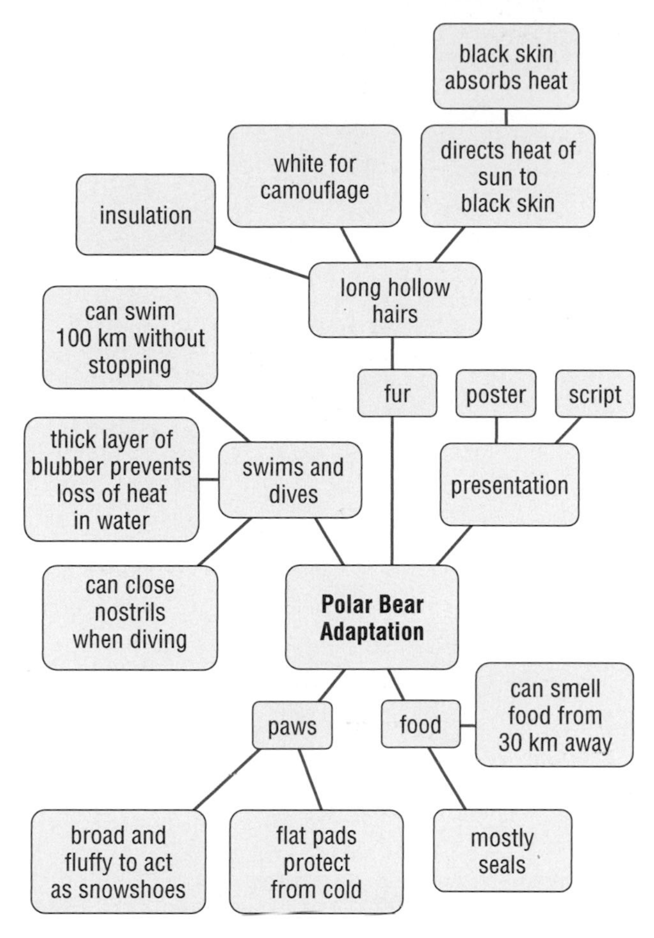

Figure S1.6 A mind map such as this one on polar bears can help you organize your thoughts and ideas. You could add sketches to your mind map to help describe your thoughts.

Practice

1. Create a Venn diagram that shows the similarities and differences between horses and dogs. Here is a list of terms that will give you some ideas:
   - have fur
   - eat meat
   - eat grass
   - are mammals
   - have hooves
   - have claws
   - larger than a person
   - usually smaller than a person
2. Draw a flow chart that shows in detail the process of making a breakfast of your choice.
3. Draw a network tree that shows the living and non-living things that make up your classroom. Here are some terms that you can use for ideas:
   - living things
   - plants
   - people
   - students
   - teacher
   - non-living things
   - furniture
   - desks
   - chairs
   - writing tools
   - chalk
   - pens
   - pencils
4. Create a time line that shows key events in your life.
5. Make a Know-Wonder-Learn chart for *one* of the following topics.
   - adaptations of dolphins
   - where electrical energy comes from
   - the planet Mars
6. Brainstorm about the things you plan to do on your summer vacation. Make a mind map as you brainstorm.
7. What graphic organizer would you use to do each of the following? (You do not need to create the graphic organizer.)
   - **(a)** keep track of brainstorming about life in Antarctica
   - **(b)** show how communication technologies have changed over time
   - **(c)** show the similarities and differences between the planets Earth and Venus
   - **(d)** plan and record your study of deep-ocean environments

CARE AND USE OF THE MICROSCOPE

This year, you might be learning how to use a microscope for the first time. This SkillPower will help you learn about the parts of the microscope and how to use it properly.

Parts of a Microscope

The photographs in Figure S2.1 show three slightly different microscopes. The difference between microscopes A and B is the part of the microscope that moves when you focus it. Microscope C has a mirror as a light source instead of an electric lamp. You will probably be using a microscope similar to one of these. Examine the next page to identify the parts of a microscope.

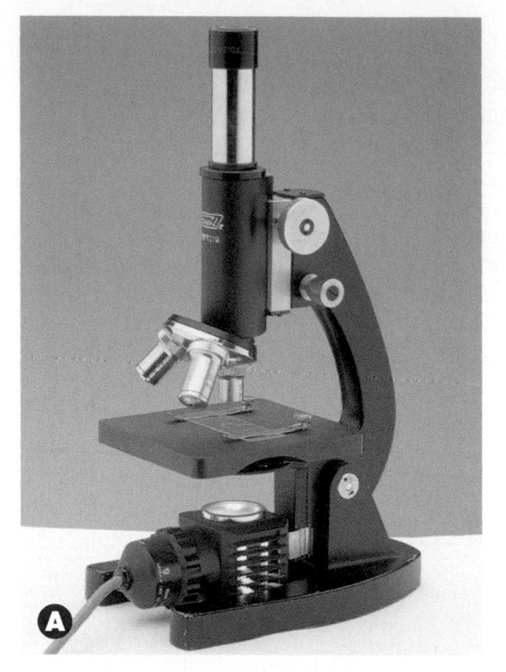

(A) Monocular, compound microscope with a movable barrel.

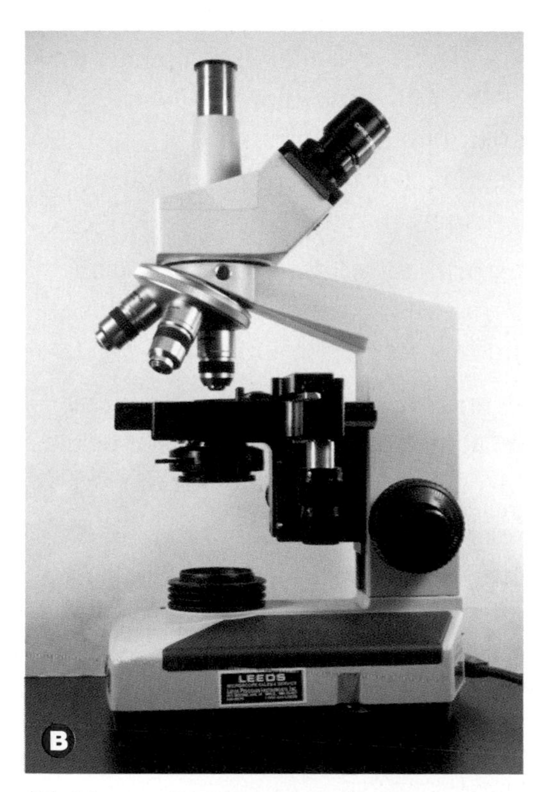

(B) Monocular, compound microscope with a movable stage.

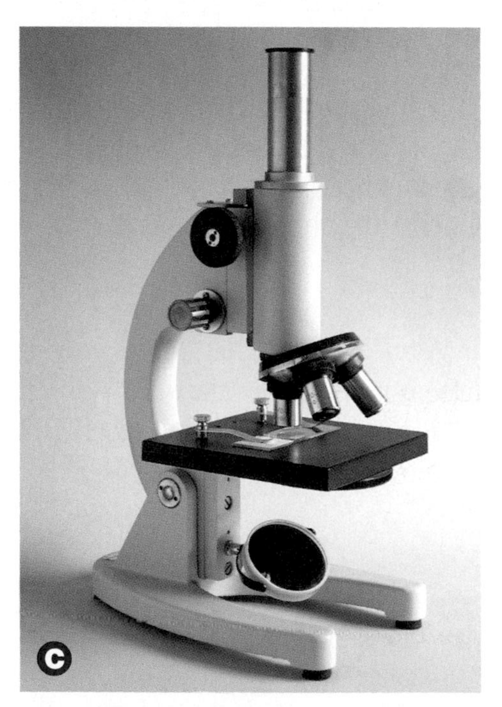

(C) Monocular, compound microscope with a mirror instead of a lamp.

Figure S2.1 When you focus microscope A, the entire barrel, including the eyepiece, tube, and objective lenses, moves up and down. When you focus microscope B, only the stage moves up and down. Microscope C is similar to microscope A. However, this microscope has a mirror instead of a lamp.

A. Light Source

The **light source** in microscopes A and B is an electric lamp that shines on the sample that you are viewing. Microscope C has a mirror instead of a lamp. **CAUTION:** Never use sunlight as the light source for the mirror.

B. Stage

The **stage** supports the objects or microscope slides. Stage clips are attached to the stage to hold the slide in place. A hole in the centre of the stage allows light to pass through the sample on the slide.

C. Objective lenses

The microscopes shown here have three **objective lenses**. Each objective lens has a different magnifying power. Some microscopes will have objective lenses with powers of 4×, 10×, and 40×. Others might have objective lenses with powers of 10×, 40×, and 100×. The three lenses are often called the low-power, medium-power, and high-power objective lenses.

D. Tube

The **tube** (also called the body tube) holds the ocular lens (see E) and the objective lens the correct distance apart. You cannot change this distance.

E. Eyepiece

You look into the **eyepiece** when viewing an object with a microscope. The eyepiece has a lens called the **ocular lens** that magnifies the sample. A lens is a piece of glass that is curved on both sides. Most eyepieces have a label that says 10×. This label means that if you used the eyepiece alone to look at an object, the object would look 10 times larger than it really is. This value is called the *magnifying power* of the lens.

F. Coarse-Adjustment Knob

The **coarse-adjustment knob** moves the barrel or stage of the microscope up and down to focus the sample. Use the coarse-adjustment knob only with the low-power objective lens.

G. Fine-Adjustment Knob

The **fine-adjustment knob** moves the barrel or stage much smaller distances than does the coarse-adjustment knob. You focus with the fine-adjustment knob when using the medium-power and high-power objective lenses. Notice that on the microscope with the moveable stage, the fine-adjustment knob is a smaller knob that sticks out from the centre of the larger coarse-adjustment knob.

H. Arm

The **arm** of the microscope holds all of the parts together in the correct position.

I. Condenser Lens

The **condenser lens** focuses the light from the light source so it passes through the sample. The condenser lens does not magnify the sample. Some compound microscopes do not have a condenser lens.

J. Diaphragm

The **diaphragm** is on the bottom of the stage. You can adjust it to control the amount of light that travels upward through the hole in the stage and through the sample you are viewing.

K. Revolving Nosepiece

The **revolving nosepiece** is a disk that holds the objective lenses. When you want to change to a different objective lens, you rotate the disk.

Using the Microscope

1. When you take a microscope from its storage shelf, always hold it with both hands. Hold the arm in one hand and the base in the other hand. Place the microscope on a sturdy table or desk. Plug the electric cord into an electrical socket.

   CAUTION: Be sure that your hands are dry when handling the electrical cord. When you unplug the cord from the socket, pull from the plug and not from the cord.
2. Microscopes should always be stored with the low-power objective lens (the shortest lens) in position. This means that the shortest lens should be pointing directly down toward the stage of the microscope. If your microscope is not in this position, rotate the nosepiece until the shortest lens is pointing down.
3. If your microscope has a movable barrel, turn the coarse-adjustment knob to move the barrel up. Leave a large distance between the objective lens and the stage. If your microscope has a movable stage, turn the coarse-adjustment knob to move the stage down as far as it will go.
4. You are now ready to view a prepared slide with your microscope.

Viewing a Slide

Reading the instructions for viewing a slide will be easier to understand if you are carrying out the steps as you read. Under your teacher's guidance, carry out the following steps.

1. Obtain a prepared slide from your teacher. Do not touch the middle of the slide with your fingers or you might see your fingerprint instead of the object you are viewing.

   CAUTION: Be very careful when handling slides. The glass breaks easily. The edges are sharp and can cut your fingers.
2. Plug in the electric cord and turn on the switch. If your microscope has a mirror, direct light from a *lamp* (not sunlight) onto the mirror. Adjust the mirror so light reflects through the hole in the stage.
3. Be sure that the diaphragm is clicked into place and will allow light to go upward through the hole in the stage.
4. Move the stage clips out of the way of the centre of the stage. Place the prepared slide on the microscope stage. Place the part of the slide that you want to view in the centre of the opening where the light shines through. Lift the clips slightly and put them on the slide to hold it in place.
5. Look through the eyepiece. Very slowly, turn the coarse-adjustment knob until the object that you are viewing comes into focus. Adjust the knob until you can see the clearest image possible. Sketch what you see.
6. Very slowly, move the slide from side to side. Describe what you see happening in the microscope when you move the slide from left to right.

7. Very slowly, move the slide forward (away from you) and then back again. Describe what you see when you move the slide forward.

8. Next you will look at the slide with the medium-power objective lens. Examine the nosepiece, as shown in Figure S2.2. Decide which way to turn the nosepiece to move the medium-power objective lens into place. Carefully turn the nosepiece until the medium-power lens clicks into place.

Figure S2.2 Always look at the nosepiece when you move it to change objective lenses.

9. Look through the eyepiece. Very slowly turn the fine-adjustment knob until the object is clearly in focus.

   CAUTION: Never turn the coarse-adjustment knob when you are looking through the medium-power objective lens. The lens could touch the slide, which could damage the lens and break the slide.

10. Sketch what you see. How does this sketch compare to the sketch that you made in step 5?

11. Before you remove the slide from the stage of the microscope, return the nosepiece to the low-power objective. Adjust the coarse-adjustment knob to increase the distance between the stage and the objective lens. Carefully lift the stage clips and move them away from the slide. Return the slide to the teacher. Return the microscope to its storage shelf.

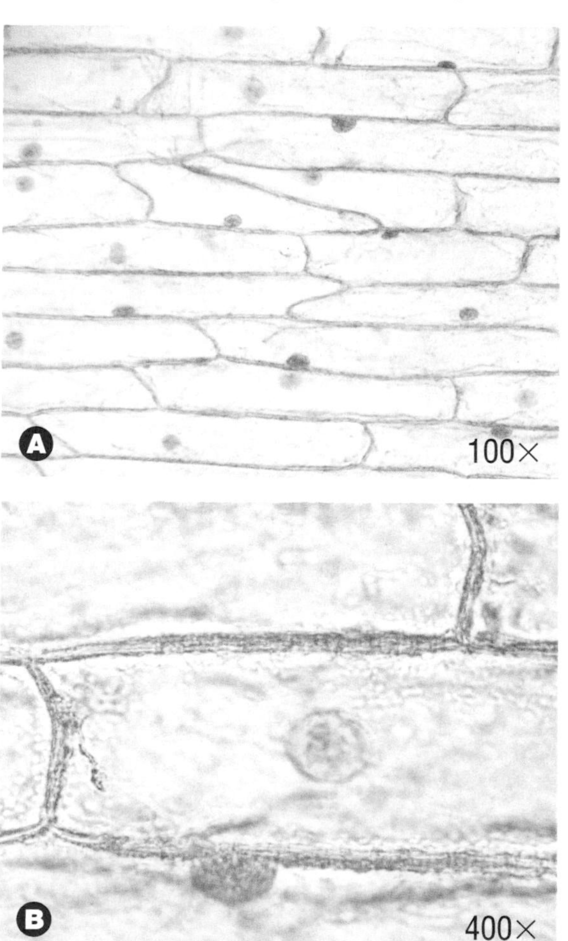

Figure S2.3 How do the images that you sketched in step 5 and step 10 compare to these images?

Examples of photographs taken through a microscope on low-power and medium-power are shown in Figure S2.3. The sample is taken from an onion bulb. The label 100× means that the image you see is 100 times as large as the actual object. The 400× indicates that the image you see is 400 times as large as the actual object.

Preparing a Wet Mount

Sometimes you will want to look at fresh samples such as plant tissue, hair, or fibres. To do so, you will need to prepare a sample called a **wet mount**. Read the steps below that tell you how to make a wet mount. Then complete the Practice that follows.

1. If your sample is dry, use tweezers to place it on a clean, dry microscope slide as shown in Figure S2.4.

Figure S2.4 Handle dry samples with tweezers.

2. Next, use a medicine dropper to place one or two drops of tap water on the sample as shown.

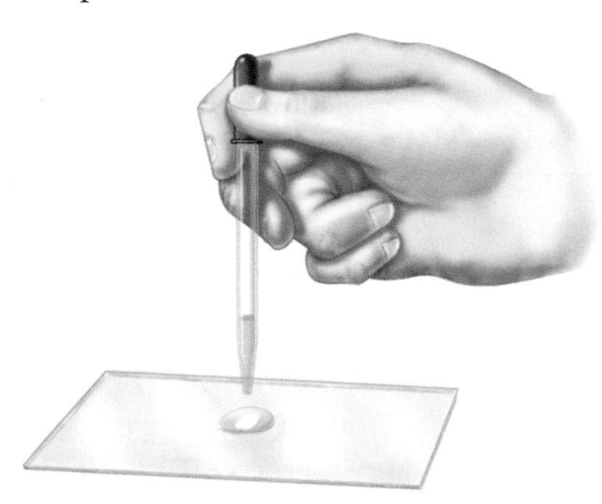

Figure S2.5 Samples must be wet in order to see them clearly with a microscope.

3. If your sample is a liquid such as pond water, use a medicine dropper to place one or two drops of your sample on the slide.
4. Hold a cover slip so that one side of the cover slip touches the slide beside the sample as shown in Figure S2.6.

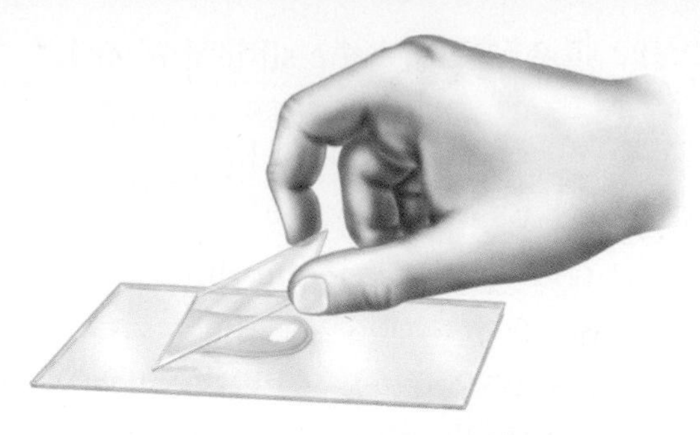

Figure S2.6 Touch one edge of the cover slip to the slide. Then slowly lower the other side. This motion will push out any air bubbles that might form. Air bubbles under the cover slip will spoil the image.

5. Very slowly, lower the cover slip until it rests on the water. If there is too much water and some water comes out from under the cover slip, touch the water with the edge of a paper towel as shown in Figure S2.7. The paper towel will draw the extra water away.

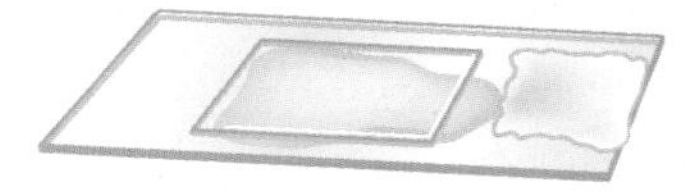

Figure S2.7 If you do not remove the extra water, the cover slip will move around and make it difficult to observe your sample.

Practice

1. Choose a sample that you wish to view with a microscope. Here are some suggestions.
   - a fibre from clothing or thread
   - a hair
   - a small fragment of onion skin (the very thin layer between the thick layers of onion)
   - a drop of pond water
2. Prepare a wet mount of your sample by following the steps above.
3. Examine your wet mount with a microscope by carrying out the same steps that you followed when viewing the prepared slide.

Skill POWER 3

THE METRIC SYSTEM AND MEASUREMENT

You have probably seen signs on highways that tell you the distance to nearby towns or cities. For example, a sign might say, "Kamloops 78 km." You probably know that "km" means kilometres. When you shop at a grocery store, you might pick up a bag of potatoes that is labelled "5 kg." The "kg" means kilograms. Did you notice that both terms started with *kilo-*? The meaning of the prefix *kilo-* that is used in front of metres and in front of grams stands for 1000. A kilometre is 1000 m and a kilogram is 1000 g. The prefix *kilo-* is part of the metric system.

Metric System

The **metric system** is based on powers of 10. A unit, such as a metre or a gram, is multiplied or divided by 10, 100, 1000, or more. Prefixes tell you how to multiply or divide the unit. For example, the prefix *kilo-* tells you that a unit of measurement is multiplied by 1000. The number 1000 is a power of 10 because it is 10 multiplied by itself three times.

$$1 \text{ km} = 1000 \text{ m}$$
$$1000 = 10 \times 10 \times 10$$

Some prefixes tell you that you need to divide a unit by a power of 10. For example, a centimetre is 1 m divided by 100. When you divide a number by a power of 10, you can also write the result as a decimal.

$$1 \text{ cm} = \frac{1}{100} \text{ m} = \frac{1}{10 \times 10} \text{ m} = 0.01 \text{ m}$$

The following table shows units that can be used with metric prefixes.

Table S3.1 Common Units

Quantity	Unit	Example
length	metre (m)	A metre stick is 1 m long.
mass	gram (g)	A paper clip has a mass of about 1 g.
volume	litre (L)	A large orange juice carton has a volume of 1 L.
time	second (s)	The heart of a large dog beats about once every 1 s.

Note that seconds, minutes and hours are not part of the metric system. But metric prefixes are still used with seconds. The following table shows you some of the common prefixes and what they mean.

Table S3.2 Metric Prefixes

Prefixes	Symbol	Relationship to the Base Unit
giga-	G	1 000 000 000
mega-	M	1 000 000
kilo-	k	1000
hecto-	h	100
deca-	da	10
—	—	1
deci-	d	0.1
centi-	c	0.01
milli-	m	0.001
micro-	μ	0.000 001
nano-	n	0.000 000 001

Practice

1. How many grams (g) are there in 1 kg?
2. How many millimetres (mm) are there in 1 m?
3. One cm is how many metres (m)?
4. How many nanoseconds (ns) are there in 1 s?

Measuring

When you are doing science experiments, you will make many measurements. Practising the skills discussed here will help you get better results from your experiments.

Measuring Distance or Length

When you measure the length of an object or the distance between two points, you will usually use a ruler or tape measure. Many rulers and tape measures are marked as shown in Figure S3.1. Some rulers have inches on one side and centimetres on the other. Always use the side with centimetres.

On most rulers, the starting line or zero line is not at the end of the ruler. Therefore, you need to be sure to place the zero line at one end of the object you are measuring.

Examine Figure S3.1. Notice that the zero line is at the end of the rectangle. Now notice that the other end of the rectangle lies on the seventh short line beyond the longer line labelled 5.

Therefore, the rectangle is 5.7 cm long. If the end of the object that you are measuring does not lie exactly on one of the lines on the ruler, examine it to see which line is closest. Identify the closest line and use that value. Practise using a ruler by completing the following Practice questions.

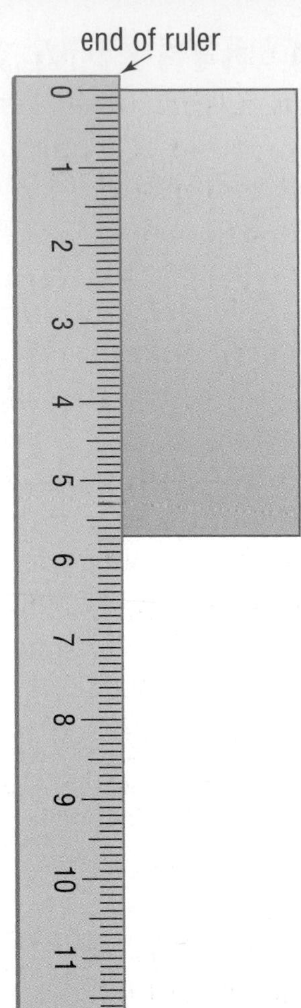

Figure S3.1 Identify the zero mark on each ruler that you use. Place the zero line on one end of the object that you are measuring. Find the line that is closest to the other end of the object. Identify the number represented by that line.

Practice

1. Measure each side of the figures shown below.

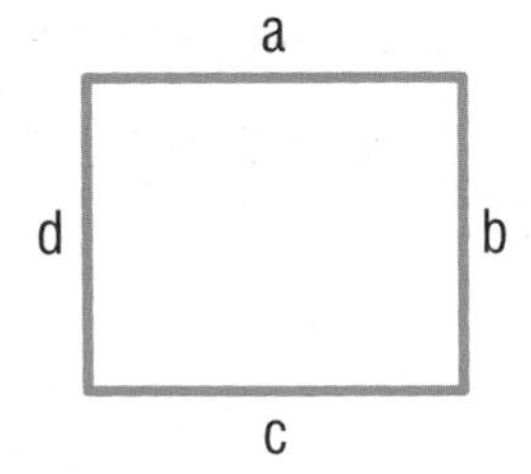

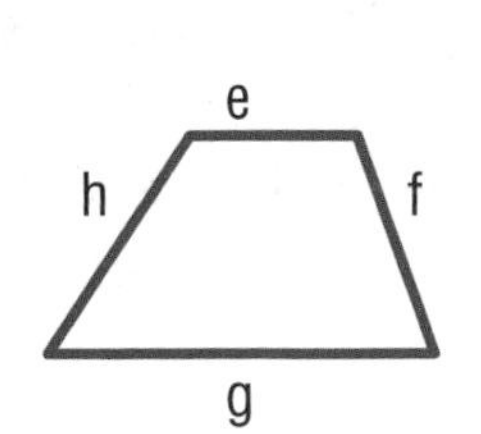

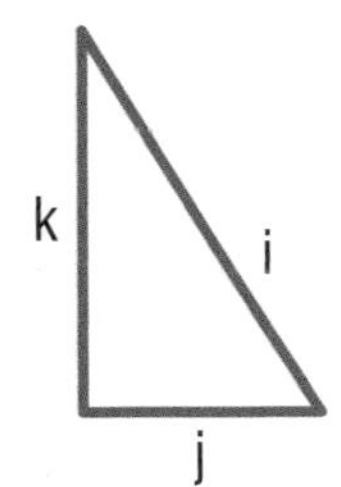

2. Measure the width of your textbook.

Measuring the Volume of a Liquid

When you measure liquids, you will often use a container called a graduated cylinder. A graduated cylinder is a tall, narrow glass or plastic container that has markings on the side. The markings usually show you the volumes of the liquid in millilitres. Graduated cylinders that measure small volumes will often have markings for tenths of a millilitre as well.

To measure the volume of liquid in a graduated cylinder, place the cylinder on a flat surface. Bend down so that your eye is level with the surface of the liquid in the graduated cylinder as shown in the figure. You will notice that the surface of the liquid is curved. The curved surface is called the **meniscus**. Find the line on the cylinder that is nearest to the *bottom* of the meniscus. This line shows the volume of the liquid.

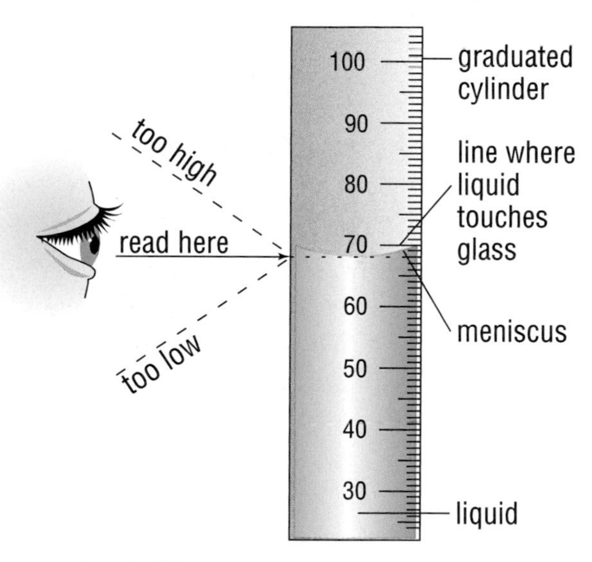

Figure S3.2 The volume of the liquid in this graduated cylinder is 68.0 mL.

Practice

Read the volume indicated by each graduated cylinder below.

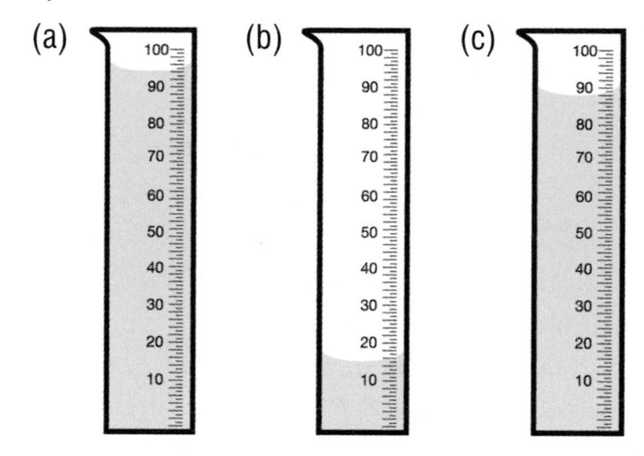

Measuring Temperature

When you need to measure the temperature of a liquid for an experiment, you will probably use a thermometer similar to the one shown in Figure S3.3. Laboratory thermometers measure temperature in degrees Celsius. The scale on a typical laboratory thermometer goes from −20°C to +110°C.

Figure S3.3 This laboratory thermometer contains dyed alcohol. Some thermometers contain mercury, a toxic heavy metal.

Tips for Using a Thermometer

- Handle thermometers with great care. The glass can break easily and cut your hands.
- Never stir a solution with a thermometer.
- Do not let the bulb of the thermometer touch the walls of the container when you are measuring the temperature of a solution.
- If your thermometer should break, do not touch it. Tell your teacher immediately.

Not all thermometers look like the thermometer shown in Figure S3.3. In the past, doctors and nurses would take a person's temperature by placing the bulb of a small thermometer under the person's tongue. Medical thermometers have changed a great deal. For example, your doctor probably uses an ear thermometer similar to the one shown in Figure S3.4.

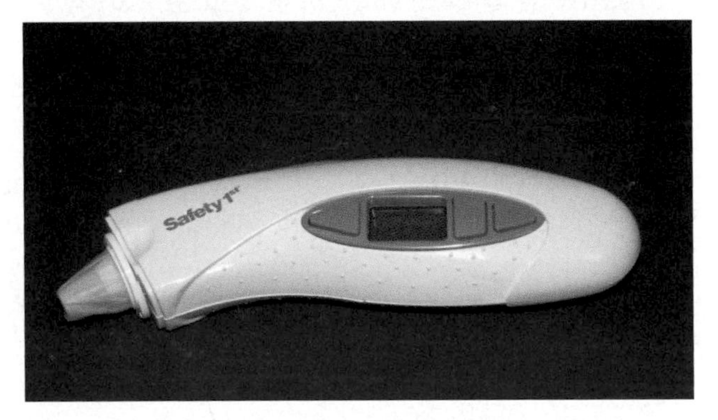

Figure S3.4 Although this ear thermometer does not look at all like the laboratory thermometer, they both measure temperature.

Practice

1. Your teacher will provide you with beakers of water at different temperatures. Several students will measure the temperature of the water in each beaker.
2. Record your measured values of temperature in a table similar to the one shown below.
3. Then compare your measured values. If your results do not agree, discuss possible reasons why the results differ.

Table S3.3 Temperature Data

	Student			
Beaker	**1**	**2**	**3**	**4**
A				
B				
C				
D				

ORGANIZING AND COMMUNICATING SCIENTIFIC DATA

In many experiments, you will be making measurements and writing down numbers. These numbers are called numerical data. Creating a table is a good way to record your data. Also, organizing your data into a table makes it easier to draw a graph from the data. Graphs are an excellent tool for analyzing and communicating information.

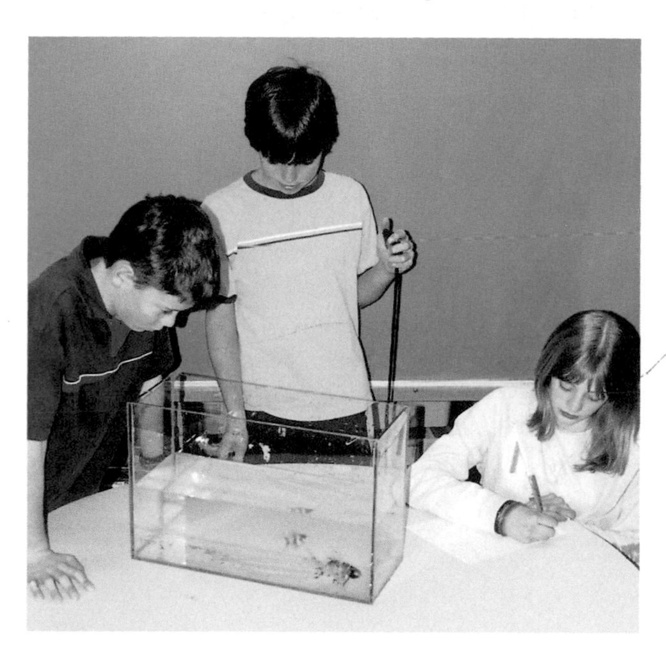

Figure S4.1 Sometimes it is difficult to do an activity, take measurements, and record the numbers alone. Working in groups makes science investigations easier and more fun.

Making a Data Table

Before doing your experiment, prepare a data table to record your observations. Preparing a data table can help you organize your experiment. For example, suppose you are going to measure the rate of growth of a plant. You decide to measure the height of the plant every week, for six weeks. Your data table for this experiment might look like Table S4.1 shown here.

Tips for Making Tables

- Always put the numbers that you have chosen, such as number of weeks, in the first column. Put measured quantities in the other column or columns to the right.
- You can print the units that you will use in the measurements in the box with the heading.
- Always give your data table a title.

Table S4.1 Plant Height versus Time

Time Since Planting (weeks)	Plant Height (cm)
0	
1	
2	
3	
4	
5	
6	

At the end of six weeks of making observations, your table might look like Table S4.2.

Table S4.2 Plant Height versus Time

Time Since Planting (weeks)	Plant Height (cm)
0	0
1	0.9
2	2.0
3	2.7
4	3.6
5	4.8
6	5.5

Now you have a completed data table. You are ready to analyze your data. A common way to analyze data is to graph it. The next section will help you improve your graphing skills. Before you read about graphing, practise your table-making skills by completing the following Practice questions. Keep the data table that you make so you can graph the data later.

Practice

1. You and a friend see some frog eggs at a pond that you visit often. You record your observations by writing the notes on the right.
   (a) Read the notes.
   (b) Decide on the headings for the columns in your chart.
   (c) Use the data from the notes to fill in the data table.
2. Measure the temperature outside your school every day for two weeks. Record your observations in a data table.

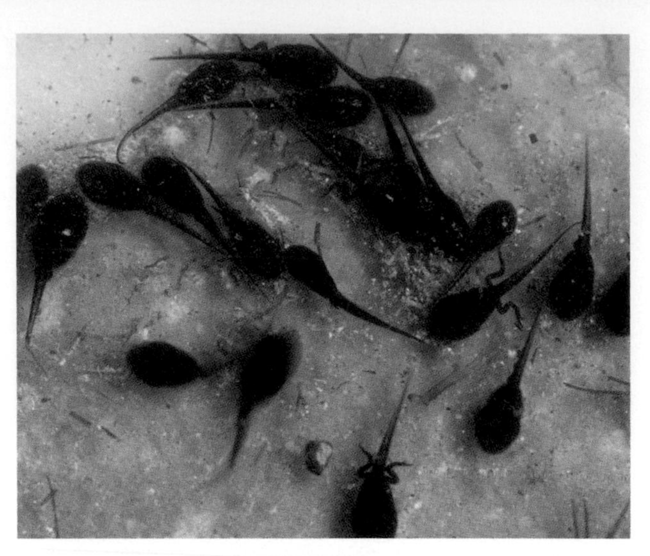

Figure S4.2 How many days after a frog lays eggs do the eggs develop into tadpoles?

Tadpole Watching

One day we saw frog eggs in the pond. We were pretty sure that they had not been there the day before. We checked them every day to see when they would hatch.

- On the first through the 14th days, we did not see any tadpoles.
- On the 15th day, some of the eggs looked like there were wiggly tadpoles inside but none were swimming around in the water.
- On the 16th day, we saw three tadpoles swimming.
- On the 17th day, there were seven tadpoles swimming in the water.
- On the 18th day, we counted 13 tadpoles.
- On the 19th day, we saw 23 tadpoles.
- On the 20th day, it was really hard to count them, but we are pretty sure there were 35 tadpoles.
- On the 21st day, we thought we could count 41 tadpoles.
- On the 22nd day, there were probably 45 tadpoles.
- On the 23rd day, we think there were 47 tadpoles.

We counted for three more days. We were pretty sure there were still 47 tadpoles

Drawing a Line Graph

After you have collected data, you need to analyze it. You often need to know how a variable such as time affects another variable such as plant height. (You can learn more about variables in SkillPower 5.)

- The variable that might affect another variable is called the **independent variable**. The experimenter can choose or manipulate the independent variable.
- The variable that is influenced or affected by the independent variable is called the **dependent variable**.

The experimenter chooses the independent variable and then observes the dependent variable to find out how it changes.

Look back at Table S4.2. The data shows that the plants grew taller over time.

- The experimenter chose the times to measure the plants so the time that had passed was the independent variable.
- Since the height of the plants *depended* on the amount of time that had passed since planting the seeds, the height was the dependent variable.

When you are plotting data on a graph, you put the independent variable on the horizontal axis and the dependent variable on the vertical axis. The following steps will guide you through the process of making a line graph.

Getting Started

1. Obtain a piece of graph paper and draw horizontal (side to side) and vertical (up and down) lines. These lines are called the *horizontal axis* and the *vertical axis*.
2. Use as much of the graph paper as possible for your graph.
3. Make the scale on each axis a little larger than the range of your data. For example, the *time* data goes from 0 to 6 weeks so the range is 6 weeks. Therefore, make seven weeks the last value on your time axis. Your *height* data goes from zero to 5.5 cm. Therefore, make 6 cm the last value on your *height* axis.
4. Label each axis and include units.

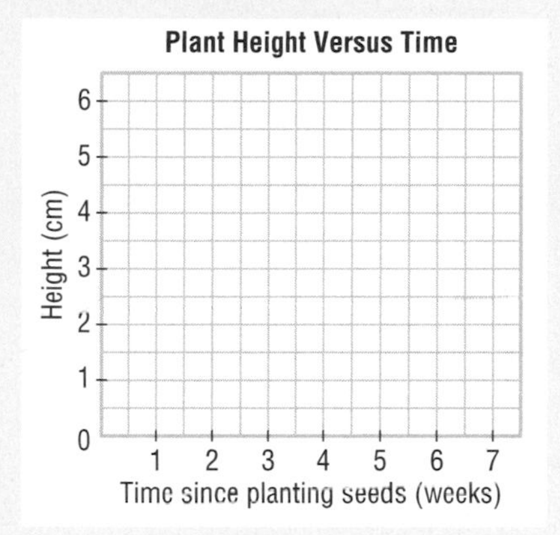

Plotting the Points

5. Start plotting your points. Time zero is the week in which you planted the seeds. Since the plants had not started to grow, the height is 0 cm. Make a point at the zero point on your graph. This point is called the origin.
6. To plot the next point, go to the right until you reach one week. Then go up until you reach a height of 0.9 cm. This value is just slightly below 1 cm. Make a dot at this point.

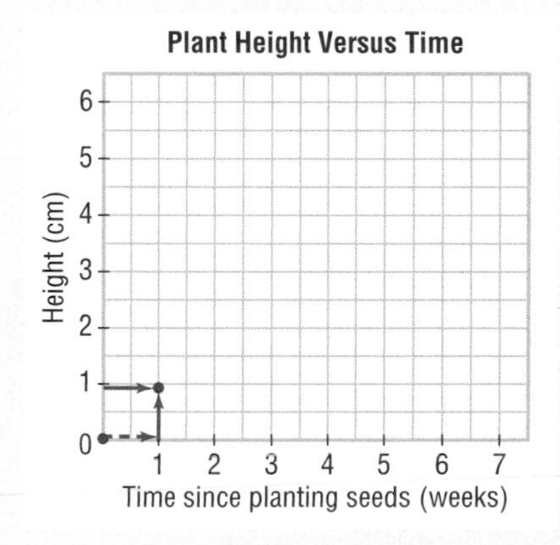

7. From the data table, you can see that the next point is at two weeks and 2.0 cm. Continue to the right on your horizontal axis until you reach two weeks. Go up until you reach the line representing 2.0 cm. Plot a point.

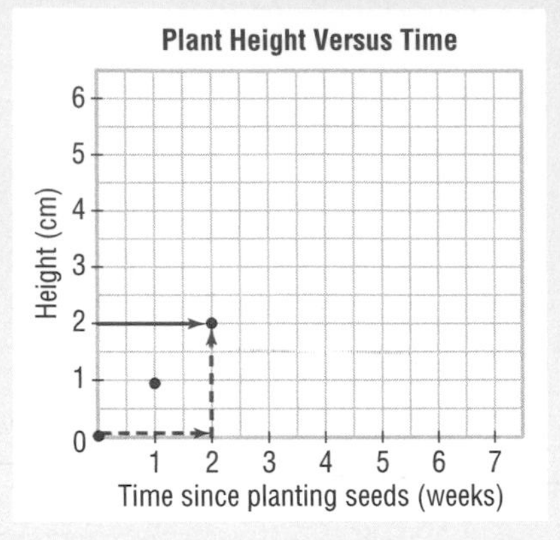

8. Plot all of the points listed on the data table. If you place a ruler on the graph, you will see that the points do not lie on a perfectly straight line. *Do not connect the points with short straight lines.*

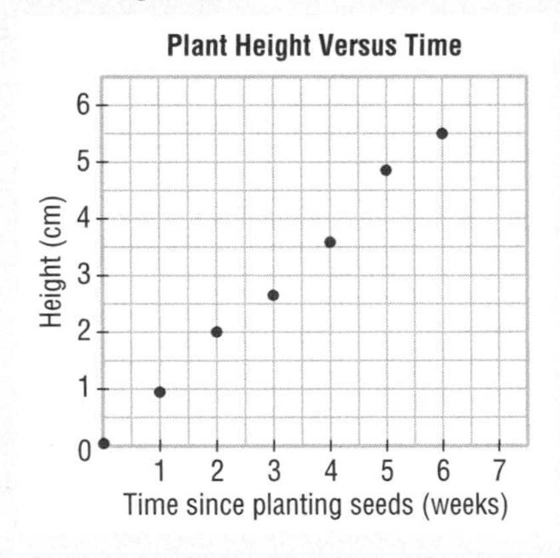

Drawing the Line

9. Place a ruler on the graph and move it around until you find a place where the ruler is close to all of the points. Try to leave about the same number of points above the line as below. Draw a straight line with the ruler. This line is called the "line of best fit."

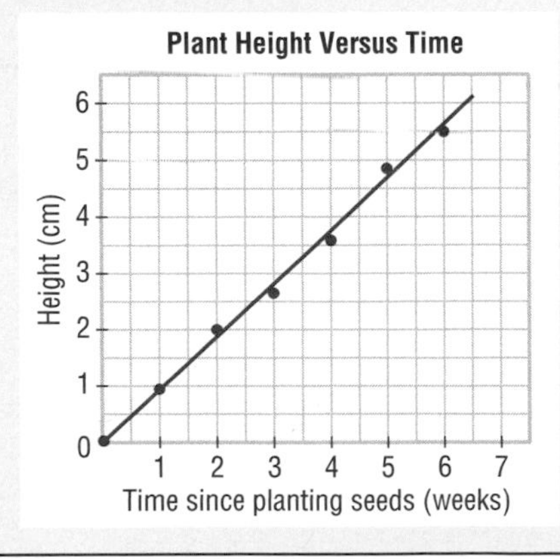

Using Line Graphs

After you have completed a line graph, you can use it to estimate values on the graph. You can estimate the height of the plant at any time during the growth period. For example, what was the height of the plant three and one half weeks after planting the seeds? To estimate this height, carry out the following steps and study the graph.

- Find 3.5 weeks on the *time* axis.
- Draw an arrow up to the line and put a mark on the graph.
- Draw another arrow from the mark over to the *height* axis.
- Read the height at the point at which the arrow touches the axis.
- Even though you did not measure the plant at 3.5 weeks, you can estimate that the plant was probably about 3.3 cm high 3.5 weeks after the seed was planted.

Plant Height Versus Time

Figure S4.3 You can use the finished graph to estimate.

Curved Line Graphs

How would you draw the line if your data points looked like those in Figure S4.4? There is no place that you could draw a straight line and come near all of the data points. In this situation, you would draw a curved line like the graph in Figure S4.5. First, sketch a smooth curved line lightly with a pencil. When you are satisfied that the line comes close enough to all of the data points, make the line darker.

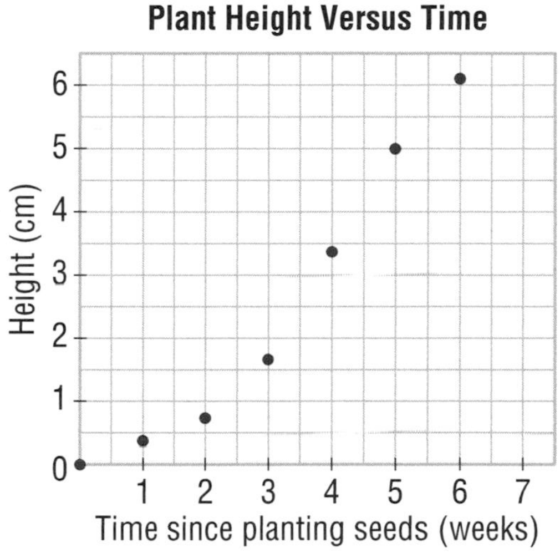

Plant Height Versus Time

Height (cm)

Time since planting seeds (weeks)

Figure S4.4 These data points do not fit on a straight line.

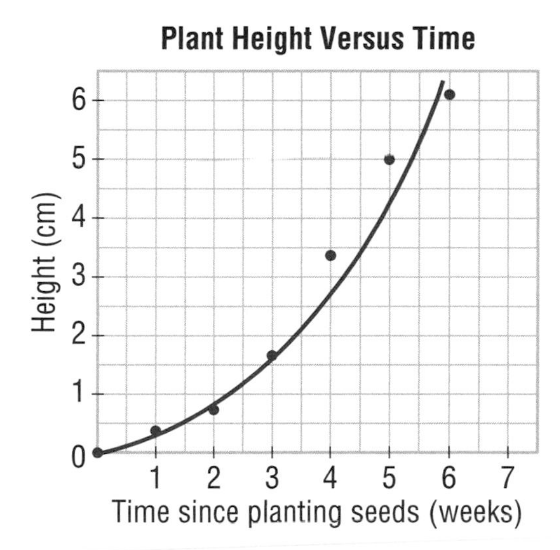

Figure S4.5 When data points do not fit a straight line, draw a smooth curved line near all of the data points.

Interpreting Line Graphs

When you examine the line graph that you made from your data, you can see how the dependent variable changes when your independent variable changes. In the graph in Figure S4.3, you can see that the plant grows steadily. It grows by the same amount every day. The graph in Figure S4.5 is curved showing that the plants grew slowly at first and then began to grow faster after three days.

Graphs do not always show how something changes with time. For example, Figure S4.6 shows how water temperature affects how active guppies are. (The experimenter decided to measure activity by counting the number of darting movements guppies made every 5 min.)

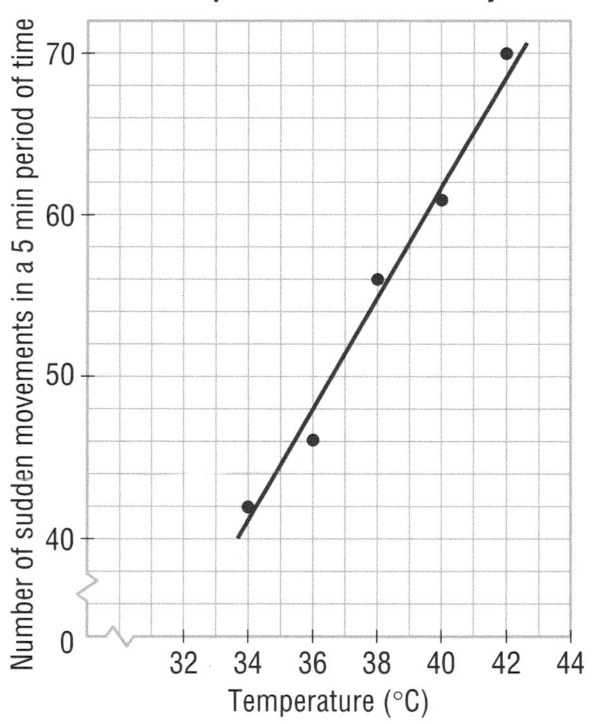

Figure S4.6 This graph shows that when the temperature of the water increases, the activity of the guppies increases.

Practice

1. Turn to the first Practice question in SkillPower 4. Find the data table that you made to record data about tadpoles.
   - **(a)** Draw a graph with days (the independent variable) on the horizontal axis and the number of tadpoles (the dependent variable) on the vertical axis.
   - **(b)** Describe what the graph tells you about the number of tadpoles that hatched from the eggs each day.
2. Suppose you conduct an experiment to find out whether extra recycling bins will encourage more students in your school to recycle. Make a graph to analyze the results of this experiment.

Table S4.3 Students Using Recycling Bins

Number of extra bins	Number of students using recycling bins
0	28
1	35
2	46
3	53
4	63

Making Bar Graphs

Sometimes you will need to compare numerical values that describe objects that fit in different groups. For example, suppose you are comparing the masses of the four largest moons of Jupiter. Figure S4.7 shows how a bar graph can help you do this. The sizes of the bars make it easy to see the differences and similarities in the masses of the moons. You can see that the heaviest moon, Ganymede, is three times as heavy as the smallest moon, Europa.

The following points will help you make your own bar graphs. Refer to the bar graph in Figure S4.7 as you study these directions.

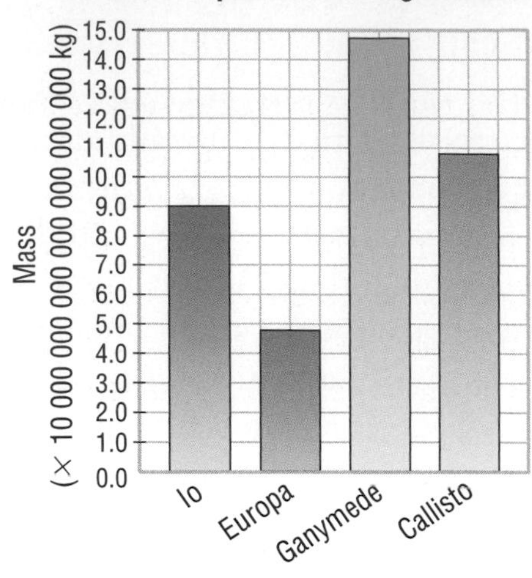

Figure S4.7 Notice that the numerical values for mass on the vertical axis must all be multiplied by 10 000 000 000 000 000 000 to give the correct mass in kilograms.

Tips for Making Bar Graphs

- Use as much of the graph paper as possible.
- Use a ruler to draw your axes.
- Make the scale of your each axis just a little larger than the largest value that you measured or counted.
- Make all of the bars the same width.
- Make the space between the bars about half as wide as the bars.
- Label the bars.
- Label your vertical axis.
- Make the bars different colours for easy viewing.
- Give your bar graph a title.

Practice

1. Use the data in the following table to make a bar graph for the populations of some species of deer in British Columbia.

Table S4.4 Deer Populations in British Columbia

Species of Deer	Population (animals) (in thousands)
Moose	175
Elk	40
Mule Deer	125
White-Tailed Deer	75

DESIGNING AND CONDUCTING EXPERIMENTS

When you think of scientific experiments, you probably think about a scientist in a laboratory. But you do not need to be a professional scientist ito do a science experiment. You do not even need a laboratory. You can do a science experiment anywhere as long as you do it properly.

Figure S5.1 People sometimes think that all scientific experiments are done in laboratories by people in white coats.

A proper experiment is often called a **fair test**. To perform a fair test, you must plan and control your experiment so that only one factor affects your results. In this SkillPower, you will learn how to perform a fair test.

Figure S5.2 Why is the student saying, "That's not fair!"?

Identifying Variables

One of the first steps in designing a fair test is identifying variables. **Variables** are all of the factors that could change (vary) or be different in different parts of an experiment and affect the results. For example, in the cartoon in Figure S5.2, the student appears to be measuring the time needed for the water in the beaker to evaporate. What are the variables that might affect the time needed for the water to evaporate? Here is a list of a few of those variables. Can you think of any more?

- the volume of water in the beaker
- the temperature of the water in the beaker
- a breeze blowing over the beaker
- whether the sun or a light is shining directly on the water in the beaker
- the humidity (amount of water vapour in the air)

As you can see, even a simple experiment can have many variables.

Consider an experiment similar to one you might do in your study of life science. Suppose that you want to test two ingredients that are sometimes used in plant food. You want to see if either of the ingredients will make a certain type of plant grow faster. You will need to grow plants under the three conditions shown in Figure S5.3.

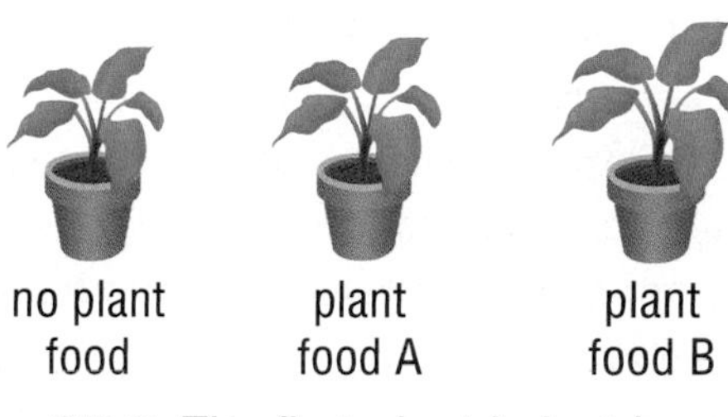

Figure S5.3 The first plant is just in soil with no plant food. Compare this plant with the other plants that are growing in soil with different plant foods.

Notice that you need to include a plant grown with no plant food. Without this plant, you would not know whether the plant foods had any effect on the rate of growth. You can compare plants grown in each of the types of plant food—A and B—to the plants grown with no added plant food.

In addition to plant food, what other variables could possibly affect the rate of growth of the plant? To perform a fair test, you must identify all of the variables that might possibly affect the rate of growth of the plants. Examine the following list of variables.

- type of soil used
- plant diseases
- type of seeds used
- insects or other pests that might harm the plants
- amount of light the plants receive
- temperature of the air and soil
- volume of water added to soil

Try to think of another variable to add to the list.

Controlling Variables

There are many variables that might affect plant height. In the experiment described above, however, you want to find out how plant food affects plant height. Plant food type is the independent variable and plant height is the dependent variable. (Turn to SkillPower 4 or the Glossary to review what these terms mean.) How can you be sure that any differences in growth rate are due to differences in plant food only? To perform a fair test, you must control all of the other variables.

Figure S5.4 It is easier to control variables such as light and water indoors. These lettuce plants are part of an experiment deep in a mine shaft near Sudbury, Ontario.

It would be much easier to control the growing conditions by growing plants indoors as shown in Figure S5.4. If the plants were indoors, you could control all of the variables except the type of plant food by carrying out the suggestions in Table S5.1. Variables that might affect the growth of the plants are listed in the first column. The second column gives a suggestion for a way to control the variable and prevent it from affecting the plant growth.

Table S5.1 Controlling Variables for Growing Plants

Variable	How to Control the Variable
type of soil	Use the same sterile potting soil for all plants.
plant diseases	Use sterile potting soil and grow the plants indoors.
seeds	Use seeds from the same variety of plant and from the same batch of seeds.
insects	Grow plants indoors where there are no insects.
light	Use the same type of grow-lights for all plants. Replace burned-out bulbs immediately.
temperature	Carefully control the room temperature. Ensure that air from windows and doors does not make the temperature in one part of the room different than other parts of the room.
water	Water all of the plants at the same time. Give them all the same volume of water from the same source of water.

Repeating Experiments

No measurement is perfect. No two plants are identical. Therefore, it is necessary to make many measurements so you can see if the results are similar. Suppose, for example, that you planted 10 seeds and measured the height of all 10 plants at the end of six weeks. How would you explain the results if your measurements were 7.2 cm, 6.9 cm, 7.1 cm, 7.5 cm, 6.8 cm, 7.2 cm, 7.4 cm, 4.1 cm, 7.3 cm, and 7.0 cm?

As you can see, one plant is much smaller than the others. Some unknown event must have affected the plant that was only 4.1 cm tall. If this 4.1 cm plant had been the only one that you observed, the results of your experiment would have been incorrect. You would not have known that most of the plants would grow much higher. You may decide that the short plant has been affected by variables that are not part of your experiment. In this case, you would discard this result.

Even though the heights of the remaining nine plants were similar, they were not identical. You get the best results if you calculate the average height of the nine plants. You can find the average by adding the heights of all of the plants then dividing the sum by the number of plants as shown below.

$$\text{average height} = \frac{\text{sum of heights of plants}}{\text{number of plants}}$$

$$\text{average height} = \frac{7.2\text{ cm} + 6.9\text{ cm} + 7.1\text{ cm} + 7.5\text{ cm} + 6.8\text{ cm} + 7.2\text{ cm} + 7.4\text{ cm} + 7.3\text{ cm} + 7.0\text{ cm}}{9}$$

$$\text{average height} = 7.2\text{ cm}$$

No matter what type of experiment you are doing, you should always do several trials and make several measurements or observations. If you make only one measurement, you cannot know if that was the one case in which something went wrong.

Making Inferences

Making an **inference** is analyzing the results of your experiment and using logic to draw a conclusion. This conclusion should be in the form of an answer to your original question. For example, suppose you asked the question, "Does ingredient A or ingredient B affect the growth of the plants?"

You planted 10 seeds in soil with no added plant food, 10 seeds in soil with plant food containing ingredient A, and 10 seeds in soil with plant food containing ingredient B. You grew them all under identical conditions. After 15 days, you measured the heights of all of the plants and found the results below.

Table S5.2 Effect of Plant Food on Average Plant Height

Conditions	Average Height of Plants
no plant food	11.2 cm
plant food with ingredient A	12.6 cm
plant food with ingredient B	16.7 cm

Your analysis of the data would be that plants grown in the presence of ingredient A were slightly taller than plants grown with no added plant food. Plants grown with ingredient B were much taller than plants grown with no plant food.

These statements only summarize your observations. You have not yet answered the original question. To answer the question, you would *infer* that ingredient A probably had a small effect on growth of the plants. However, 12.6 cm is not much taller than 11.2 cm. Therefore, you cannot be sure that ingredient A really caused the plants to grow larger. Ingredient B, however, seemed to have a very large effect on plant growth. Ingredient B caused the plants to grow 5.5 cm higher than the plants with no plant food.

Practice

1. Suppose that you want to do an experiment to find out how acid affects seed growth. You decide to experiment by watering plants with undiluted vinegar (a mild acid) and with vinegar you have diluted with water.
   - **(a)** List all of the possible variables that you think might affect whether or not a seed grows.
   - **(b)** Write out plans for the experiment. Describe the type of data that you might collect. Explain how you would analyze the results. How would you know whether the vinegar had any effect on the growth of the seeds?

2. Imagine that, while learning about science, you became interested in frog development. In a science article, you read that the amount of iodine in the water influenced the rate at which tadpoles matured into adult frogs. You decided to do an experiment with two different amounts of iodine added to the water. You wanted to find out whether the iodine had any effect on frog development and what that effect might be.
   - **(a)** List all of the possible variables that you think might affect the rate of development of tadpoles into frogs.
   - **(b)** Design an experiment that you might be able to perform on frog development. Write out plans for the experiment. Describe the type of data that you might collect. Explain how you would analyze the results. How would you know whether the iodine had any effect on the metamorphosis of tadpoles into frogs?

TECHNOLOGICAL PROBLEM SOLVING

You often see the terms "science and technology" written together. Have you ever wondered if there are differences between science and technology? How is *technological* problem solving different from *scientific* investigation?

Comparing Scientific Investigation and Technological Problem Solving

Science is a way to understand the universe and to explain natural events. A good example of scientific investigation is found in the study of extreme environments. If you studied life in the ocean depths, you might have read about life forms that live nearly 4 km below the surface of the ocean. In 1977, scientists discovered life forms such as tube worms and white crabs on the ocean floor. Figure S6.1 shows these organisms.

Figure S6.1 Scientists did not know that these red tube worms and white crabs existed before 1977. Before 1977, no technology was available that would allow people to explore the deepest parts of the ocean floor.

Researchers conducted a scientific investigation when they looked for answers to the question, "How do living things obtain energy when they live in total darkness on the ocean floor?" Until that time, scientists had believed that all forms of life obtained energy from the Sun. Then they discovered that a rare form of bacteria lives on the ocean floor. These bacteria obtain energy from chemicals released by hot water that bubbles up through cracks in the ocean floor. Tube worms and other life forms consume the bacteria.

Before scientists could find these life forms, engineers had to solve a major technological problem. Ordinary submarines were designed to go to depths of 300–400 m. These submarines would be crushed by the weight of the water before they reached 1 km below the surface of the ocean. The engineers had to build a vessel that could safely go to the deepest parts of the ocean floor.

The solution is shown in Figure S6.2. This small, scientific submarine is named *Alvin*. *Alvin* is capable of diving to a depth of 4 km.

Figure S6.2 Some of *Alvin's* systems, including the sonar, were developed by a company in Port Coquitlam, British Columbia.

Designing Alvin: Solving a Technological Problem

Technological problem solving is using available knowledge to build a device or design a process that will solve a practical problem. The available knowledge might come from common sense, trial and error, or scientific information. For example, the engineers who designed *Alvin* had to solve the following problems:

- They had to find or design materials that could withstand great pressures.
- They had to know how to seal the craft so it would not leak.
- They had to design a propeller that could turn freely even under high pressures.
- They needed to design an artificial arm on the outside of the submarine that could grasp objects and place them in a basket. When the submarine returned to the surface, they could recover and study the objects brought up from the ocean floor.
- The engineers also had to design a window so the pilot and researchers could see out into the water. The window would have to be strong enough to withstand very high pressures.

These are just a few of the technological problems that the engineers solved when they designed *Alvin*.

Guidelines for Technological Problem Solving

No two technological problems are alike. Still, a general pattern will help you solve each problem. The flow chart in Figure S6.3 shows the main steps in the process.

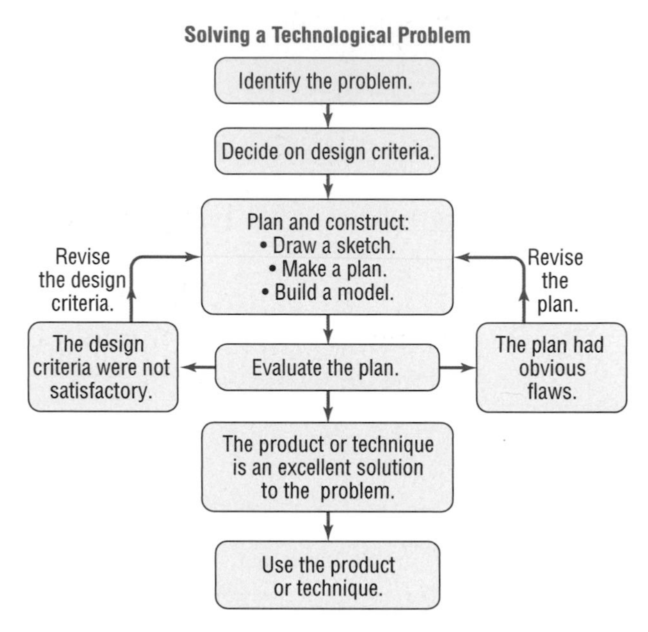

Figure S6.3 This general plan for technological problem solving can be applied to many different types of problems.

To learn how to use these guidelines, apply them to the problem on the next page. This problem is similar to designing the artificial arm on the submarine, *Alvin*, or the Canadarm on the space shuttle.

Designing A Mechanical Arm

Suppose that you are raising a very rare type of fish. The fish are sensitive to any pollution in the water. They are also frightened easily. You want to design a device that you can use to add or remove items in the aquarium. You cannot reach in with your hands because that will frighten the fish. As well, any substances, even soap, on your hands will get into the water and might harm the fish. You can add or remove items by lowering or raising them with a wire mesh basket that is supported with strings.

You need to design a device that will lift a 50 g glass, plastic, or metal object into and out of the wire mesh basket. The water is at least 25 cm deep.

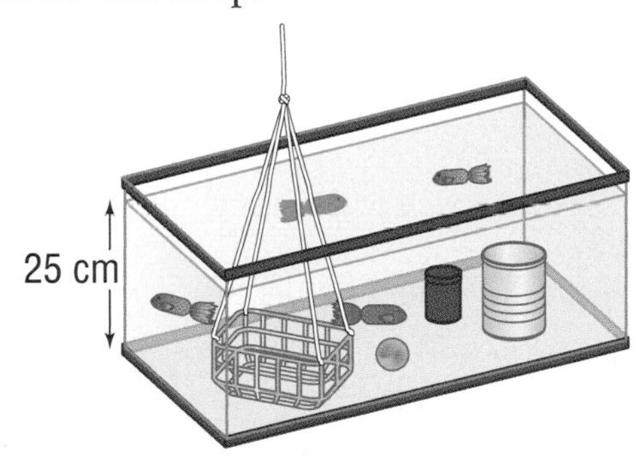

Figure S6.4 Your challenge is to design a mechanical arm. The arm must be able to pick up the items in the tank and place them in the wire mesh basket.

1. Identify the problem

Before you can start to solve a technological problem, you must clearly understand what it is that you need to do. Write out a clear statement of exactly what is needed to show that you understand the problem.

2. Decide on design criteria

List, in detail, all of the tasks that your device must be able to do. For example:

- You must be able to operate your mechanical arm from above the water. The water is 25 cm deep.
- The part of the arm that grasps objects must be able to lift 50 g objects without allowing the objects to slip when they are under water.

3. Plan and construct

Draw and label a sketch of what you think your device should look like. The device you sketch should fit the design criteria you listed. Next, write out a plan for building a working model. Collect the materials that you will need. Build the working model.

4. Evaluate the plan

Test your model. Find out if it meets your criteria.

Result 1: Your test might show you that your design criteria were not workable. *Revise your criteria.*

Result 2: Your test might show you that your plan was faulty and the device did not work. Or you might see ways to improve your device. *Revise your plan.*

Result 3: The device worked. You have solved the problem.

Practice

1. With your teacher's permission and guidance, design and build a device for grasping objects under the water in a fish tank as described above.
2. With a partner, invent a technological challenge. Exchange challenges with another pair of students. With your teacher's permission and guidance, use the guidelines provided in this SkillPower to solve the challenge.

Glossary

How to Use This Glossary

This Glossary provides the definitions of the key terms that are shown in **boldface** type in the textbook (instructional boldfaced words such as "observe," and "gather" used in investigations are not included). Other terms that are not critical to your understanding, but that you may wish to know, are also included in the Glossary. The Glossary entries show the section where you can find the boldfaced words. In addition, the Glossary entries list related words in brackets after the definition. The pronunciations of terms that are difficult to say appear in square brackets after the terms. Use the following pronunciation key to read them:

a = mask, back	ih = ice, life	uh = sun, caption
ae = same, day	i = simple, this	uhr = insert, turn
air = stare, where	o = stop, thought	yoo = cute, human
e = met, less	oh = home, loan	
ee = leaf, clean	oo = food, boot	

A

adaptations physical characteristics or behaviours that give an organism a better chance of surviving and reproducing in a particular environment (2.2) (see also *behaviour*, *organism*)

airlock an airtight chamber where the air pressure can be reduced until it matches the vacuum of space (7.3) (see also *vacuum*)

altitude elevation; how high a location is above sea level (2.1)

amphibians a class of vertebrates (3.2) (see also *classification systems*, *vertebrates*)

Animalia one of six kingdoms used to classify organisms; includes all multicellular organisms that have cells with a nucleus but do not have the protective walls found around the cells of plants and fungi; that need to eat plants or other animals to obtain food; and can move from place to place to find food, shelter, mates, and to escape from enemies (3.2) (see also *classification systems*, *kingdom*, *multicellular*, *organism*)

astronaut an explorer who travels in space (7.1) (see also *exploration*)

atmosphere the gases that surround Earth (7.2)

atoms tiny particles that make up everything (4.1)

aquatic having to do with water; used to describe water biomes (2.1) (see also *biomes*)

attract used to describe how electricity "pulls together" objects (4.2) (see also *repel*)

B

bacteria unicellular micro-organisms (3.2) (see also *micro-organisms*, *Monera*)

battery a device that turns chemical energy into electricity (4.3)

behaviour the way an organism acts (2.2) (see also *adaptations*)

behavioural adaptations habits and activities of organisms that are important for survival in a particular habitat (2.2) (see also *behavioural adaptations*, *organisms*)

biomass energy bioenergy; the energy stored in all plant and animal tissues (6.2)

biome a large land region that has a distinct climate, soil, plants, and animals (2.1) (see also *climate*)

birds a class of vertebrates (3.2) (see also *classification systems*, *vertebrates*)

booster rockets engines in a space vehicle that provide thrust during a launch (7.3) (see also *rocket*, *thrust*)

C

camouflage an organism's ability to blend into its surroundings (2.2) (see also *adaptations*, *colouration*)

Canadarm a Canadian remote manipulator or robot designed to carry out jobs normally done by astronauts during space walks (8.3) (see also *Dextre*, *robot*)

cell membrane the part of a cell that controls what enters and leaves the cell (1.2) (see also *cells*)

cells the basic units of life (1.1)

cell wall a rigid structure that provides protection and support for plant cells (1.2) (see also *cells*)

charged used to describe an atom with an imbalance of positive or negative charges (4.1) (see also *atoms*, *imbalance*)

chlorophyll a green chemical found in plants cells that traps energy from the Sun that the plant uses to make food (1.2) (see also *cells*, *chloroplasts*)

chloroplasts parts of plant cells that contain chlorophyll, which traps energy from the Sun that the plant uses to make food (1.2) (see also *cells*, *chlorophyll*)

circuit a complete pathway or loop for the flow of electricity (5.1) (see also *source*, *load*)

circuit breakers devices that prevent too much current from flowing through a circuit (6.3) (see also *circuit*, *fuses*)

classification systems ways of grouping things based on similarities (3.1)

climate the average weather patterns of an area according to records that have been kept for many years (2.1) (see also *weather*)

closed circuit a complete and unbroken pathway (circuit) for the flow of electricity from a source to a load and back to the source (5.1) (see also *circuit*, *load*, *open circuit*, *source*, *switch*)

coal a solid fossil fuel (6.1) (see also *fossil fuels*)

colouration the colour and pattern on an animal's skin, fur, or feathers (2.2) (see also *adaptations*, *camouflage*)

conductors materials that allow electrons to flow through them easily (4.3) (see also *electrons*, *insulators*)

cover slip a thin, transparent cover that can be placed on top of an object on a microscope slide to hold the object in place and keep water off the lens of the microscope (1.3) (see also *microscope*, *slide*)

cryobot a robot that can penetrate thick ice (8.1) (see also *robot*)

current the flow of electrons through a circuit within a certain amount of time; measured in units called "amperes" or "amps" (5.1) (see also *circuit*, *electrons*)

current electricity electricity that results when electrons from an imbalanced charge move through an object (4.3) (see also *electrons*, *imbalance*)

cytoplasm a clear, jelly-like material that surrounds the nucleus of a cell and holds the organelles in place (1.2) (see also *cells*, *nucleus*, *organelles*)

D

Deepworker a single-seat submersible that can explore the ocean to a depth of 600 m and can be fitted with a pair of manipulator arms that can be used to recover rockets that have fallen from their spacecrafts (8.3)

dependent variable in an experiment, a condition that is influenced or affected by the independent variable (see also *independent variable*)

Dextre a robot used to carry out jobs normally done by astronauts during space walks; part of Canada's contribution to the International Space Station (8.3) (see also *robot*)

discharge the transfer of electrons from one object to another (4.2) (see also *electrons*)

E

electric generator a machine that uses the energy of motion to produce electricity (6.1) (see also *turbines*)

electric meter a device that measures how much electricity is used in a building (6.3)

electrocution death caused by a strong electric current passing through a living organism (6.3)

electromagnet a temporary magnet (objects that can attract some metals) created by an electric current (5.3) (see also *electromagnetism*)

electromagnetic waves waves of energy, including radio waves, light waves, X-rays and microwaves (7.3)

electromagnetism used to describe magnetism (an attraction to some metals) produced by electricity (5.3) (see also *electromagnet*)

electrons small particles around the nucleus (centre) of an atom (4.1) (see *atoms, nucleus*)

endoplasmic reticulum (ER) the part of a cell that transports food, water, and waste around and out of the cell (1.2) (see also *cells*)

energy conservation used to describe how certain things can be done without electricity to save energy (6.3)

energy consumption the amount of energy we use (6.3)

energy efficiency used to describe how less energy or electricity can be used to accomplish the same task (6.3)

environment refers to the conditions that surround people and affect how they live (7.1)

estuaries places where rivers flow into the sea; one of three salt-water biomes (2.1) (see also *biomes*)

Exosuit a revolutionary suit for deep ocean exploration (8.3)

exploration travelling to discover what a place is like (7.1)

extreme anything at the end of a scale or range (7.1)

fair test a proper experiment that has been planned and controlled so that only one factor affects the results

fibre optics a technology using very long thin tubes of glass or plastic to transmit data (8.3) (see also *NEPTUNE*)

filament a thin wire with a very high resistance used in light bulbs; when a light is turned on, electrons are pushed through the filament, which resists the electrons, heats up, and produces light. (5.3) (see also *electrons, light bulb, resistance*)

fish a class of vertebrates (3.2) (see also *classification systems, vertebrates*)

flow chart a graphic organizer that shows the order of events in a process

fossil fuels sources of energy that are created in nature from the rotted body parts of ancient plants and animals (6.1)

frogbot a robot with a single leg that has a spring at its jointed "knee" and that can be used in places where wheels might not work (8.1) (see also *robot*)

fungi organisms that obtain the nutrients they need to survive by absorbing them from other organisms (3.2) (see also *organisms*)

fuses devices that prevent too much current from flowing through a circuit (6.3) (see also *circuit, circuit breakers*)

geosynchronous used to describe how higher satellites move at the same speed as Earth rotates, making one complete orbit in 24 h (8.2) (see also *satellites*)

geothermal energy renewable energy from Earth's crust (6.2) (see also renewable)

gravity a force that pulls together any two objects that have mass (7.2)

ground used to describe how an object can be connected through a conductor to the ground, or Earth, to prevent an electric charge from building or to get rid of an electric charge (4.3) (see also *conductor*)

hazard used to describe something that is dangerous, such as when electricity can flow through a closed circuit in a way that is not controlled (5.1) (see also *short circuit*)

HHPS an acronym that stands for Hazardous Household Product Symbols; symbols that appear on containers to help people use household products safely

hibernation a period of time when animals are much less active and use a lot less energy (2.2) (see also *behavioural adaptations*)

hydro hydro-electric power; power produced by falling water that spins turbines in an electric generator (6.1) (see also *electric generator, turbines*)

hydrogen fuel cell fuel cells that use the energy of a chemical reaction between hydrogen and oxygen to create electricity (6.2)

hydro-electric dam a large barricade that blocks a river to store water for the production of hydro-electric power (6.1) (see also *electric generator, hydro-electric power, turbines*)

hydro-electric power power produced by falling water that spins turbines in an electric generator (6.1) (see also *electric generator, turbines*)

imbalance refers to when the number of negative charges that orbit the nucleus of an atom is not equal to the number of positive charges in the nucleus of that atom (4.1) (see also *atoms, electrons, nucleus, protons*)

independent variable in an experiment, a condition that might affect another condition (dependent variable) (see also *dependent variable*)

inference a conclusion made by analyzing the results of an experiment and using logic

inherited characteristic a characteristic that an organism is born with (2.2) (see also *adaptations*)

insulators materials that block the flow of electrons (4.3) (see also *conductors, electrons*)

invertebrates animals that do not have backbones or any other bones (3.2) (see also *vertebrates*)

kingdom in biology, the term used for the largest groups of types of organism (see also *classification systems*)

Know-Wonder-Learn chart a table that helps you plan and record your study of different science topics; the table has the following three columns: "What I Know," "What I Wonder," and "What I Have Learned."

lander a special spacecraft used to transport rovers to Mars and designed to land safely on the planet's surface (8.1) (see also *rovers*)

latitude the distance north or south of the equator; measured in degrees north (N) or south (S) (2.1)

life-support used to describe methods of providing air, water and food using special techniques (7.1)

light bulb a device that turns electric energy into heat and light energy (5.3) (see also *load*)

load any component along an electric circuit that uses electricity; a basic component of an electric circuit (5.3) (see also *circuit, light bulb*)

magnify to make objects appear larger (1.3) (see also *microscope*)

mammals a class of vertebrates (3.2) (see also *classification systems, vertebrates*)

meniscus the slight curve on the surface of a liquid where it touches the sides of its container

metric system a system of measurement based on powers of 10

microgravity zero gravity; used to describe a condition where objects fall together at the same rate and seem to float (7.3) (see also *gravity*)

micro-organisms very small organisms that can only be seen using a microscope (1.2) (see also *microscope, organisms*)

microscope a tool that magnifies objects by bending light through a piece of curved glass (a lens) (1.3) (see also *magnify*)

migration the movement of animals from one region to another in response to a change in seasons (2.2) (see also *behavioural adaptations*)

mimicry used to describe an adaptation some animals have that makes them look like something else (their surroundings or something unpleasant like an organism that stings or tastes bad) (2.2) (see also *adaptations, colouration*)

mind map a tool that you can use for note-taking, brainstorming, planning a presentation, and many other purposes

mitochondria parts of a cell where the processes that produce energy occur (1.2) (see also *cells*)

Monera one of six kingdoms used to classify organisms (3.2) (see also *bacteria, classification systems, kingdom*)

motor a device with an electromagnet that spins within a permanent magnet and causes other components, such as wheels and fans, to turn (5.3) (see also *electromagnet*)

multicellular used to describe organisms that have many cells (1.2) (see also *cells, organisms*)

natural gas a fossil fuel that is usually found as a mixture of gases (6.1) (see also *fossil fuels*)

negative charge the charge on an electron (4.1) (see also *electrons, positive charge*)

NEPTUNE a project that will allow researchers to study large areas of the ocean continuously using powered fibre optic cables placed on the seabed off the coast of British Columbia; acronym for North-East Pacific Time-series Undersea Networked Experiments (6.1) (see also *fibre optics*)

network tree a graphic organizer that helps you see how one main idea or object breaks down into smaller pieces

non-renewable an energy source that cannot be replaced within a human lifetime (6.1) (see also *renewable*)

nuclear energy the energy released when the nucleus of an atom is split apart (6.1) (see also *atoms*, *nucleus*)

nucleus in biology, an organelle that controls all of the activities in the cell (1.2) (see also *cells*, *organelles*); in chemistry, the centre of an atom (4.1) (see also *atoms*)

oil a liquid fossil fuel (6.1) (see also *fossil fuels*)

open circuit an incomplete electric circuit that cannot carry the flow of electricity (5.1) (see also *circuit*, *closed circuit*)

orbit the path of an object in space around a larger object (7.2)

organelles parts (structures) inside a cell that do specific jobs (1.2) (see also *cells*)

organisms living things (1.1)

parallel circuit electric circuits in which electrons have more than one complete pathway to follow; each load in the circuit can be turned on or off without affecting the other (5.2) (see also *circuit*, *electrons*, *load*, *series circuit*)

payload the cargo, measuring instruments, astronauts, and equipment that a rocket carries; makes up much of the weight of a rocket (7.3) (see also *rocket*)

photosynthesis [foh-toh-SIN-thuh-sis] the process by which plants use carbon dioxide gas to make food (1.1)

physical adaptations characteristics that make organisms suited to the biomes where they live (2.2) (see also *adaptations*, *biomes*)

Plantae one of six kingdoms used to classify organisms; includes wildflowers, grasses, trees, mosses; all plants are multicellular and use photosynthesis to obtain food (3.2) (see also *classification systems*, *kingdom*)

planet a large body that orbits a star (7.2) (see also *solar system*)

positive charge the charge of the particles in the nucleus (centre) of an atom (4.1) (see also *atoms*, *negative charge*, *nucleus*)

precipitation the amount of rain or snow

Protista one of six kingdoms used to classify organisms; kingdom with the biggest variety of members; includes organisms that are unicellular or multicellular, have cells with one nucleus, usually live in moist or wet environments, and of which some make their own food using photosynthesis (3.2) (see also *classification systems*, *kingdom*)

recycling using the same material over and over again (7.1) (see also *life-support*)

renewable an energy source that can renew or replace itself (6.1) (see also *non-renewable*)

repel used to describe how electricity "pushes away" objects away from each other (4.2) (see also *attract*)

reproduction the process by which living things produce offspring (1.1)

reptiles a class of vertebrates (3.2) (see also *classification systems*, *vertebrates*)

resistance used to describe how difficult it is for electrons to pass through the load of a circuit; measured in units called "ohms" (Ω) (5.1) (see also *circuit*, *electrons*, *load*)

robot a machine or device that works automatically or by remote control (8.1)

rocket a tube with fuel at the lower end; when the fuel burns, the force of hot gases escaping from the bottom of the tub pushes the rocket upward (7.3)

rovers exploration vehicles that are controlled and operated by signals that are sent by scientists on Earth (8.1) (see also *lander*, *robot*)

satellite any object in orbit around another object in space, such as Earth or another planet (8.2)

science a way of thinking and asking questions about nature on Earth and far beyond Earth; a collection of knowledge and ideas that help people understand and explain how nature works

series circuit a single pathway for electrons to travel from a source(s) through to a load(s) and then from the load(s) back to the source(s) (5.1) (see also *circuit*, *load*, *parallel circuit*, *source*)

short circuit a closed circuit that does not have a useful load (5.1) (see also *circuit*, *hazard*, *load*)

slide a flat piece of glass that objects can be put on for viewing through a microscope (1.3) (see also *microscope*)

snakebot a flexible robot shaped like a string of sausages that has cameras at both ends and can collect data (8.1) (see also *robot*)

society used to describe all the people who live together in a certain place and at a certain time

solar energy renewable energy from the Sun (6.2) (see also *renewable*, *Sun*)

solar system used to describe all of the planets and their moons, asteroids (chunks of rock), dust, gases, and other objects that orbit the Sun (7.2) (see also *planets*, *Sun*)

source a basic component of an electric circuit that provides the "push" that causes electrons to move through a circuit (5.1) (see also *circuit*, *electrons*)

space junk refers to the satellites, parts of spacecraft, tools, and other garbage travelling in orbit around Earth at very high spends (8.4) (see also *satellites*)

species the most specific level of classification of an organism (3.1) (see also *classification systems*, *organisms*)

spiderbot a multi-legged robot that can climb over a variety of difficult surfaces (8.3) (see also *satellite*, *orbit*)

stage the name for each section of a rocket's engine; as a rocket travels towards space, each section drops off after all of its fuel is burned up (see also *rocket*)

static electricity a buildup of electric charges on two objects that have become separated from each other (4.2)

Sun a huge ball of hot gases, similar to billions of other stars that we see in the night sky (7.2) (see also *solar system*)

switch a device that closes or opens a circuit to start or stop the flow of electricity (5.1) (see also *circuit*, *closed circuit*, *open circuit*)

technological problem solving using available knowledge to build a device or design a process that will solve a practical problem

technology the practical knowledge and tools that people use to make life easier; the use of devices, methods, and scientific knowledge to solve practical problems (7.1)

telescope an instrument that uses lenses and mirrors to make distant objects appear larger and closer (7.2)

thrust a force produced by escaping gases which causes a rocket to move (7.3) (see also *rocket*)

tidal energy renewable energy of the moving water in the tides (6.2) (see also *renewable*)

time line a graphic organizer that shows you the order in which events occurred and how much time passed between events

turbines a spinning device used by an electric generator to create motion that produces electricity (6.1) (see also *electric generator*)

uncharged electrically neutral; an atom that contains an equal number of positive and negative charges (4.1) (see also *atoms*)

unicellular used to describe organisms that are made up of only one cell (1.2) (see also *cells*, *organisms*)

vacuole the part of a cell that stores water and food that the cell cannot use right away (1.2) (see also *cells*)

vacuum a region where there is very little or no air or other gases (7.3)

variables all of the factors that could change (vary) or be different in different parts of an experiment

vertebrates animals with spines (backbones) (3.2) (see also *invertebrates*)

voltage used to describe the size of the "push" from a source that forces electrons through a circuit; measured in units called "volts" (V) (5.1) (see also *circuit*, *electrons*)

weather the local conditions that change from day to day, or even hour to hour (2.1)

wet mount a fresh sample prepared on a slide for viewing under a microscope (see also *microscope*, *slide*)

WHMIS an acronym that stands for Workplace Hazardous Materials Information System; safety symbols used throughout Canada to identify dangerous materials that are found in all workplaces

wind energy the renewable energy of moving air (6.2) (see also *renewable*)

Index

The page numbers in **boldface** type indicate the pages where the terms are defined. Terms that occur in investigations (inv) and activities (act) are also indicated.

A
Aboriginal technology, xviii–xxiii
aboriginal technology for explorers, 197*act*
action-reaction forces, 209
adaptations, **47**–56
  animal voices, 52–53*act*
  behavioural, **50**
  camouflage, 54, 55*inv*
  physical, **48**, 49*act*
  symbiosis, 51
airlock, **211**
altitude, **36**
Anik satellites, 240
Animalia (animals), **80**–82
Antarctica, 188, 220
Aristotle, 63
astronaut, **192**
atmosphere, **200**
atoms, **99**
  electric charge, **100**–102
  model, 101*act*
attract (electric), **105**, 108–109

B
bacteria, 70–73
  ancient bacteria, 70, 71
  "true" bacteria, 70, 72*act*
battery, **121**
behaviour, **47**
behavioural adaptations, **50**
biomass energy (bioenergy), **165**
biomes
  38, **39**–43, 44*act*
  aquatic biomes, **42**–43
  land biomes, 38–41
  people in biomes, 41*act*
birds, **82**
booster rockets, **210**
boreal forest, 38

C
camouflage, **54**, 55*inv*
Canadarm, **249**
Canadian technologies, 248–249, 250*inv*, 251–253, 254–255*act*
carbon dioxide, 7
cell city, 20–21
cell membrane, **17**
cell wall, **17**
cells
  basic units of life, 7, **15**–22
  model of a cell, 18*act*
  one or many?, 19
  parts (organelles), 16–17, 18*act*
  size, 19
charged, **100**
  see also electric charges
chlorophyll, **17**
chloroplasts, **17**
circuit, **128**
  closed or open, **129**
  parallel *see* parallel circuit
  series *see* series circuit
  short circuit, 130
  switch, 129, 133*inv*
circuit breakers, **170**
classification keys, 67*act*, 68*act*
classification systems, **62**–69
  Aboriginal peoples, 66
  scientists, 62–66
  six kingdoms of life, 70–84
climate change, 51
climates, **36**, 37*inv*
closed circuit, **129**
clothing for extreme environments, 196
coal, **156**
coastal rain forest, 60
colouration, **54**
communicating
  with a robot on Mars, 234–235*inv*
  with spacecraft, 216–217
conductors, **118**
coral reefs, 43
cover slip, 26
cryobot, **236**
current, **134**
  creating flow, 120*act*, 121
  observing flow, 116–117*inv*
  parallel and series circuits, 137*act*
current electricity, **115**–122
  conductors and insulators, **118**
  grounding, **119**
  using current electricity, 127*act*
  see also circuit
cytoplasm, **16**

D
Deepworker, **252**
dependent variable, 288
desert, 39, 41
Dextre, **248**
diesel submarine, 259
discharge, **111**
distribution lines, 169
diversity of life in BC, 8*act*, 44–45

E
electric charges, 97*act*, 98–103
  attraction and repulsion, **105**, 108–109
  charged atoms, **100**, 101*act*, 102
  imbalance of charges, **100**
  positive or negative, 99–101
  static electricity, 106–107*inv*
  uncharged, **99**
electric generator, **152**
electric meter, **171**
electrical energy source, 153*act*
  burning fossil fuels, 156–157, 158
  hydro-electric power, 154, 155*inv*
  new technologies, 160–168
  nuclear energy, 157, 158
electricity
  current electricity, **115**–122
  generating, 152–159, 167*inv*
  hazard, 129, 130*act*
  movement of electrons, 101–102
  safety, 132, 171
  series circuits, 128–135
  static electricity, **104**–114
  transforming, 143–146
  transmission and distribution, 169
electrocution, **171**
electromagnet, **144**, 145*inv*
electromagnetic waves, **216**
electromagnetism, **144**
electrons, **99**–102
  transfer produces electricity, 102, 104–105
endoplasmic reticulum (ER), **17**

36090

energy conservation, **172**, 173*inv*
energy consumption, **172**
energy efficiency, **172**
energy rating, 172
environment and living things, 34–59
adaptations, 47–56
climates and biomes, 36–44
definition of environment, **188**
extreme environments, **188**
estuaries, **42**, 43
Exosuit, **252**
exploration, **192**
exploring BC, 187*act*
extremes
on land, 190
in ocean, 191

F

fair test, 293
fibre optics, **253**
filament, **143**
fish, **82**
flashlight, 131*act*
flow chart, 272
food for astronauts, 192, 193*act*
fossil fuels, **156**–157
"free fall," 214
frogbot, **236**
fungi, **77**, 78–79*inv*
fuses, **170**, 170–171*act*

G

generating electricity, 152–159
geosynchronous orbit, **241**
geothermal energy, **164**
global-positioning system (GPS), 241
Goddard, Robert, 209, 210
grassland, 39
graphic organizers, 272
gravity, **201**
and orbits, 202*inv*
grounding, **119**
growth, 9, 12–13*inv*

H

hand lens, 23, 24*act*
hazard
of electricity, 129, 130*act*
of space travel, 258
heat-exchange systems, 164
hibernation, **50**
Hooke, Robert, 16
Hubble space telescope, 244
hybrid vehicle, 259
hydro-electric dam, **154**
hydro-electric power (hydro), **154**
model station, 155*inv*
hydrogen fuel cell, **166**

I

imbalance of charges, **100**
independent variable, 288
inference, 295
inherited chacteristic, **47**
insulators, **118**
International Space Station (ISS), 226, 244, 245
invertebrates, **81**

K

kingdom (in science), 63
six kingdoms of life, 70–82, 83*act*, 84
Know-Wonder-Learn chart, 274

L

lakes and ponds, 43
lander, **228**, 229
latitude, **36**
Leeuwenhoek, Anton van, 28
life in space, 220
life-support systems, **194**
light bulb, **143**, 144
lightning, 112, 113*act*
Linnaeus, Carolus, 64, 65
living things
to Aboriginal peoples, 6
differences from non-living things, 5*act*
diversity in BC, 8*act*, 44–45
exchange of gases, 7
growth, 9, 12–13*inv*
made of cells, 7, 15
reproduction, 10
response to stimuli, 11, 12–13*inv*
to scientists, 7–11
and their environment, 34–59
use of energy, 8–9
viewing microscopic organisms, 23–30
see also cells; organisms (living things)
loads, electric, 128, 133*inv*

M

magnetic field, 144
magnify, 23
mammals, **82**
meniscus, 284
metric system, 281
micro-hydro turbines, 154
micro-organisms, **19**
bacteria, 70–73
in extreme environments, 220
protists, 73, 74–75*inv*
viewing *see* microscopes
microgravity, **214**, 215*act*
microgravity vibration isolation mount (MIM), 251
microscopes, 25–29
investigations, 25, 27, 74–75
parts, 26
practice, 29*act*
then and now, 28
migration, **50**
mimicry, **54**
mind map, 274
mitochondria, **16**
model
of a cell, 18*act*
of an atom, 101*act*
extreme environment, 189*act*
Monera (bacteria), 70–73
motor, electric, 144
multicellular, **19**

N

natural gas, **156**
negative charge, **99**–101
NEPTUNE project, **253**, 254
network tree, 273
non-living things, 5*act*
non-renewable energy source, **158**
nuclear energy, **157**
nuclear fission, 166
nuclear fusion, 166
nuclear submarine, 259
nucleus, **16, 99**

O

observing the sky, 199–207
Oersted, Hans, 144
oil, **156**
open circuit, **129**
orbit, **200**
geosynchronous orbit, **241**
and gravity, 202*inv*
organelles, **16**–17, 18*act*
organisms (living things)
adaptations, **47**–56
characteristics, 7–14
classification systems, 60–69
six kingdoms of life, 70–82, 83*act*, 84
see also living things
oxygen, 7, 192

P

parallel circuit, **136**, 137*act*
building, 138*inv*, 140*inv*
compared with series circuit, 141
current in, 137*act*
flow of electricity, 138*inv*
sources in parallel, 139
payload of a rocket, **210**
photoelectric cell, 161
photosynthesis, **7**
phylum, 64
physical adaptations, **48**, 49*act*
planets, **203**, 204–205*inv*
Plantae (plants), **75**, 76*act*
positive charge, **99**–101
power, measured in watts, 172
precipitation, **36**
Protista (protists), 73, 74–75*inv*

R

radio waves, 216–217
recycling, **194**
renewable energy source, **158**
new technologies, 160–168
repel (electric), **105**, 108–109
reproduction, **10**
reptiles, **82**
resistance, **134**, 143
respiration, 7
rivers and streams, 42
robots, **232***act*, 233
communicating with a Mars robot, 234–235*inv*
designing a robot, 237*act*
types of robots, 236, 238
rocket, **208***act*, 209–210
rovers, **228**, 229, 233
"run-of-river" generators, 154, 155

S

Safety, viii–xi
safety with electricity, 132, 171
satellites, 240–244
communications satellites, **240**–241
observation satellites, 242, 243*act*
timeline, 246*act*
science, **xiii**
scientific names, 65–66
seashores, 43
series circuit, 128–135
closed and open circuits, **129**
compared with parallel circuit, 141
current in, 134, 137*act*
a single pathway, **131**
sources and loads, 132–133*inv*
shelter in extreme environments, 196–197
short circuit, **130**
slide, 26, 27*inv*
snakebot, **238**
society, **xvi**
solar energy, **161**
solar system, **203**, 204–205*inv*
sources (electric) **128**, 132–133*inv*
connected in parallel, 139
connected in series, 131, 132–133*inv*
space arm, 248–249, 250*inv*
space junk, **258**
space program, 230–231*inv*
space report, 218*inv*
space stations, 226, 244, 245
spacesuit, 211, 212–213*inv*, 260*inv*
species, 64
spiderbot, **238**
stage of rocket engine, **210**
static electricity, **104**–114
discharge, **111**
use of, 109, 110*act*
stems in plants, 76*act*
stimuli, 11, 12–13*inv*
Sun, **206**
solar energy, 161
structure of the Sun, 206
switch, **129**, 133*inv*
symbiosis, 51

T

technology, **xiii**, **188**
aboriginal, for explorers, 197*act*
benefits of space technology, 258
extreme environments, 195, 227*act*
producing electricity, 160–168
testing, 257–259, 260*inv*, 261*inv*
using responsibly, 261*inv*
technological problem solving, 298
telescope, **199**
temperate forest, 38, 40
thermal plasma analyzer, 251
thrust, **209**
tidal energy, **163**
time line, 273
timeline for satellites, 246*act*
transformers, 169
transmission lines, 169
tropical rain forest, 38
tundra, 38
turbine, **152**

U

uncharged, **99**
unicellular, **19**

V

vacuole, **17**
vacuum of space, **211**
variables, 293
vascular tissues, 75, 76*act*
venn diagram, 272
Venus, 221*inv*, 222
VENUS test network, 253, 254
vertebrates, **81**, 82
voltage, **134**, 137
Voyager spacecraft, 219

W

wastes, nuclear, 157
water for explorers, 192
watt (W), 172
weather, **36**
weather satellites, 242, 243*act*
wet mount, 280
wind energy, **162**
wind turbines, 162
Workplace Hazardous Materials Information System, **xi**

Photo Credits

viii Erin Carr; ix Erin Carr; **x** Erin Carr; **xii top** Jim Watt/ MaxxImages.com; **centre**, CORBIS CANADA; **bottom**, Johnny Johnson/MaxxImages.com; **xiii left**, Allan and Sandy Carey/IVY IMAGES; **right** Larry Williams/CORBIS CANADA; **xiv top**, Sam Ogden/Science Photo Library; **bottom** Pascal Goetgheluck/Science Photo Library; **xvi** Gina Gallant; **xviii**, Jacqueline Windh Photography; **xix top** O.Bierwagon /IVY IMAGES; **bottom**, Nancy Greene; **xxi top** Bill Ivy/IVY IMAGES; **bottom**, Ernie Stelzer; **xxii top** S. McCutcheon/IVY IMAGES, **bottom** IVY IMAGES; **xxiii left** Bill Lowry/ IVY IMAGES, **top right**, Mabel Joe; **bottom right**, Gunter Marx Photography/CORBIS CANADA; **Unit 1 Opener**, Bill Ivy/IVY IMAGES; **Chapter 1 Opener**, Aubrey Lang/VALAN PHOTOS; **p5**, Edmond Van Hoorick/Getty Images; **p6 left**, Celia Nord; **p7 top**, Dr. Linda Stannard/ UCT/Photo Researchers, Inc., Doc White/IVY IMAGES; **p8**, © OSF/MaxxImages.com; **p9**, **top**, PhotoLink/Getty Images, Alexis Rosenfeld/Photo Researchers, Inc., Suzanne and Joseph Collins/Photo Researchers, Inc.; **p10 top**, Brandon D. Cole/CORBIS CANADA, **centre left**, © 2005 Lon Lauber/AlaskaStock.com, **centre right**, Stephen Dalton/ Photo Researchers, Inc., **bottom left**, Bruce A. Iverson, **right**, Jonelle Weaver/Getty Images; **p11**, V. Wilkinson/VALAN PHOTOS, **inset**, Gail Jankus/ Photo Researchers, Inc.; **p12**, Ian Crysler; **p15**, Jeff Rotman, Photo Researchers, Inc., **inset**, © James A. Cosgrove; **p19**, Biophoto Associates/Photo Researchers, Inc.; **p22**, Bruce A. Iverson; **p23**, Sehn-Yong Shin; **p25**, A-E, Bruce A. Iverson, F, Eye of Science/Photo Researchers, Inc.; **p31**, V. Wilkinson/VALAN PHOTOS; **Chapter 2 Opener**, Albert Kuhnigk/VALAN PHOTOS, **inset**, Fred Bruemmer/VALAN PHOTOS; **p35**, Al Harvey/ www.slidefarm.com; **p40**, Alex L. Fradkin/Getty Images; **p41**, V. Last, Geographical Visual Aids; **p42 top left**, T.E. Adams/ Visuals Unlimited, Wayne Lankinen/VALAN PHOTOS, **top right**, Al Harvey/www.slidefarm.com, Ken Lucas/Visuals Unlimited; **p43 top left**, Al Harvey/www.slidefarm.com, Bill Lowry/IVY IMAGES, inset, Ernie Stelzer, **top right**, W. Towriss/IVY IMAGES, Frank & Joyce Burek/Getty Images; **p44 left**, Jeannie R. Kemp/VALAN PHOTOS, W. Towriss/IVY IMAGES; **p45 top left**, S.C. Fried/Photo Researchers, Inc., Parks Canada, **top right**, Allan & Sandy Carey/ Getty Images, J.A. Wilkinson/VALAN PHOTOS; **p47**, James R. Page/VALAN PHOTOS; **p48 top**, used by permission from the Canadian Forest Service, Northern Forestry Centre, **bottom left**, V. Wilkinson/VALAN PHOTOS, PhotoLink/ Getty Images; **p49 top**, Ryan McVay/ Getty Images, W. Lankinen/IVY IMAGES, Wayne Lankinen/ VALAN PHOTOS, Stephen J. Kraseman/VALAN PHOTOS; **p50**, StockTrek/ Getty Images; **p51 top**, Gilbert Grant/ Photo Researchers, Inc., J. Mathias/IVY IMAGES; **p54**, W. Lankinen/IVY IMAGES, G. McDaniel/Photo Researchers, Inc., **centre left**, Dr. John Alcock/Visuals Unlimited, Bruce A. Iverson, **bottom**, Valerie Hodgson/Visuals Unlimited; **p55**, Erin Carr; **p57**, Neil G. McDaniel/Photo Researchers, Inc.; **Chapter 3 Opener**, Peter Lane Taylor/Visuals Unlimited; **p61 top**, David Tanaka, Ken Lucas/Visuals Unlimited; **p62**, Jim Zipp/Photo Researchers, Inc.; **p63**, J.A. Wilkinson/ VALAN PHOTOS; **p64**, Suzanne L. Collins/Photo Researchers, Inc., **inset**, Michael Patrick O'Neill/Photo Researchers, Inc.; **p66, top**, A-B, Allan and Sandy Carey/IVY IMAGES, Tom and Pat Leeson/Photo Researchers, Inc.; **p67 top**, V. Last, Geographical Visual Aids, **bottom left**, Stephen J. Kraseman/ VALAN PHOTOS, Ralph A. Reinhold/VALAN PHOTOS; **p68**, Erin Carr; **p71**, BSIP/Photo Researchers, Inc.; **p72**, Biophoto Associates/Photo Researchers, Inc.; **p73 left**, Robert C. Simpson/VALAN PHOTOS, Al Harvey /www.slidefarm.com, Bruce A. Iverson, Gary Retherford/ Photo Researchers, Inc.; **p76**, Erin Carr; **p77**, Herman H. Geithorn/VALAN PHOTOS; **p78**, Erin Carr; **p80 top**, V. Wilkinson/VALAN PHOTOS, W. Lynch/IVY IMAGES, John Cancalosi/VALAN PHOTOS; **p81 top**, Bill Ivy/IVY IMAGES, Lynn and Donna Rogers/IVY IMAGES; **p82 top left**, Rondi/Tani Church/Photo Researchers, Inc., Brandon D. Cole/Visuals Unlimited, Tom and Pat Leeson/Photo Researchers, Inc., **centre**, Ken Cole/VALAN PHOTOS, **top right**, Bill Ivy/IVY IMAGES, Dennis Schmidt/VALAN PHOTOS, James D. Markov/VALAN PHOTOS; **p83**, Erin Carr; **p85 top**, Ken Cole/VALAN PHOTOS, Michael Patrick O'Neill/Photo Researchers, Inc.; **p88 banner**, © 2005 Jeff Schultz/AlaskaStock.com, Calvin W. Hall/Alaska Stock.com, © 2005 Clark J. Mishler/AlaskaStock.com, Ernie Stelzer; **p88**, Barbara Wilson; **p89**, Bill Ivy/IVY IMAGES; **p90**, Jane Watson; **p91**, J. Kraseman/VALAN PHOTOS; **p93**, Bill Ivy/IVY IMAGES; **Unit 2 Opener**, Al Harvey/ www.slidefarm.com; Chapter 4 Opener, Gregory Ochcki/ Photo Researchers, Inc.; **p97**, Brian Sprout/MaxxImages.com; **p98**, Eastcott:Momatiuk/VALAN PHOTOS; **p99**, Iggi Ruggieri/Stockfood/MaxxImages.com; **p100**, Layne Kennedy/CORBIS CANADA; **p104**, David Tanaka; **p106**, Erin Carr; **p108**, Bill Ivy/IVY IMAGES; **p109**, Teck Cominco Metals, Ltd. Trail Operations; **p110**, Erin Carr; **p112 top**, Ernie Stelzer, **bottom left**, Chuck Doswell/Visuals Unlimited, Science Source/Photo Researchers, Inc.; **p115**, Ben S. Kwiatkowski/Fundamental Photographs, NYC; **p118**, James R. Page/VALAN PHOTOS; **p119**, O. Bierwagon/IVY IMAGES; **p123**, Chuck Doswell/Visuals Unlimited; **Chapter 5 Opener**, Raymond Gehman/CORBIS CANADA; **p127**, Ryan McVay/ Getty Images; **p128**, Bill Lowry/IVY IMAGES; **p134**, Bruce A. Iverson; **p136**, Bryan & Cherry Alexander/Photo Researchers, Inc.; **p141**, Dick Hemingway; **p143**, G. Benoit/IVY IMAGES; **p144**, Dick Hemingway; **p149**, Alex Bartel/Photo Researchers, Inc.; **Chapter 6 Opener**, NASA; **p151**, Dick Hemingway; **p152**, photo courtesy of B.C. Hydro, Power House @ Stave Falls; **p153**, O. Bierwagon/ IVY IMAGES; **p154**, photo courtesy of WINDSTREAM POWER SYSTEMS; **p155** Erin Carr; **p156**, Marc Asnin/CORBIS/SABA/CORBIS CANADA; **p157 top**, O. Bierwagon/IVY IMAGES, V. Last, Geographical Visual Aids; **p158 top left**, James R. Page/ VALAN PHOTOS, Phil Norton/VALAN PHOTOS, **right**, Roger Russmeyer/CORBIS CANADA; **p160 top left**, J. DeVisser/IVY IMAGES, Al Harvey/www.slidefarm.com, **right**, Tom Parkin/VALAN PHOTOS; **p161 left**, CORBIS CANADA, Gunter Marx Photography/CORBIS CANADA; **p162 left**, photo courtesy of Toronto Hydro Energy Inc., Holberg Wind Energy Project; **p163 left**, O. Bierwagon/ IVY IMAGES, Wolfgang Kaehler/CORBIS CANADA; **p164 top**, Krafft/Photo Researchers, Inc., **bottom left**, Charles O'Rear/CORBIS CANADA, photo courtesy of Ministry of Forests, British Columbia; **p165**, V. Last, Geographical Visual Aids; **p166 top left**, Science Source/ Photo Researchers, Inc., **right**, Reuters/CORBIS CANADA, Marnie Dobell; **p169 top**, O. Bierwagon/IVY IMAGES, Dick Hemingway; **p169**, Erin Carr; **p170 top**, Dick Hemingway, J.A. Wilkinson/VALAN PHOTOS; **p171**, David Tanaka; **p172**, David Tanaka; **p175**, Rob Melnychuk/ Getty Images; **p177 left**, Steve Maslowski/Visuals Unlimited,

Mario Palumbo/Manitoba Hydro; **p178 banners**, Bill Ivy/IVY IMAGES, J. Kraseman/VALAN PHOTOS, © 2005 Jeff Schultz/AlaskaStock.com; **p178**, Jean Issac; **p179**, Al Harvey/www.slideshow.com; **p180**, Taras Atleo; **Unit 3 Opener**, Julian Baum/Science Photo Library; **Chapter 7 Opener**, Hot Shots/IVY IMAGES; **p187**, Francis Lepine/VALAN PHOTOS; **p188**, G. Wiltsie/Spectrum Stock/IVY IMAGES; **p190 top**, Adam Jones/Photo Researchers, Inc., **bottom left**, Jeremy Bishop/Photo Researchers, Inc., J. DeVisser/IVY IMAGES; **p191 top**, Dr. Paul A. Zahl/Photo Researchers, Inc., **bottom left**, Science VU/Visuals Unlimited, John Chumack/Photo Researchers, Inc.; **p193**, O. Bierwagon/IVY IMAGES; **p195 top left**, S. McCutcheon/IVY IMAGES, **top centre**, Bill Ivy/IVY IMAGES, **top right**, F. Bruemmer/IVY IMAGES, **bottom left**, Denis Roy/VALAN PHOTOS, Paul Nicklen/National Geographic Society, **inset**, J. Zoiner/Hot Shots/IVY IMAGES; **p196 top**, F. Bruemmer/IVY IMAGES, Kit Kittle/CORBIS CANADA, **right**, StockTrek/Getty Images; **p197**, M. Leger/IVY IMAGES; **p200**, StockTrek/Getty Images; **p203**, NASA; **p204** Erin Carr; **p208**, Hot Shots/IVY IMAGES; **p209 right**, Science Source/Photo Researchers, Inc.; **p210**, PhotoLink/Getty Images; **p211**, NASA; **p214**, Canadian Space Agency; **p215**, NASA/Getty Images; **p215**, Erin Carr; **p216**, © Andrew Gray; **p217 left**, J. Passchoff/Visuals Unlimited, Prof. Justin Jonas, Rhodes University; **p219 left**, NASA, Richard Terrile/JPL/NASA; **p220 top**, Galileo Project/JPL/NASA, George F. Mobley/National Geographic Society; **p221**, NASA/Science Source/Photo Researchers, Inc.; **Chapter 8 Opener**, NASA; **p228 top**, JPL/NASA, **bottom**, Kennedy Space Center, NASA; **p232 left**, T. Mihok/IVY IMAGES, David Parker/Photo Researchers, Inc.; **p236**, JPL/NASA; **p237**, NASA; **p238**, JPL/NASA; **p239**, Ames Research Center/NASA; **p241**, Bill Ivy/IVY IMAGES; **p242**, Digital Globe; **p243**, NOAA; **p244**, NASA; **p248**, images courtesy MD Robotics and Canadian Space Agency 2003; **p249 top**, NASA, **bottom**, Canadian Space Agency; **p251 top**, Canadian Space Agency, **bottom left**, Aurbis Technical Centre Saskatoon, NASA; **p252**, Nuytco Research Ltd., www.nuytco.com; **p253 top**, image provided courtesy the Neptune Project (www.neptune.washington.edu) and CEV (Center for Environmental Visualization), M. Keller/IVY IMAGES; **p257 left**, photo NASA Haughton Mars Project/Pascal Lee, Canadian Space Agency; **p259 top**, CP (Andrew Vaughan), G. Fritz/IVY IMAGES; **p265**, JPL/NASA; **p266 banner**, Kluntaxa Nation Archives, Kluntaxa Nation Archives, Halle Flygare/VALAN PHOTOS, Kluntaxa Nation Archives/Canadian Museum of Civilization; **p266**, Wilfred Jacobs; **p267**, Judith Nickol, Coaldale, Alberta; **p268**, Mark Fricker; **p269 top** and **bottom**, NASA; **p272 top**, John Fowler/VALAN PHOTOS; **second from top**; Bill Ivy/IVY IMAGES; **third from top**, Bill Ivy/IVY IMAGES; **bottom**, Bill Ivy/IVY IMAGES; **p276 top**, Karen and Betty Collins/Visuals Unlimited; **bottom**, Bruce A. Iverson; **p279**, Bruce A. Iverson; **p284**, RDF/Visuals Unlimited; **p285**, Erin Carr; **p286**, Bill Ivy/IVY IMAGES; **p292**, H. Ehricht/Hot Shots/IVY IMAGES; **p295 left**, Science VU/VISUALS UNLIMITED; **right**, Ralph White/CORBIS CANADA.

Illustrations Credit

Technical illustrations provided by ArtPlus Limited. Cartoons provided by Jean Morin and Steve Attoe. **p5**, From *Life Science*, Glencoe/McGraw-Hill; **p16–17** From *science.connect 1*, McGraw-Hill Ryerson Ltd., **p26** From *SciencePower 9*, McGraw-Hill Ryerson Ltd; **p28**, From *SciencePower8*, McGraw-Hill Ryerson Ltd., **p33** From *science.connect 1*, McGraw-Hill Ryerson Ltd.; **p36** From *science.connect 1*, McGraw-Hill Ryerson Ltd.; **p38–39**, From *SciencePower 7*, McGraw-Hill Ryerson Ltd., **p40**, From *SciencePower 7*, McGraw-Hill Ryerson Ltd., **p65**, From *Science*, McGraw-Hill; **p70**, From *Science Voyages*, McGraw-Hill; **p76**, From *Science*, McGraw-Hill; **p80**, From *Biology*, 6th edition, McGraw-Hill; **p111**, From *Science*, McGraw-Hill; **p156**, From *SciencePower9*, McGraw-Hill Ryerson Ltd.; **p163**, From *SciencePower 9*, McGraw-Hill Ryerson Ltd.; **p194 top**, From *Science*, McGraw-Hill; **p203**, From *Science*, McGraw-Hill; **p206**, From *Science*, McGraw-Hill; **p241 top**, From *ScienceFocus 9*, McGraw-Hill Ryerson Ltd.; **bottom**, From *SciencePower 9*, McGraw-Hill Ryerson Ltd.; **p280 left and top right**, From *ScienceFocus 9*, McGraw-Hill Ryerson Ltd.; **p283**, From *BC Science 7*, McGraw-Hill Ryerson Ltd.; **p291 left top and bottom**, From *ScienceFocus 7*, McGraw-Hill Ryerson Ltd.; **right**, From *BC Science 7*, McGraw-Hill Ryerson Ltd.

Text Credit

p52, Voices of the Wild: An Animal Sensagoria used by permission of the author, David Bouchard.